Future Perspectives in Nanostructured Materials Preparation, Characteristics and Applications

Future Perspectives in Nanostructured Materials Preparation, Characteristics and Applications

Editor

Thomas Dippong

Basel • Beijing • Wuhan • Barcelona • Belgrade • Novi Sad • Cluj • Manchester

Editor
Thomas Dippong
Chemistry and Biology
Technical University of
Cluj Napoca
Baia Mare, Romania

Editorial Office
MDPI
St. Alban-Anlage 66
4052 Basel, Switzerland

This is a reprint of articles from the Special Issue published online in the open access journal *International Journal of Molecular Sciences* (ISSN 1422-0067) (available at: https://www.mdpi.com/journal/ijms/special_issues/Z6E1S5674T).

For citation purposes, cite each article independently as indicated on the article page online and as indicated below:

Lastname, A.A.; Lastname, B.B. Article Title. *Journal Name* **Year**, *Volume Number*, Page Range.

ISBN 978-3-0365-9227-5 (Hbk)
ISBN 978-3-0365-9226-8 (PDF)
doi.org/10.3390/books978-3-0365-9226-8

© 2023 by the authors. Articles in this book are Open Access and distributed under the Creative Commons Attribution (CC BY) license. The book as a whole is distributed by MDPI under the terms and conditions of the Creative Commons Attribution-NonCommercial-NoDerivs (CC BY-NC-ND) license.

Contents

About the Editor vii

Preface ix

Thomas Dippong, Erika Andrea Levei, Ioan Petean, Iosif Grigore Deac, Raluca Anca Mereu and Oana Cadar
Screening of Mono-, Di- and Trivalent Cationic Dopants for the Enhancement of
Thermal Behavior, Kinetics, Structural, Morphological, Surface and Magnetic Properties of
$CoFe_2O_4$-SiO_2 Nanocomposites
Reprinted from: *Int. J. Mol. Sci.* **2023**, 24, 9703, doi:10.3390/ijms24119703 1

Chinmaya Mutalik, Muhammad Saukani, Muhamad Khafid, Dyah Ika Krisnawati, Widodo, Rofik Darmayanti, et al.
Gold-Based Nanostructures for Antibacterial Application
Reprinted from: *Int. J. Mol. Sci.* **2023**, 24, 10006, doi:10.3390/ijms241210006 27

Dania Racolta, Constantin Andronache, Maria Balasoiu, Leonard Mihaly-Cozmuta, Vadim Sikolenko, Oleg Orelovich, et al.
Influence of the Structure on Magnetic Properties of Calcium-Phosphate Systems Doped with
Iron and Vanadium Ions
Reprinted from: *Int. J. Mol. Sci.* **2023**, 24, 7366, doi:10.3390/ijms24087366 49

Marco Zannotti, Andrea Rossi, Marco Minicucci, Stefano Ferraro, Laura Petetta and Rita Giovannetti
Water Decontamination from Cr(VI) by Transparent Silica Xerogel Monolith
Reprinted from: *Int. J. Mol. Sci.* **2023**, 24, 7430, doi:10.3390/ijms24087430 59

Rawan E. Elbshary, Ayman A. Gouda, Ragaa El Sheikh, Mohammed S. Alqahtani, Mohamed Y. Hanfi, Bahig M. Atia, et al.
Recovery of W(VI) from Wolframite Ore Using New Synthetic Schiff Base Derivative
Reprinted from: *Int. J. Mol. Sci.* **2023**, 24, 7423, doi:10.3390/ijms24087423 73

Jing Tang, Shengyu Feng and Dengxu Wang
Facile Synthesis of Sulfur-Containing Functionalized Disiloxanes with Nonconventional
Fluorescence by Thiol–Epoxy Click Reaction
Reprinted from: *Int. J. Mol. Sci.* **2023**, 24, 7785, doi:10.3390/ijms24097785 99

Jiaxin Li, Yachao Yan, Yingzhi Chen, Qinglin Fang, Muhammad Irfan Hussain and Lu-Ning Wang
Flexible Curcumin-Loaded Zn-MOF Hydrogel for Long-Term Drug Release and Antibacterial
Activities
Reprinted from: *Int. J. Mol. Sci.* **2023**, 24, 11439, doi:10.3390/ijms241411439 113

Taimei Cai, Huijie Chen, Lihua Yao and Hailong Peng
3D Hierarchical Porous and N-Doped Carbonized Microspheres Derived from Chitin for
Remarkable Adsorption of Congo Red in Aqueous Solution
Reprinted from: *Int. J. Mol. Sci.* **2023**, 24, 684, doi:10.3390/ijms24010684 127

Bethan K. Charlton, Dillon H. Downie, Isaac Noman, Pedro Urbano Alves, Charlotte J. Eling and Nicolas Laurand
Surface Functionalisation of Self-Assembled Quantum Dot Microlasers with a DNA Aptamer
Reprinted from: *Int. J. Mol. Sci.* **2023**, 24, 14416, doi:10.3390/ijms241914416 143

David Ruiz Arce, Shima Jazavandi Ghamsari, Artur Erbe, and Enrique Samano
Metallic Nanowires Self-Assembled in Quasi-Circular Nanomolds Templated by DNA Origami
Reprinted from: *Int. J. Mol. Sci.* **2023**, *24*, 13549, doi:10.3390/ijms241713549 . . . . **155**

Alina Ioana Ardelean, Madalina Florina Dragomir, Marioara Moldovan, Codruta Sarosi, Gertrud Alexandra Paltinean, Emoke Pall, et al.
In Vitro Study of Composite Cements on Mesenchymal Stem Cells of Palatal Origin
Reprinted from: *Int. J. Mol. Sci.* **2023**, *24*, 10911, doi:10.3390/ijms241310911 . . . . **171**

Alexandra L. Ivlieva, Elena N. Petritskaya, Dmitriy A. Rogatkin, Inga Zinicovscaia, Nikita Yushin and Dmitrii Grozdov
Impact of Chronic Oral Administration of Gold Nanoparticles on Cognitive Abilities of Mice
Reprinted from: *Int. J. Mol. Sci.* **2023**, *24*, 8962, doi:10.3390/ijms24108962 . . . . **185**

Yun-Hsiu Tseng, Tien-Li Ma, Dun-Heng Tan, An-Jey A. Su, Kia M. Washington, Chun-Chieh Wang, et al.
Injectable Hydrogel Guides Neurons Growth with Specific Directionality
Reprinted from: *Int. J. Mol. Sci.* **2023**, *24*, 7952, doi:10.3390/ijms24097952 . . . . **197**

About the Editor

Thomas Dippong

Thomas Dippong (Associate Professor at the Technical University of Cluj Napoca) is a chemical engineer with a Ph.D. in chemistry. His current research activities are related to the characterization of nanoparticles for various applications as part of an ongoing research project in partnership with the Technical University of Cluj-Napoca within the field of ferrite embedded in silica matrix. He is an expert in analytical chemistry, organic/inorganic chemistry, thermal treatment, instrumental analysis, and synthesis of nanomaterials. Dr Dippong has published 140 peer-reviewed publications (84 papers in high-ranked scientific ISI-Thomson journals (46 Q1, 14 Q2, and 24 Q3) and 56 in other national and international journals), with 1741 citations and an h-index of 36 (WoS). He has given 35 lectures at international conferences (the 14th ICTAC in Brazil, 5th CEEC-TAC in Roma, JTACC in Budapest, ESTAC in Brasov, etc.). He has also published 2 books with international publishing houses and 15 books with national publishing houses. In the last 3 consecutive years (2021–2023), Dr Thomas Dippong was included in the prestigious list of World's Top 2% researchers. He has been a contract manager for many projects and is currently an active member of four projects. Dr. Dippong has reviewed 440 scientific articles for 91 ISI-Thomson journals. He has been a Guest Editor for four Special Issues published by three prestigious Q1 ISI-Thomson journals.

Preface

The characteristics of nanoparticles, such as their size, shape, surface morphology, crystal structure, grain boundaries, and composition, can be tuned to match the intended use. Thus, the design of nanostructured materials with a wide range of applications in the engineering, chemistry, physics, ceramics, biotechnology, biomedicine, and environmental fields represents an active area of research. Nanostructured materials are now widely used in research and industry, and their particular properties enable advanced physicochemical, electrical, thermal, catalytic, coloristic, optical, and magnetic applications.

The present Special Issue aims to provide a comprehensive overview of original research articles, communications, and reviews that focus on the development, characterization, and/or applications of nanoparticles and their composites. This Special Issue focuses on the following topics: synthesis routes for nanoparticles and nanocomposites; preparation and characteristics of nanoparticles, nanotubes, nanowires, or nanofibers; thermal behavior of nanomaterials and their composites; structural characterization using XRD, FT-IR, Mossbauer, or XPS; surface characterization via BET; morphological characterization using TEM, SEM, or AFM; magnetic applications of nanostructured materials; adsorption or photocatalytic properties of nanostructured materials used in the environmental field; application of nanocomposites for water decontamination: removal of heavy metals, dyes, pesticides, and other pollutants; pigments based on materials used in glazes; materials for high-performance batteries; nanosensors, catalysis and photocatalysis, oxidation processes, and antimicrobial activities; and medical applications of nanomaterial.

Thomas Dippong
Editor

Article

Screening of Mono-, Di- and Trivalent Cationic Dopants for the Enhancement of Thermal Behavior, Kinetics, Structural, Morphological, Surface and Magnetic Properties of $CoFe_2O_4$-SiO_2 Nanocomposites

Thomas Dippong [1], Erika Andrea Levei [2], Ioan Petean [3], Iosif Grigore Deac [4], Raluca Anca Mereu [3] and Oana Cadar [2,*]

[1] Faculty of Science, Technical University of Cluj-Napoca, 76 Victoriei Street, 430122 Baia Mare, Romania
[2] INCDO-INOE 2000, Research Institute for Analytical Instrumentation, 67 Donath Street, 400293 Cluj-Napoca, Romania
[3] Faculty of Chemistry and Chemical Engineering, Babes-Bolyai University, 11 Arany Janos Street, 400084 Cluj-Napoca, Romania; raluca.mereu@ubbcluj.ro (R.A.M.)
[4] Faculty of Physics, Babes-Bolyai University, 1 Kogalniceanu Street, 400084 Cluj-Napoca, Romania
* Correspondence: oana.cadar@icia.ro

Abstract: $CoFe_2O_4$ is a promising functional material for various applications. The impact of doping with different cations (Ag^+, Na^+, Ca^{2+}, Cd^{2+}, and La^{3+}) on the structural, thermal, kinetics, morphological, surface, and magnetic properties of $CoFe_2O_4$ nanoparticles synthesized via the sol-gel method and calcined at 400, 700 and 1000 °C is investigated. The thermal behavior of reactants during the synthesis process reveals the formation of metallic succinates up to 200 °C and their decomposition into metal oxides that further react and form the ferrites. The rate constant of succinates' decomposition into ferrites calculated using the isotherms at 150, 200, 250, and 300 °C decrease with increasing temperature and depend on the doping cation. By calcination at low temperatures, single-phase ferrites with low crystallinity were observed, while at 1000 °C, the well-crystallized ferrites were accompanied by crystalline phases of the silica matrix (cristobalite and quartz). The atomic force microscopy images reveal spherical ferrite particles covered by an amorphous phase, the particle size, powder surface area, and coating thickness contingent on the doping ion and calcination temperature. The structural parameters estimated via X-ray diffraction (crystallite size, relative crystallinity, lattice parameter, unit cell volume, hopping length, density) and the magnetic parameters (saturation magnetization, remanent magnetization, magnetic moment per formula unit, coercivity, and anisotropy constant) depend on the doping ion and calcination temperature.

Keywords: cobalt ferrite; silica matrix; doping; calcination; magnetic behavior

1. Introduction

Due to their potential use in various technological fields such as gas sensors, catalysis, magnetic imaging, high-density recording media, ferrofluids, microwave devices, magnetic data storage, and transformer cores, spinel ferrites have attracted considerable attention [1–3]. Depending on the cation distribution on A and B sites, spinel ferrites display ferrimagnetic, antiferromagnetic, spin-glass, or paramagnetic behavior [1]. Among them, cobalt ferrite ($CoFe_2O_4$) presents superparamagnetic behavior, high chemical stability, good mechanical hardness, moderate saturation magnetization (M_S), high coercivity (H_C), positive anisotropy constant (K), significant electrical resistance, low eddy current losses and low production costs [4–7]. $CoFe_2O_4$ is a promising candidate for magnetic resonance imaging, drug delivery, tomography, tissue imaging ferrofluids, recording heads, magnetic sensors, microwave devices, magnetic refrigeration, catalysis, pigments, gas detectors, environmental remediation, etc. [4,8–11].

Citation: Dippong, T.; Levei, E.A.; Petean, I.; Deac, I.G.; Mereu, R.A.; Cadar, O. Screening of Mono-, Di- and Trivalent Cationic Dopants for the Enhancement of Thermal Behavior, Kinetics, Structural, Morphological, Surface and Magnetic Properties of $CoFe_2O_4$-SiO_2 Nanocomposites. *Int. J. Mol. Sci.* **2023**, *24*, 9703. https://doi.org/10.3390/ijms24119703

Academic Editors: Raphaël Schneider and Christian M. Julien

Received: 7 May 2023
Revised: 28 May 2023
Accepted: 1 June 2023
Published: 2 June 2023

Copyright: © 2023 by the authors. Licensee MDPI, Basel, Switzerland. This article is an open access article distributed under the terms and conditions of the Creative Commons Attribution (CC BY) license (https://creativecommons.org/licenses/by/4.0/).

By doping a small amount of an external ion in the composition of $CoFe_2O_4$, the physicochemical properties can be tailored without undesired phase transformation [3]. The properties of $CoFe_2O_4$ can be easily tailored as they depend on the synthesis method, composition, doping ions, and particle size distribution [4,10,11]. The properties of $CoFe_2O_4$ doped with divalent (Mg^{2+}, Cu^{2+}, Ni^{2+}, Zn^{2+}, Mn^{2+}) and trivalent (La^{3+}, Ru^{3+}, Gd^{3+}, Al^{3+}) ions were extensively studied [12–18], while the doping with monovalent ions received less attention. The most common monovalent ion for ferrite doping is Ag^+ due to its antibacterial activity against different pathogens [19]. Doping $CoFe_2O_4$ with non-magnetic Na^+ ions resulted in good chemical stability but decreased magnetic properties [20]. Non-magnetic metal ions such as Cd^{2+} are particularly important because they do not display a magnetic moment but may change the magnetic properties by disturbing the magnetic moment's equilibrium in the ferrite composition [3]. The photocatalytic and magneto-optical properties of transition metal-doped $CoFe_2O_4$ have received extensive research for future use in electronics, telecommunication, environmental and biomedical applications, catalysis, etc. [21–24]. The properties of $CoFe_2O_4$, especially those related to magnetic behavior, conductivity, and catalytic activity, are enhanced by doping with noble metal ions [25]. Additionally, Ag nanoparticles play a non-magnetic spacer role that allows the tuning of the dipolar magnetic interactions between the magnetic nanoparticles [5]. The Ag doping in $CoFe_2O_4$ provides good physical properties and antimicrobial activity, but its preparation requires complex (i.e., surface linkers, reducing and protecting agents) and time-consuming procedures that make it inappropriate for large-scale production [19,26,27]. So, a simple route for its large-scale preparation is still being developed. There is an ongoing debate on spinel ferrites doping with La^{3+}, as some studies reported that the effect on coercivity and superexchange occurs most probably due to the location of La^{3+} ions on the sample's surface or interstitial sites. However, the most often proposed mechanism is the partial replacement of Fe^{3+} by La^{3+} in B sites, resulting in unit cell expansion, increased structural distortion, smaller particle size, and higher density than of undoped samples [6,8,9]. Due to the large difference between the ionic radii of the dopant and the substituted cations, doping with La^{3+} is energetically unfavorable [12]. $CaFe_2O_4$ has attracted significant attention in various applications due to its excellent magnetic properties, narrow band gap and, consequently, easy absorption of photons in the visible light range [28]. Though, to our knowledge, no studies report the synthesis and characterization of Ca doped $CoFe_2O_4$.

The properties of ferrite nanoparticles are greatly sensitive to their elemental composition, crystal structure, dopants, synthesis method, and preparation conditions [1]. Various methods were developed for the preparation of undoped $CoFe_2O_4$ and M-doped $CoFe_2O_4$, such as co-precipitation, sol-gel, thermal decomposition, microwave-assisted, laser pyrolysis, mechanical size reduction, gas phase synthesis, thermal hydrolysis, pulse laser deposition, microemulsion, solvothermal, sonochemical, mechanical–chemical processing, ball milling, hydrothermal synthesis, auto-combustion, etc., with each method having their advantages and shortcomings [4,8,9,21–24]. Of these, the chemical methods are susceptible to nanoparticle agglomeration; self-agglomeration is one of the key impediments in the obtaining of magnetic nanoparticles and can be diminished by their embedding into inorganic or organic matrices or by using surfactants to cover the nanoparticle surface [11].

The sol-gel route has the advantages of simplicity, flexibility, low cost, good control over the structure and properties, and production of nanosized ferrites with trustable and reproducible physical properties [11,21–23]. To stabilize and diminish nanoparticle agglomeration in order to obtain single-phase ferrites, the nanoparticles can be coated by a uniform and stable ultrathin layer or dispersed in a non-magnetic matrix. The sol-gel method involves mixing reactants with tetraethylorthosilicate (TEOS), forming strong networks with moderate reactivity, allowing the incorporation of various organic and inorganic molecules and short gelation time [21–23]. In this regard, forming an inactive SiO_2 coating on the surface of oxide systems could prevent their agglomeration and improve their chemical stability, with the SiO_2 acting as a physical barrier that control the attraction between the magnetic nanoparticles [21–23,29,30]. Moreover, embedding magnetic ferrites into the SiO_2 matrix gained considerable attention due to the possibility of controlling the particle size and minimizing the surface roughness and spin disorder, thus enhancing the magnetic properties of the obtained nanocomposites [22,23]. Additionally, the non-magnetic SiO_2 matrix possesses a high surface area and does not affect the magnetic behavior or electric properties of the $CoFe_2O_4$ nanoparticles due to its lower dielectric constant [29].

In the present work, we investigate the changes in the structural, morphological, surface, and magnetic properties of undoped and doped $CoFe_2O_4$ with transition monovalent (Ag^+, $Ag_{0.1}Co_{0.95}Fe_2O_4$; Na^+, $Na_{0.1}Co_{0.95}Fe_2O_4$), divalent (Ca^{2+}, $Ca_{0.1}Co_{0.9}Fe_2O_4$; Cd, $Cd_{0.1}Co_{0.9}Fe_2O_4$) and trivalent (La^{3+}, $La_{0.1}CoFe_{1.9}O_4$) metal ions embedded in the SiO_2 matrix obtained through the sol-gel route, followed by calcination at various temperatures.

The novelties of this paper consist of the following:

(i) A study of the influence of the SiO_2 matrix and dopant ion on forming metallic succinates and their decomposition into ferrites embedded in the SiO_2 matrix.

(ii) A comparative study of the impact of monovalent, divalent, and trivalent ion doping of $CoFe_2O_4$ embedded in the SiO_2 matrix on the morphological, magnetic and structural properties in order to find new strategies to increase their potential for existing and new possible applications.

(iii) Filling the gap in the existing literature on the effect of Na^+, Ag^+, Cd^{2+}, La^{3+}, and Ca^{2+} ion doping on the physicochemical properties of $CoFe_2O_4$ embedded in the SiO_2 matrix; the embedding of $CoFe_2O_4$ nanoparticles in the non-toxic, inert SiO_2 matrix via the sol-gel method allows the control of the particle growth, reduces the nanoparticle agglomeration and enhances the chemical stability and magnetic guidance [11].

(iv) The elucidation of the thermal behavior under isothermal and non-isothermal conditions and the formation kinetics of undoped and doped $CoFe_2O_4$.

(v) Obtaining single crystalline ferrites at low calcination temperatures (400 and 700 °C) by doping $CoFe_2O_4$ with mono-, di- and trivalent ions.

2. Results and Discussion

2.1. Thermal Stability of $CoFe_2O_4$ and Role of Doping Metals

Figure 1 shows the TG and DTA curves for the gels dried at 40 °C. The DTA curves (Figure 1) of the gels dried at 40 °C show the following three processes: (i) the loss of residual water (physically adsorbed water and moisture) indicated by the endothermic peak with the maximum at 33.6–35.5 °C, with a mass loss of 5.3–7% between 25 and 110 °C; (ii) the formation of metal succinates indicated by two endothermic effects at 143.6–145 °C (formation of Co, Ag, Na, Ca, and Cd succinates) and 177–184 °C (formation of Fe and La succinates) with a mass loss of 14.4–16.3% and (iii) the decomposition of metal succinates to metal oxides and the formation of ferrites shown by two exothermic effects at 250–257 °C (mass loss of 18.8–20.9%), corresponding to the oxidative decomposition of the succinates to Ag, Na, Co, Ca, and Cd oxides and at 283–294 °C (mass loss of 10.1–13.2%), consistent with the formation of Fe and La oxides [21,22,29,30]. The two-stage (144–145 °C and 177–184 °C) evolution of the redox reaction between the metal nitrates and 1,4-butanediol (1,4-BD) is due to the free aqueous, stronger acid $[Fe(H_2O)_6]^{3+}$ ion than $[Co(H_2O)_6]^{2+}$ ion [21]. As a result, the Fe succinate is formed at higher temperatures [21]. The corresponding mass loss on the TG curve is due to the loss of crystallization water from the nitrates and volatile products (H_2O, NO_2), resulting in the redox reaction [21]. An additional mass loss (0.1–1.6%) can also be observed on some TG curves between 900 and 1000 °C. This effect is the most visible in the undoped $CoFe_2O_4$. The highest total mass loss is observed for $CoFe_2O_4$ (57.6%), while in the doped $CoFe_2O_4$, the total mass loss is slightly lower (53.4–54.6%). The SiO_2 matrix undergoes various transformations during the thermal process, which makes the processes' delimitation ascribed to the formation and decomposition of succinate precursors difficult [21–23,29,30].

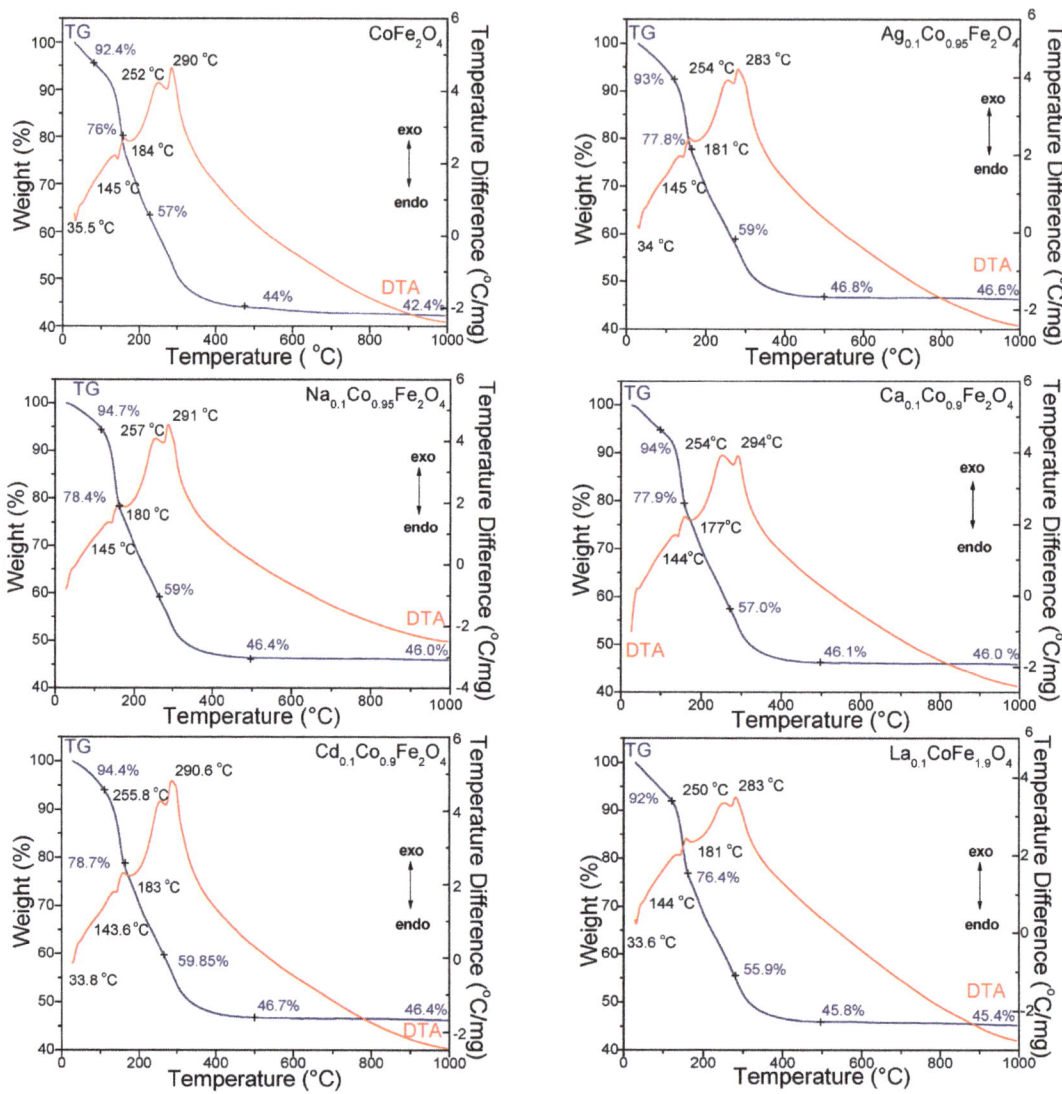

Figure 1. TG and DTA of $CoFe_2O_4$, $Ag_{0.1}Co_{0.95}Fe_2O_4$, $Na_{0.1}Co_{0.95}Fe_2O_4$, $Ca_{0.1}Co_{0.9}Fe_2O_4$, $Cd_{0.1}Co_{0.9}Fe_2O_4$ and $La_{0.1}CoFe_{1.9}O_4$ gels heated at 40 °C.

The decomposition of metal succinates under isothermal conditions (150, 200, 250, and 300 °C) is presented in Figure 2. In all cases, a rapid mass loss in the first 10 min, followed by a slow mass loss of up to 120 min, is observed. The mass loss is lower at 150 and 200 °C due to the incomplete formation of metal succinates and is comparable at 250 and 300 °C, confirming the formation of ferrites around 250 °C. $Ag_{0.1}Co_{0.95}Fe_2O_4$ presents the lowest mass loss on the 150 and 200 °C isotherms, while $Na_{0.1}Co_{0.95}Fe_2O_4$ presents the lowest mass loss on the 250 and 300 °C isotherms. The highest mass loss appears on the 150 °C isotherm for $CoFe_2O_4$, on the 200 °C and 250 °C isotherms for $Cd_{0.1}Co_{0.9}Fe_2O_4$, and the 300 °C isotherm for $Ca_{0.1}Co_{0.9}Fe_2O_4$.

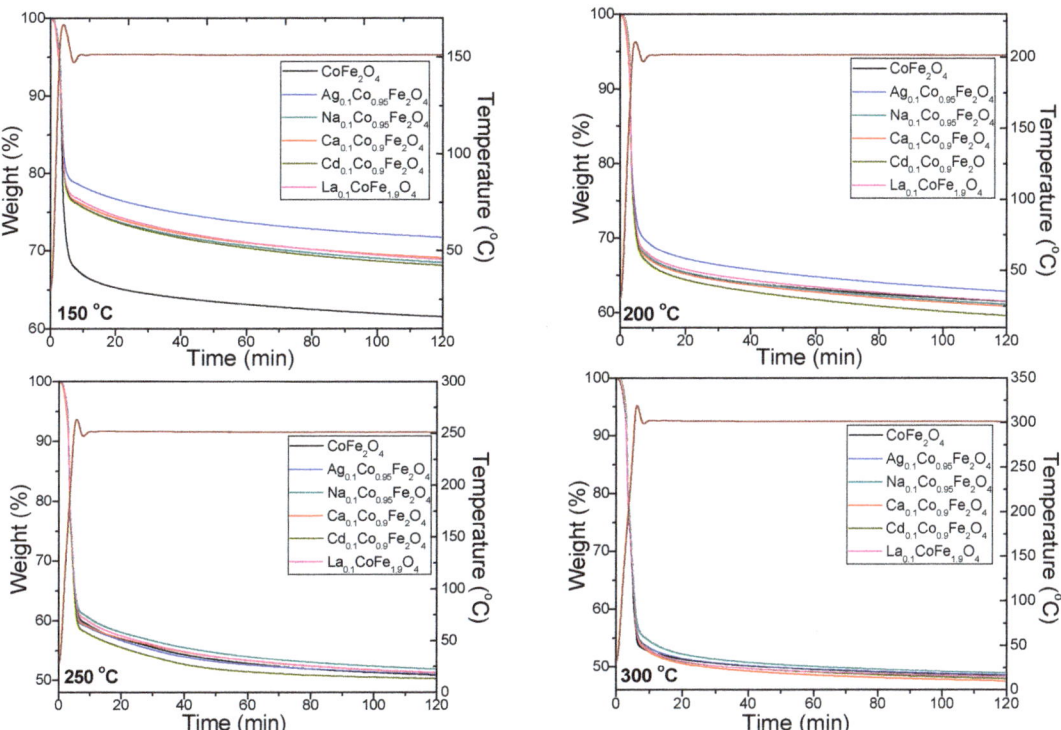

Figure 2. TG isotherms of $CoFe_2O_4$, $Ag_{0.1}Co_{0.95}Fe_2O_4$, $Na_{0.1}Co_{0.95}Fe_2O_4$, $Ca_{0.1}Co_{0.9}Fe_2O_4$, $Cd_{0.1}Co_{0.9}Fe_2O_4$ and $La_{0.1}CoFe_{1.9}O_4$, gels.

2.2. Kinetics of Doped and Undoped $CoFe_2O_4$ Formation

The rate constant (k) was calculated using the isotherms recorded at 150, 200, 250 and 300 °C according to the following first-order kinetic equation, Equation (1):

$$\frac{dx}{dt} = k(x_o - x)^{2/3} \qquad (1)$$

where dx/dt is the reaction rate, x_o is the initial mass (mg), x is the mass (mg) at time t, t is the time and k is the rate constant defined as $k = A\, e^{(-E/RT)}$, where A is the pre-exponential factor, E is the activation energy, R is the ideal gas constant, and T is the temperature [31]. The integration of Equation (1) leads to Equation (2) as follows:

$$(x_o - x)^{1/3} = (x_o)^{1/3} - kt \qquad (2)$$

Each isotherm's rate constant (k) value was computed using ten values in the 0–10 min range, where the highest mass loss occurs. A higher k indicates a faster reaction [31,32]. The k value increases with the increase in the calcination temperature. At 150 and 200 °C, the average k values of the doped ferrites are lower than that of the undoped $CoFe_2O_4$, except for $Ca_{0.1}Co_{0.9}Fe_2O_4$. At 250 and 300 °C, the average k value is higher for $Ag_{0.1}Co_{0.95}Fe_2O_4$, $Ca_{0.1}Co_{0.9}Fe_2O_4$, and $La_{0.1}CoFe_{1.9}O_4$, and lower for $Na_{0.1}Co_{0.95}Fe_2O_4$ and $Cd_{0.1}Co_{0.9}Fe_2O_4$ compared with the average k value of the undoped $CoFe_2O_4$ (Table 1).

Table 1. Rate constant values (k) and activation energies (E_a) of $CoFe_2O_4$, $Ag_{0.1}Co_{0.95}Fe_2O_4$, $Na_{0.1}Co_{0.95}Fe_2O_4$, $Ca_{0.1}Co_{0.9}Fe_2O_4$, $Cd_{0.1}Co_{0.9}Fe_2O_4$ and $La_{0.1}CoFe_{1.9}O_4$ gels heated at 150, 200, 250 and 300 °C.

Gels	k (s^{-1})				E_a (kJ/mol)
	150 °C	200 °C	250 °C	300 °C	
$CoFe_2O_4$	0.223	0.285	0.330	0.365	1.217
$Ag_{0.1}Co_{0.95}Fe_2O_4$	0.196	0.271	0.336	0.391	1.709
$Na_{0.1}Co_{0.95}Fe_2O_4$	0.205	0.256	0.293	0.320	1.121
$Ca_{0.1}Co_{0.9}Fe_2O_4$	0.222	0.300	0.359	0.404	1.501
$Cd_{0.1}Co_{0.9}Fe_2O_4$	0.184	0.251	0.303	0.341	1.549
$La_{0.1}CoFe_{1.9}O_4$	0.194	0.275	0.340	0.391	1.743

The activation energy (E_a) and the pre-exponential factor (A) for the formation of doped and undoped $CoFe_2O_4$ were calculated by plotting the logarithm of the rate constant (k) versus the inverse temperature, log k vs. $1/T$ (Figure 3). The activation energy (E_a) was calculated according to the Arrhenius equation (Equation (3)) [32] as follows:

$$\log k = \log A - \frac{E_a}{2.303R} \cdot \frac{1}{T} \quad (3)$$

where A is the pre-exponential factor, E_a is the activation energy (J/mol), R is the ideal gas constant (8.314 J/mol·K) and T is the temperature (K). The E_a value of the undoped $CoFe_2O_4$ (1.217 kJ/mol) increases by doping with Ag^+, Ca^{2+}, Cd^{2+} and La^{3+}, and decreases by doping with Na^+ (Table 1).

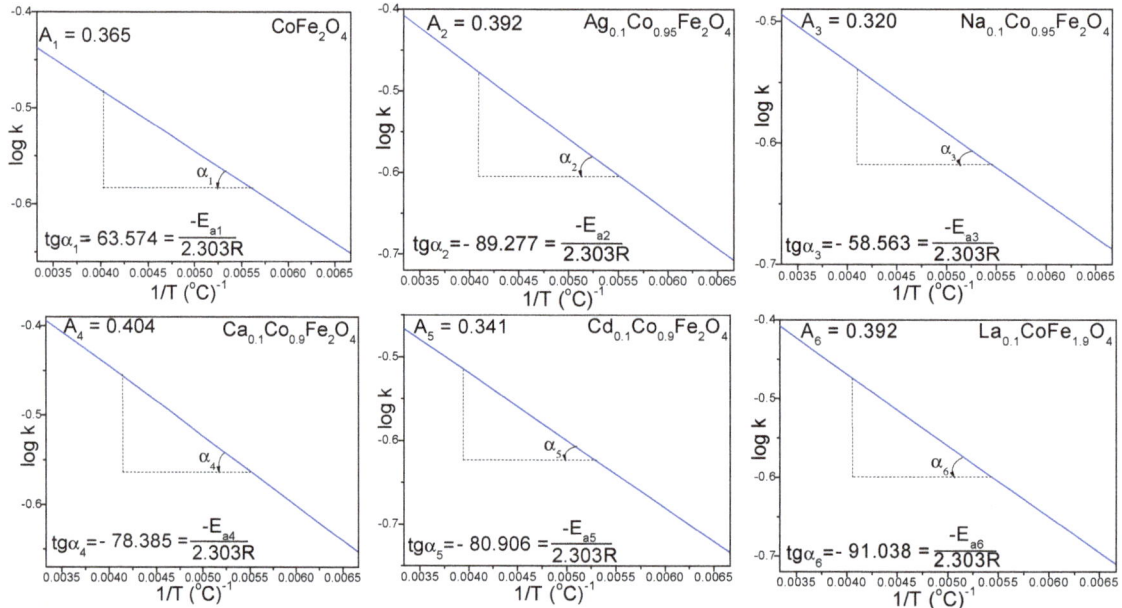

Figure 3. The log k vs. $1/T$ plots and pre-exponential factors (A) of $CoFe_2O_4$, $Ag_{0.1}Co_{0.95}Fe_2O_4$, $Na_{0.1}Co_{0.95}Fe_2O_4$, $Ca_{0.1}Co_{0.9}Fe_2O_4$, $Cd_{0.1}Co_{0.9}Fe_2O_4$ and $La_{0.1}CoFe_{1.9}O_4$ gels heated at 150, 200, 250 and 300 °C.

2.3. FT-IR Spectroscopy

The FT-IR spectra (Figure 4) of the gels heated at 40 °C show an intense band around 1380 cm^{-1}, which is characteristic to nitrate groups, two bands at 2953 and 2862 cm^{-1}, which are specific to C–H bond vibrations and a band at 3200 cm^{-1}, which is attributed to intermolecular hydrogen bonds in 1,4-BD [22,23]. These bands are not remarked in the spectra of the gels heated at 200 °C, confirming the decomposition of metal nitrates and the formation of the metal succinates, respectively.

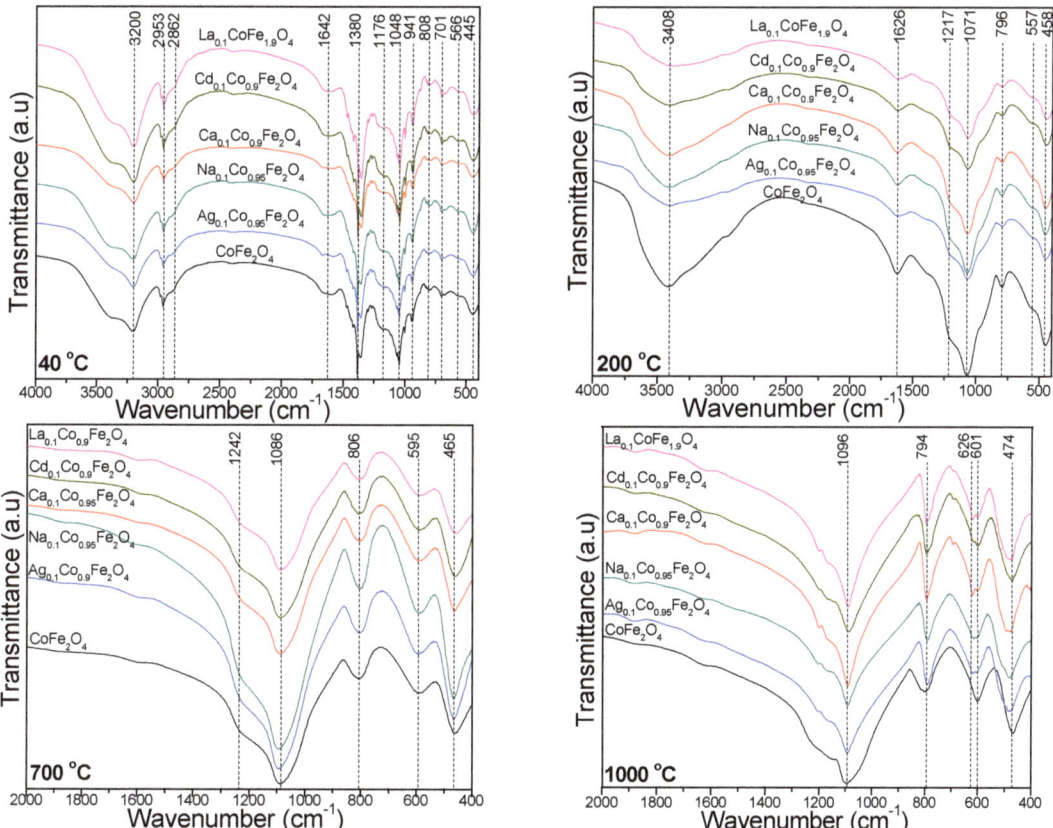

Figure 4. FT-IR spectra of gels heated at 40 and 200 °C and calcined at 700 and 1000 °C.

The vibrations of the OH groups in 1,4-BD and the adsorbed molecular water appear at 1642 cm^{-1}, while the band at 950–944 cm^{-1} is attributed to the deformation vibration of Si-OH that occurs during the hydrolysis of the –Si(OCH$_2$CH$_3$)$_4$ groups in TEOS [21–23]. The band at 808 cm^{-1} is attributed to the stretching vibration of the Si–O chains in the SiO$_4$ tetrahedron, the band at 1048 cm^{-1} is attributed to the stretching vibration of the Si–O–Si bonds, while the shoulder at 1176 cm^{-1} is attributed to the stretching vibration of the Si–O bonds in the SiO$_2$ matrix [22,23]. As shown in the XRD data, in the samples calcined at low temperatures (400, 700 °C) the SiO$_2$ is amorphous, while in the samples calcined at 1000 °C, the SiO$_2$ (quartz and cristobalite) is crystalline. In general, macro-sized silica is crystalline, and nano-sized silica is amorphous.

At 200 °C, the band around 1626 cm^{-1} is attributed to the vibration of the COO$^-$ groups, indicating the formation of a chelated complex through the coordination of succinates by metal ions. The disappearance of this band in the FT-IR spectra of the samples calcined at 700 and 1000 °C suggests that the formed cobalt ferrite nanoparticles have no residual organic compounds. The band at 3408 cm^{-1} is attributed to O–H stretching and intermolecular hydrogen bonds in metal succinates [22,23]. The samples heated at 40 and 200 °C show that the absorption band around 560 cm^{-1} is assigned to the stretching vibrations of tetrahedral M–O bonds and cyclic Si–O–Si structures, while the band at around 450 cm^{-1} is attributed to the octahedral M–O and Si–O bond vibrations. The presence of the SiO$_2$ matrix is indicated by the symmetric stretching and bending vibration of the Si–O–Si chains at around 800 cm^{-1} and the stretching vibration of the Si–O–Si bonds at around 1071 cm^{-1}, with a shoulder at around 1217 cm^{-1} [22,23]. The FT-IR spectra of the gels calcined at 700 and 1000 °C (Figure 4) show the specific bands of the SiO$_2$ matrix as follows: Si–O–Si bond stretching vibration (1086–1096 cm^{-1} and the shoulder at 1242 cm^{-1}), Si–O chain vibration in the SiO$_4$ tetrahedron (794–806 cm^{-1}), Si–O bond vibration (465–474 cm^{-1}) and Si–O–Si cyclic structure vibration (595–601 cm^{-1}) [22,23]. At 1000 °C, the supplementary band at 626 cm^{-1} is attributed to cristobalite. This band does not appear for the CoFe$_2$O$_4$ calcined at 1000 °C, supporting the obtained XRD results. The stretching vibrations of the Ag–O, Na–O, Ca–O, Co–O, Cd–O, and La–O bonds are slightly shifted toward lower wavenumbers (595–601 cm^{-1}), while the Fe–O bond vibration is shifted toward slightly higher wavenumbers (465–474 cm^{-1}) [21–24]. The shift of these bands' positions in samples with different doping cations and different calcination temperatures is a consequence of the distortion of the lattice due to different M–O distances [33]. The bands at 794–806 cm^{-1} are attributed to the O–Fe–O and Fe–OH bond vibrations, while the band at 465–474 cm^{-1} is ascribed to the Fe–O linkage [24].

2.4. X-ray Diffraction of Undoped and Doped CoFe$_2$O$_4$

As the oxidic phases at low temperatures are poorly crystalline or amorphous, the desired surface properties and crystallinity can be achieved by tailoring the calcination conditions. In addition, the reactivity of the amorphous SiO$_2$ allows its participation in various chemical transformations.

The XRD patterns of the gels calcined at 400, 700, and 1000 °C are presented in Figure 5. At 400 and 700 °C, the diffraction peaks matching with the reflection planes of (220), (311), (222), (400), (422), (511), and (440) confirm the presence of the pure, low-crystallized CoFe$_2$O$_4$ (JCPDS #00-022-1086) phase with a cubic spinel structure (space group *Fd3m*) [22–24]. At low temperatures (400 and 700 °C), doping ions did not produce any secondary impurity-associated reflections, and the spinel crystal structure of the produced gels was maintained. The absence of secondary phases points toward the successful insertion of doping ions. The broad peak at 2θ = 20–40° reveals the low crystallization of gels calcined at 400 and 700 °C. At 1000 °C, the undoped ferrite displays the single, well-crystallized CoFe$_2$O$_4$ phase, which is accompanied by cristobalite (JCPDS #89-3434) for the Na- and Ag-doped ferrites and by cristobalite and quartz (JCPDS #85-0457) for the Ca-, La- and Cd-doped ferrites. Calcination also led to a slight shift in the 2θ peak position, small changes in the peak width, and higher crystallite sizes [22,23]. The Cd$_{0.1}$Co$_{0.9}$Fe$_2$O$_4$ gel displays the lowest intensity diffraction peaks, indicating the lowest crystallization compared to the other gels. The increase in the diffraction peaks' intensity with the calcination temperature indicates the increase in the crystallinity degree and crystallite size [22,23]. The presence of La^{3+} ions did not generate any secondary impurity-related reflections, and the spinel crystal structure of the CoFe$_2$O$_4$ was maintained [6]. Oppositely, Mansour et al. reported the appearance of LaFeO$_3$ as a secondary phase due to the diffusion of some La^{3+} ions to the grain boundaries that react with Fe to form LaFeO$_3$ [8]. The XRD patterns are influenced not only by the calcination temperature and doping ions but also by the crystallite size, lattice strain, and defects [22].

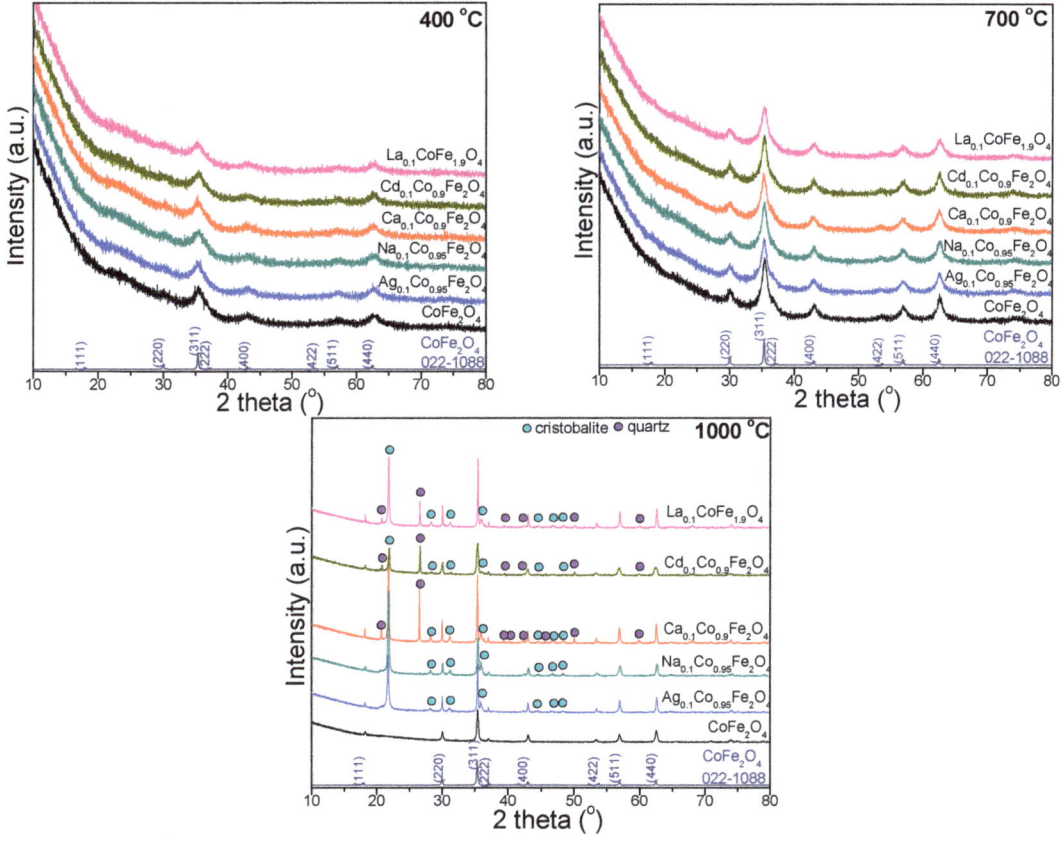

Figure 5. XRD patterns of gels calcined at 400, 700 and 1000 °C.

The structural parameters, namely, the crystallite size (D_{XRD}), degree of crystallinity (DC), lattice parameter (a), unit cell volume (V), distance between the magnetic ions and the hopping length in the A (d_A) and B (d_B) sites, physical density (d_p), X-ray density (d_{XRD}) and porosity (P), of the gels calcined at 400, 700 and 1000 °C determined by XRD are presented in Table 2. The D_{XRD} increases with the calcination temperature since at high temperatures (1000 °C), the crystallite agglomeration without subsequent recrystallization led to the formation of a single crystal rather than a polycrystal structure [22]. Mariosi et al. reported an increase in the crystallite size from 4.5 nm to 4.9 nm for $CoLa_{0.05}Fe_{1.95}O_4$ for a higher La^{3+} content ($CoLa_{0.1}Fe_{1.9}O_4$) [6].

The DC is the ratio of the area of crystalline peaks over the total area of the diffractogram. A possible reason for the slight reduction in D_{XRD} by Cd^{2+} doping could be the local temperature increase to release the latent energy on the surface. This process leads to a strike in the crystal growth and lowers the concentration of ferrites in the vicinity [3]. The D_{XRD} increases for the doped ferrites, except in the case of Cd^{2+} doping (Table 2). The observed expansion of the unit cell and the high structural distortion of the doped $CoFe_2O_4$ compared to the undoped $CoFe_2O_4$ are attributed to the difference in the ionic radii of the host and dopant ions, as well as to the change in the cation distribution that occurs due to the Ag^+, Na^+, Ca^{2+}, Cd^{2+}, and La^{3+} doping in the spinel structure [1,4,6,27]. A possible explanation for the largest unit cell value of $La_{0.1}CoFe_{1.9}O_4$ is the much larger ionic radius of La^{3+} (1.216 Å) than of Fe^{3+} (0.65 Å), with the unit cell expansion via doping with La^{3+} taking place to compensate for the crystal deformation; accordingly, Shang et al. stated that the replacement of Fe^{3+} ions by La^{3+} ions result in a higher potential barrier for the formation of a spinel ferrite crystal structure [9]. The lattice parameter of the undoped and doped $CoFe_2O_4$ gels increases with the calcination temperature, which is also ascribed to the expansion of the unit cell [27]. A possible explanation for the difference between the theoretical and experimental values could be the assumption that the ions are rigid hard spheres [22]. The obtained results and the absence of supplementary phases in the XRD patterns indicate that the doping ions are incorporated into the $CoFe_2O_4$ structure [5,6]. The increase in the molecular weight is more significant than the increase in the V (Table 2); however, the molecular weight is more influenced by the increase in the V [22]. The decrease in the unit cell volume is expected with the introduction of smaller-sized monovalent (Ag^+, Na^+) ions in the crystal lattice. The d_A and d_B of the gels calcined at 700 and 1000 °C are higher for the doped $CoFe_2O_4$ than the undoped $CoFe_2O_4$ and display a decreasing trend for the monovalent dopant (Ag^+, Na^+) ion and an increasing trend for the trivalent dopant (La^{3+}) ion (Table 2). The lower value of the d_p (Table 2) of the undoped $CoFe_2O_4$ compared to the doped $CoFe_2O_4$ could be attributed to the pore formation through the synthesis processes [22]. The variation of the d_p caused by small fluctuations in the lattice constant is probably due to the changes in the cation distribution among the A and B sites. The P (Table 2) of the doped $CoFe_2O_4$ is lower than that of the undoped $CoFe_2O_4$. Additionally, the rapid densification during the calcination and the growth of irregular shape grains decrease the porosity at higher calcination temperatures. The decrease in the P with the increase in the d_p may result from the different grain sizes [22]. In conclusion, the increase in the D_{XRD}, DC, a, V, d_A, d_B, and d_p, and the decrease in the d_{XRD} and P at higher calcination temperatures are observed.

2.5. Elemental Composition of Undoped and Doped $CoFe_2O_4$

To investigate the elemental composition and verify the stoichiometric amount of each element in the undoped and doped $CoFe_2O_4$, the M/Co/Fe molar ratio was determined using inductively coupled plasma optical emission spectrometry (ICP-OES) after microwave digestion (Table 2). The best fit between the experimental and theoretical data is observed for the gels calcined at 1000 °C.

2.6. Morphology and Surface Parameters of Undoped and Doped $CoFe_2O_4$

The thermal treatment concomitantly enables the crystalline phase formation and grain growth with a relative coalescence between the particles. The coalescence can be attributed to the physical attraction forces between the small particles or the bonding bridges between the particles, especially at high calcination temperatures [34,35]. As the microscopic examination of gels failed due to the agglomeration of small particles into clusters, the powders were dispersed into deionized water under intense stirring to break up the powder clusters and release the free particles into dispersion. The water dispersion also prevents the particles' re-agglomeration and allows their transfer onto a solid substrate as a thin film through adsorption [36,37]. This method is usually used to prepare thin films of noble metal nanoparticles directly from the mother solution [38,39], but it was also successfully used for the Ni- and Mn-doped ferrites [29,30]. The obtained thin films were subjected to atomic force microscopy (AFM) (Figure 6). Mansour et al. suggested a coalescence phenomenon, resulting in large particle size of the obtained La-doped $CoFe_2O_4$ nanoparticles [8].

The undoped $CoFe_2O_4$, after calcination at 400 °C, displays small spherical particles of around 25 nm (Figure 6a), which increase to 30 nm with the calcination temperature at 700 °C (Figure 6b), and to 40 nm at 1000 °C (Figure 6c). The polycrystalline particles formed at 400 °C consist mainly of low-crystalized $CoFe_2O_4$ mixed with amorphous material. At 700 °C, the growth of $CoFe_2O_4$ crystallites and the presence of amorphous matter are remarked. For the gels calcined at 1000 °C, the crystallite size estimated via XRD (37.2 nm) is comparable with the particle size observed using AFM (around 40 nm). The undoped $CoFe_2O_4$ calcined at 1000 °C is mono-crystalline, as indicated by the spherical particle shape with slightly square corners (Figure 6c). These results are in good agreement with the data in the literature [40,41].

Table 2. Structural parameters determined via XRD and M/Co/Fe molar ratio of gels calcined at 400, 700 and 1000 °C.

Parameter	Temp (°C)	$CoFe_2O_4$	$Ag_{0.1}Co_{0.95}Fe_2O_4$	$Na_{0.1}Co_{0.95}Fe_2O_4$	$Ca_{0.1}Co_{0.9}Fe_2O_4$	$Cd_{0.1}Co_{0.9}Fe_2O_4$	$La_{0.1}CoFe_{1.9}O_4$
D_{XRD} (nm)	400	13.2	14.0	13.6	14.6	12.3	15.5
	700	23.4	24.7	23.7	25.2	22.5	26.1
	1000	37.2	65.1	47.6	73.7	35.0	81.8
DC (%)	400	58.8	62.7	64.2	63.7	64.8	60.2
	700	71.7	65.2	69.6	67.8	72.9	75.4
	1000	85.7	86.9	87.8	92.5	88.9	90.6
a (Å)	400	8.391	8.380	8.377	8.374	8.400	8.408
	700	8.395	8.400	8.390	8.388	8.412	8.411
	1000	8.403	8.410	8.400	8.397	8.422	8.423
V (Å^3)	400	590.8	588.5	587.8	587.2	592.7	594.4
	700	591.6	592.7	590.6	590.2	595.2	595.0
	1000	593.3	594.8	592.7	592.1	597.4	597.6
d_A (Å)	400	3.633	3.629	3.627	3.626	3.637	3.641
	700	3.635	3.637	3.633	3.632	3.643	3.642
	1000	3.639	3.642	3.637	3.636	3.647	3.647
d_B (Å)	400	2.967	2.963	2.962	2.961	2.970	2.973
	700	2.968	2.970	2.966	2.966	2.974	2.974
	1000	2.971	2.973	2.970	2.969	2.978	2.978
d_p (g/cm^3)	400	4.433	4.805	4.472	4.445	4.755	4.685
	700	4.453	4.833	4.494	4.462	4.771	4.706
	1000	4.471	4.861	4.517	4.491	4.794	4.721
d_{XRD} (g/cm^3)	400	5.276	5.473	5.288	5.265	5.445	5.429
	700	5.269	5.435	5.263	5.239	5.422	5.424
	1000	5.253	5.415	5.244	5.222	5.402	5.400
P (%)	400	16.0	12.2	15.4	15.6	12.7	13.7
	700	15.5	11.1	14.6	14.8	12.0	13.2
	1000	14.9	10.2	13.9	14.0	11.3	12.6
M/Co/Fe molar ratio	400	0/0.95/1.97	0.08/0.91/1.95	0.08/0.91/1.98	0.08/0.86/1.96	0.08/0.84/1.96	0.07/0.94/1.85
	700	0/0.96/1.97	0.08/0.92/1.97	0.08/0.92/1.96	0.08/0.86/1.97	0.08/0.86/1.96	0.07/0.97/1.86
	1000	0/0.99/2.01	0.09/0.94/1.98	0.11/0.96/1.98	0.11/0.89/2.01	0.09/0.89/2.02	0.09/0.98/1.89

The Ag-doped cobalt ferrite calcined at 400 °C is low crystalline, with crystallites of 14 nm mixed with amorphous matter into small spherical particles of around 27 nm (Figure 6d). Increasing the calcination temperature to 700 °C leads to better-developed crystallites of 24.7 nm and increases the particle size to about 31 nm. The development of the crystalline phase and the reduction in the amorphous component determines a significant alteration of the particle shape, which becomes spheroidal (Figure 6e). The particle shape evolving tendency continues by calcination at 1000 °C, with an increase in size to 70 nm and a crystalline core of 65.1 nm (Figure 6f). The presence of secondary phases prevents the development of cubic shape features, which remains spherical, as reported by Prabagar et al. [42]. Mahajan et al. also reported a higher Ag-doped $CoFe_2O_4$ crystallite size than the undoped $CoFe_2O_4$ [43].

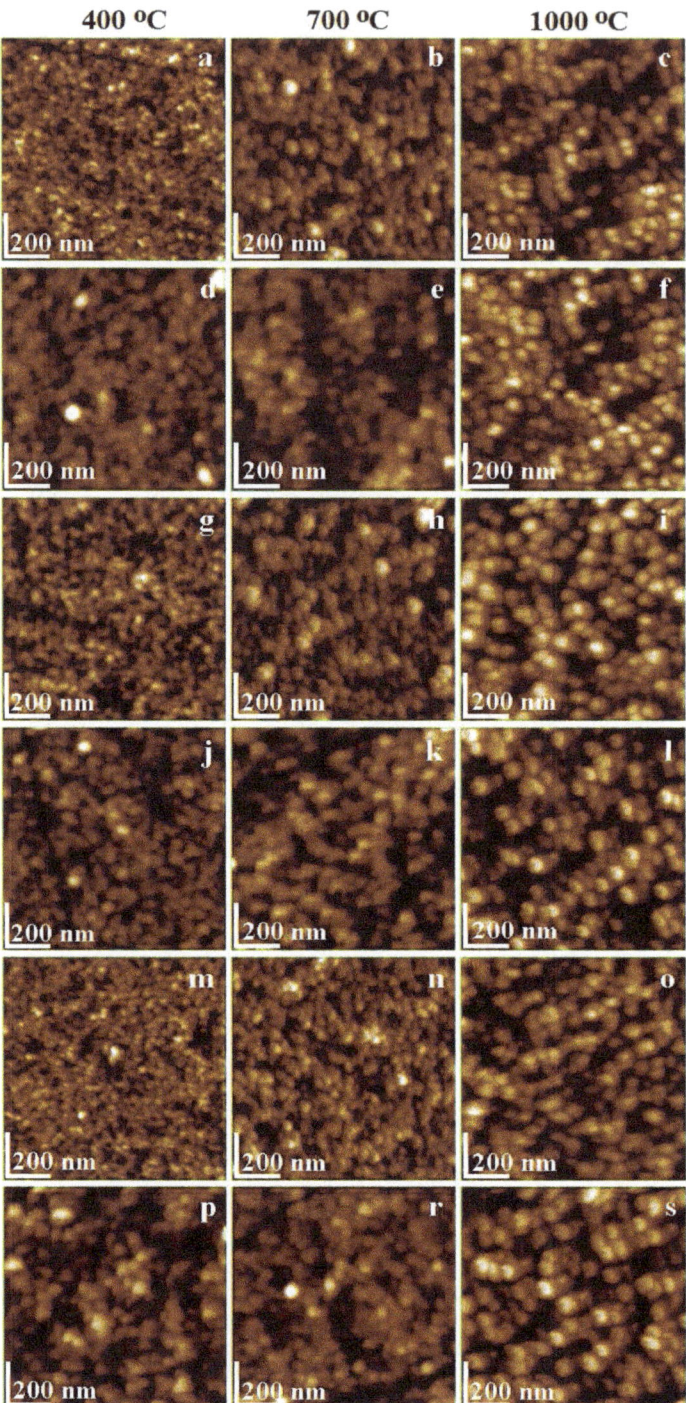

Figure 6. AFM topographic images of CoFe$_2$O$_4$ (**a–c**), Ag$_{0.1}$Co$_{0.95}$Fe$_2$O$_4$ (**d–f**), Na$_{0.1}$Co$_{0.95}$Fe$_2$O$_4$ (**g–i**), Ca$_{0.1}$Co$_{0.9}$Fe$_2$O$_4$ (**j–l**), Cd$_{0.1}$Co$_{0.9}$Fe$_2$O$_4$ (**m–o**) and La$_{0.1}$CoFe$_{1.9}$O$_4$ (**p–s**) gels calcined at 400, 700 and 1000 °C.

The Na-doped cobalt ferrite shows no significant modification of the crystallite size for the gels calcined at 400 °C and 700 °C (Table 2). Like the undoped $CoFe_2O_4$, the amorphous matter between the crystallites leads to slightly larger spherical particles of 24 nm at 400 °C, and 35 nm at 800 °C (Figure 6g,h). Surprisingly, after calcination at 1000 °C, well-structured particles with a 52 nm diameter and a ferrite core of 47.6 nm covered with some traces of cristobalite are remarked (Figure 6i). Due to the presence of the crystalline core, the particle shape becomes spheroidal. The observed size and shape are in good agreement with the data in the literature [44].

The Ca-doped cobalt ferrite, calcined at low temperatures, exhibits spherical particles of around 28 nm (Figure 6j), formed by a 14.6 nm ferrite crystallite core coated with amorphous material. Kumar and Kar also reported crystallites of 10 nm for this composition by calcination at 550 °C for 2 h [45]. Higher calcination temperature at 700 °C facilitates the crystal growth, leading to crystallites of 25.2 nm mixed with some amorphous matter, which generates particles of about 33 nm. Predominately spherical particles are accompanied by several right corners formed on the most representative particles (Figure 6k). Calcination at 1000 °C generates well-formed $Ca_{0.1}Co_{0.9}Fe_2O_4$ crystallites of around 73.7 nm. Due to the traces of cristobalite and quartz crystalline phases, the crystallites become the core of the particles of about 75 nm (Figure 6l). The particle has cubic shapes with rounded edges, which is in good agreement with the data in the literature [46]. Only a few blunted cristobalite and quartz particles are observed around ferrites.

The Cd-doped cobalt ferrite gels calcined at 400 and 700 °C present small spherical particles of around 20 and 28 nm, respectively, containing small ferritic crystallites and amorphous material (Figure 6m,n). The heterogeneous nanostructure of $Cd_{0.1}CoFe_{1.9}O_4$ was also evidenced by Shakil et al. [47]. The calcination at 1000 °C leads to particles of around 40 nm (Figure 6o) having a ferrite core of 35 nm, which is in good agreement with the less developed XRD peaks. Particle sizes of 30–50 nm were previously reported for a similar composition [3].

The La-doped cobalt ferrite has a significant influence on the particle size and shape. The calcination at 400 °C results in spherical particles of around 30 nm containing crystallites of 15.5 nm mixed with amorphous material (Figure 6p). A possible explanation could be the higher value of the La atomic radius (2.50 Å) compared to that of Fe (1.26 Å) [48]. This effect is enhanced by calcination at 700 °C, resulting in particles of 38 nm with a crystalline core of 26.1 nm (Figure 6r), covered with amorphous material.

Figure 6n shows several bigger particles that might indicate that the crystallization process is in progress, but most particles do not have enough time to reach a larger size. By calcination at 1000 °C, particles of around 90 nm, with a ferrite crystalline core of 81.8 nm covered by traces of cristobalite and quartz (Figure 6s), are formed. The spherical particle features a specific aspect derived from a cubic crystallite core, which is in good agreement with the data in the literature [49,50]. Some small particles of about 40–50 nm belong to the secondary phases. The particle adsorption onto the solid substrate develops thin films with specific topographic characteristics, which depend on the morphological aspects correlated with their density on the surface [51,52].

The tridimensional profiles of the undoped $CoFe_2O_4$ and the Ca-, Na- and Cd-doped $CoFe_2O_4$ reveal that the particles resulting after calcination at 400 °C build uniform films of well-individualized nanoparticles (Figure 7a,d,p). Interestingly, the doping with Ag and La leads to a slightly irregular film due to the occurrence of local heights (Figure 7g,m), probably due to the irregular adsorption generated by local influences related to the bigger crystallite sizes. The particles obtained by calcination at 700 °C generally display a complex topography of the thin film with randomly spotted bigger particles, which may contain a more developed crystalline core for all the doped $CoFe_2O_4$ gels, except for Cd (Figure 7e,h,k,n). The undoped ferrite and $Cd_{0.1}Co_{0.9}Fe_2O_4$ calcined at 700 °C form uniform and smooth films (Figure 7b,r).

The surface roughness (Rg) values are presented in Table 3. The particles formed by calcination at 1000 °C are well individualized and present crystalline topographic aspects and form relatively uniform thin films with no signs of coalescence tendency (Figure 7c,f,i,l,o,s). The observed topographic aspects may be helpful in the further processing of the obtained particles as the main ingredient for dedicated thin film preparation.

Table 3. Morphological parameters of gels calcined at 400, 700 and 1000 °C determined using AFM.

Parameter	Temp. (°C)	$CoFe_2O_4$	$Ag_{0.1}Co_{0.95}Fe_2O_4$	$Na_{0.1}Co_{0.95}Fe_2O_4$	$Ca_{0.1}Co_{0.9}Fe_2O_4$	$Cd_{0.1}Co_{0.9}Fe_2O_4$	$La_{0.1}CoFe_{1.9}O_4$
D_{AFM} (nm)	400	25	27	24	28	20	30
	700	30	31	35	33	28	38
	1000	40	70	52	75	39	90
D_{TEM} (nm)	1000	38	67	50	76	37	84
Height (nm)	400	12	13	10	11	15	9
	700	19	7	25	8	15	12
	1000	41	42	45	37	63	36
Rg (nm)	400	0.81	0.95	0.97	0.77	1.10	1.04
	700	1.95	0.83	2.42	0.75	1.50	1.10
	1000	5.80	6.16	6.21	5.35	7.82	5.58
Powder surface area (nm^2)	400	1017	1013	1016	1015	1018	1012
	700	1028	1021	1032	1023	1024	1026
	1000	1083	1095	1094	1062	1160	1069

The AFM investigation allows for the effective area of topographic features at a precise scanned area to be measured. Hence, the thin films obtained through adsorption from aqueous dispersion are uniform and compact (Figure 6), allowing the measurement of powder surface area. The particle number and diameter influence the variation of the obtained values (Table 3) since a large number of bigger particles leads to a large powder area, while a small number of particles with a small diameter leads to a small powder area [22,23]. Therefore, low calcination temperatures generate thin films with a small surface area by spreading the secondary phases among ferrite particles. The well-crystallized ferrites obtained after calcination at 1000 °C form thin films with a significantly larger powder surface area.

Since the AFM topographic images reveal the exterior aspect of the particles, in order to obtain information on the internal structure of the particles, the transmission electron microscopy (TEM) images were recorded on the gels calcined at 1000 °C (Figure 8), considering that the ferrite crystallites are better developed at this temperature. The crystallites appear in dark grey shades surrounded by a lighter gray hollow, indicating a dense ferrite core covered by a thin layer of SiO_2. Some particles are associated in clusters of about 90–100 nm, but these clusters are not observed in the AFM images due to their sedimentation in the aqueous dispersion before transferring the nanoparticles onto the glass slide.

Figure 7. Tridimensional profiles of $CoFe_2O_4$ (**a–c**), $Ag_{0.1}Co_{0.95}Fe_2O_4$ (**d–f**), $Na_{0.1}Co_{0.95}Fe_2O_4$ (**g–i**), $Ca_{0.1}Co_{0.9}Fe_2O_4$ (**j–l**), $Cd_{0.1}Co_{0.9}Fe_2O_4$ (**m–o**) and $La_{0.1}CoFe_{1.9}O_4$ (**p–s**) gels calcined at 400, 700 and 1000 °C.

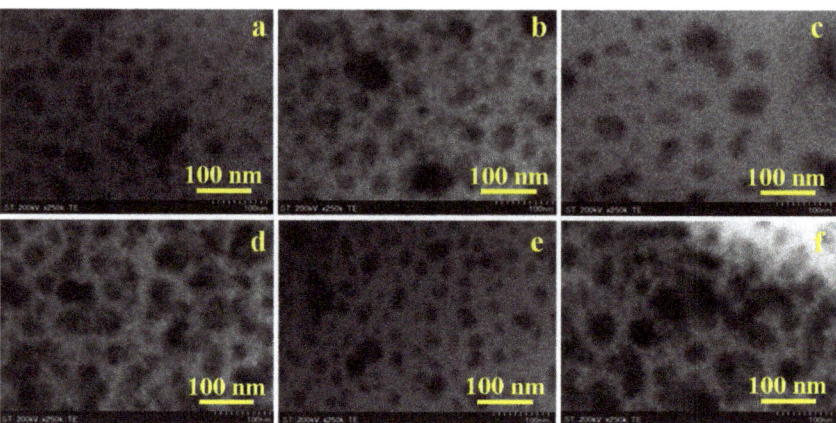

Figure 8. TEM images of (**a**) $CoFe_2O_4$, (**b**) $Ag_{0.1}Co_{0.95}Fe_2O_4$, (**c**) $Na_{0.1}Co_{0.95}Fe_2O_4$, (**d**) $Ca_{0.1}Co_{0.9}Fe_2O_4$, (**e**) $Cd_{0.1}Co_{0.9}Fe_2O_4$ and (**f**) $La_{0.1}CoFe_{1.9}O_4$ gels calcined at 1000 °C.

The undoped $CoFe_2O_4$ (Figure 8a) displays mainly fine particles of about 38 nm in diameter, which is in agreement with the AFM, but slightly bigger than the average value estimated by the XRD, probably due to the presence of the cristobalite and quartz exterior layer. A particle cluster of about 95 nm is observed on the central side of Figure 8a. Small and homogenously distributed particles result when the nucleation rate exceeds the growth rate, but the small particles tend to agglomerate into bigger structures. A possible explanation for the agglomeration tendency of small particles could be the interaction between the magnetic ions, van der Waals forces at the particle surface, and interfacial surface tensions [21,22,29,53,54]. The volume expansion and the internal energy produced during calcination may also lead to particle growth. The SiO_2 matrix reduces the number of particles that interact with each other and, thus, reduces particle agglomeration [10,11,18,42,53,54].

$Ag_{0.1}Co_{0.95}Fe_2O_4$ presents spherical particles of about 67 nm (Figure 8b), which is very close to the value observed by the AFM. The dark core of the particles observed in the TEM images is very close to the average crystallite size estimated via XRD, while the light hollow surrounding the dark core suggests the ferrite coating by a thin layer of cristobalite and quartz, a fact also sustained by the spherical particle shape. The TEM images of $Na_{0.1}Co_{0.95}Fe_2O_4$ (Figure 8c) show a less dense distribution of spherical particles with a diameter of about 50 nm, which is in good accordance with the AFM. The intense gray shade indicates the presence of a 47.6 nm ferrite core inside the particles, which is in accordance with the crystallite diameter determined by the XRD data. Oppositely, the TEM image shows a compact structure of well-developed $Ca_{0.1}Co_{0.9}Fe_2O_4$ particles with a diameter of about 76 nm (Figure 8d). The particle size is slightly bigger than that observed by the AFM. The dark core corresponds to the ferrite crystallite evidenced by XRD, while the lighter halo on the exterior is attributed to the cristobalite and quartz layer. Figure 8e reveals a compact and uniform package of fine $Cd_{0.1}Co_{0.9}Fe_2O_4$ spherical particles of about 37 nm, which is in good agreement with the AFM. The core is darker due to the ferrite crystallite presence, and the lighter exterior shade corresponds to the crystalline SiO_2 layer. $La_{0.1}CoFe_{1.9}O_4$ presents big spherical particles of about 84 nm (Figure 8f), which is in good agreement with the AFM. The dark core corresponds to the ferrite, evidenced by XRD, and the outer halo to the cristobalite and quartz layer. The presence of some crystallite clusters indicates a local powder agglomeration.

2.7. Magnetic Properties of Undoped and Doped $CoFe_2O_4$

The magnetic hysteresis loops, $M(\mu_0H)$, and the magnetization first derivatives ($dM/d(\mu_0H)$) of the gels calcined at 700 °C (Figure 9) and 1000 °C (Figure 10) indicate a typical ferromagnetic behavior. The derivative of the hysteresis loops (total susceptibility) is the local slope of the M–H curve. For the gels calcined at 700 °C, a single maximum in the $dM/d(\mu_0H)$ vs. the μ_0H curve, close to the coercive field, consistent with a single magnetic phase, is observed. These behaviors suggest crystalline samples with a single magnetic phase [21–23]. The magnetic hysteresis loops indicate moderate coercivity due to the coalescence of the particles accompanied by their magnetic coupling and improved magnetization. Although the magnetization first derivative $dM/d(\mu_0H)$ of the undoped $CoFe_2O_4$ calcined at 1000 °C shows two maxima (a more intense and better differentiated maximum next to a less intense one), one on each side of the coercivity, these two magnetic phases are magnetically coupled inside of the particle along their magnetic moments [21–23]. The doping effect of the monovalent (Ag^+ and Na^+) ions supports the formation of the two magnetic phases (an intense peak and one as a shoulder merged with the other for $Ag_{0.1}Co_{0.95}Fe_2O_4$, and a broader maximum peak suggesting the merging of the two maxima, characteristic of the two magnetic phases for $Na_{0.1}Co_{0.95}Fe_2O_4$). Oppositely, the doping with the divalent (Ca^{2+} and Cd^{2+}) and trivalent (La^{3+}) ions improves the magnetic properties, leading to the formation of a single magnetic phase characterized by a single maximum, which is very intense and sharp on the $dM/d(\mu_0H)$ vs. the μ_0H curve.

Due to the change in the magnetocrystalline anisotropy or the particle sizes by doping $CoFe_2O_4$ with non-magnetic ions, the values of M_S, remnant magnetization (M_r), H_C, magnetic moment per formula unit (n_B), and K are higher than those of the undoped $CoFe_2O_4$, with few exceptions [1,4,26]. For the gels calcined at 1000 °C, the peak heights and their horizontal shifts are associated with the strength of the magnetic phases, with the broader peaks indicating a large particle size distribution accompanied by a large H_C [21–23]. A significant increase in H_C is observed on the hysteresis loops of the gels calcined at 1000 °C (they are much broader) compared to those at 700 °C. For the gels calcined at 1000 °C, the doping effect of the monovalent metals (Ag^+, Na^+) increases the already large H_C of the undoped $CoFe_2O_4$, while the doping effect of the divalent (Ca^{2+}, Cd^{2+}) and trivalent (La^{3+}) ions leads to a decrease in H_C, which is observable on the hysteresis loops of the gels calcined at 1000 °C.

The magnetic parameters M_s, M_R, H_c, n_B and K values determined using the hysteresis loops and $M(H)$ curves are presented in Table 4. Generally, the M_s for the spinel ferrites is dictated by the superexchange interactions between the A and B site cations. The M_s decreases with the increase in the crystallite size due to the larger number of surface defects [55].

Table 4. Magnetic parameters of gels calcined at 700 and 1000 °C.

Parameter	Temp (°C)	$CoFe_2O_4$	$Ag_{0.1}Co_{0.95}Fe_2O_4$	$Na_{0.1}Co_{0.95}Fe_2O_4$	$Ca_{0.1}Co_{0.9}Fe_2O_4$	$Cd_{0.1}Co_{0.9}Fe_2O_4$	$La_{0.1}CoFe_{1.9}O_4$
M_S (emu/g)	700	30.0	25.4	27.7	26.7	26.2	26.5
	1000	31.5	29.0	31.2	39.4	36.3	36.6
M_R (emu/g)	700	3.5	7.7	7.9	2.7	3.0	1.8
	1000	13.4	13.8	11.3	14.5	17.0	15.1
H_c (Oe)	700	600	530	360	410	440	385
	1000	1750	1850	1760	840	1070	1300
n_B	700	0.935	0.815	0.857	0.821	0.831	0.821
	1000	0.977	0.814	0.965	0.917	1.151	1.175
$K \cdot 10^3$ (erg/cm^3)	700	1.13	0.84	0.63	0.68	0.72	0.64
	1000	3.46	2.90	3.45	2.08	2.44	2.99

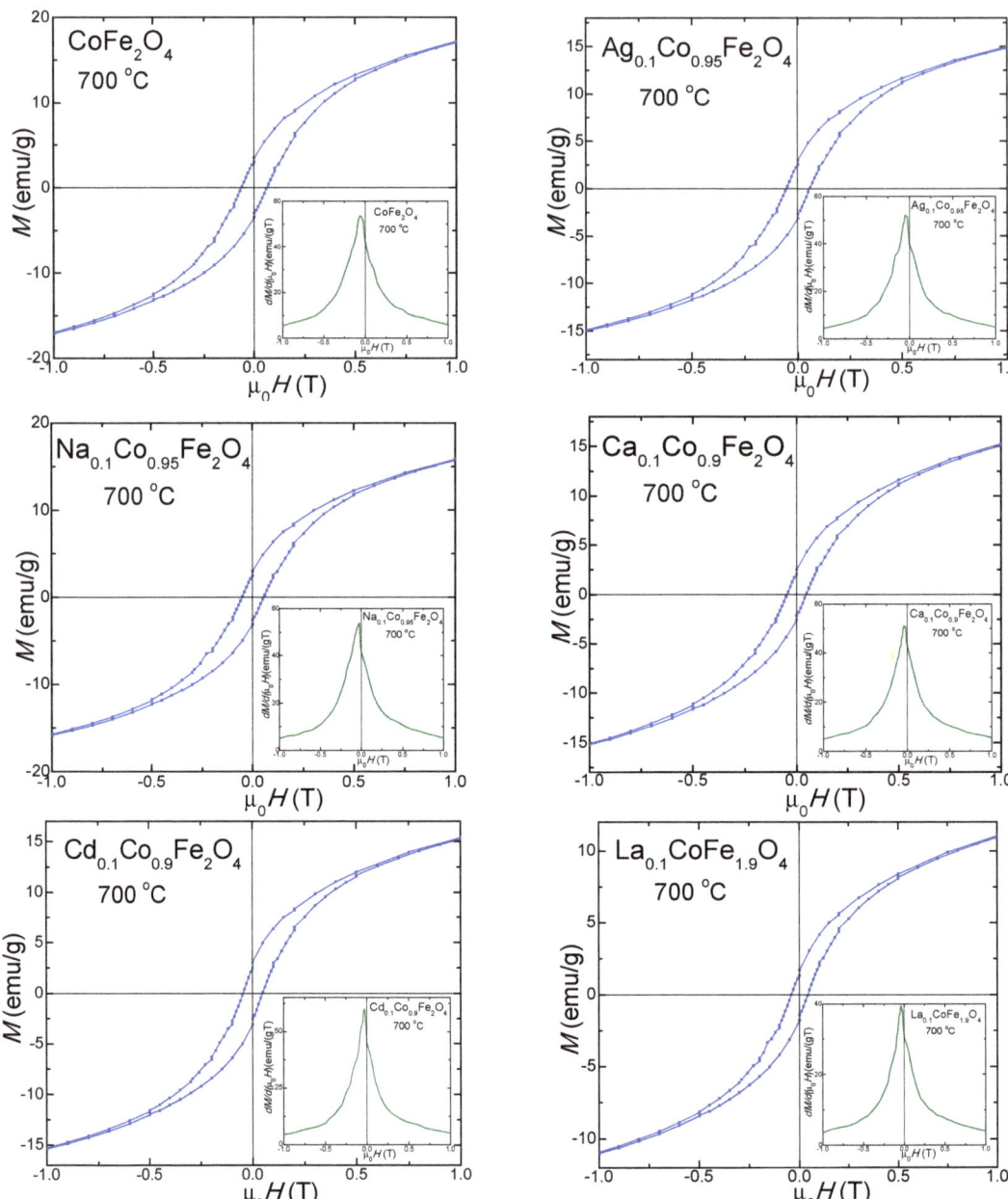

Figure 9. Magnetic hysteresis loops of gels calcined at 700 °C.

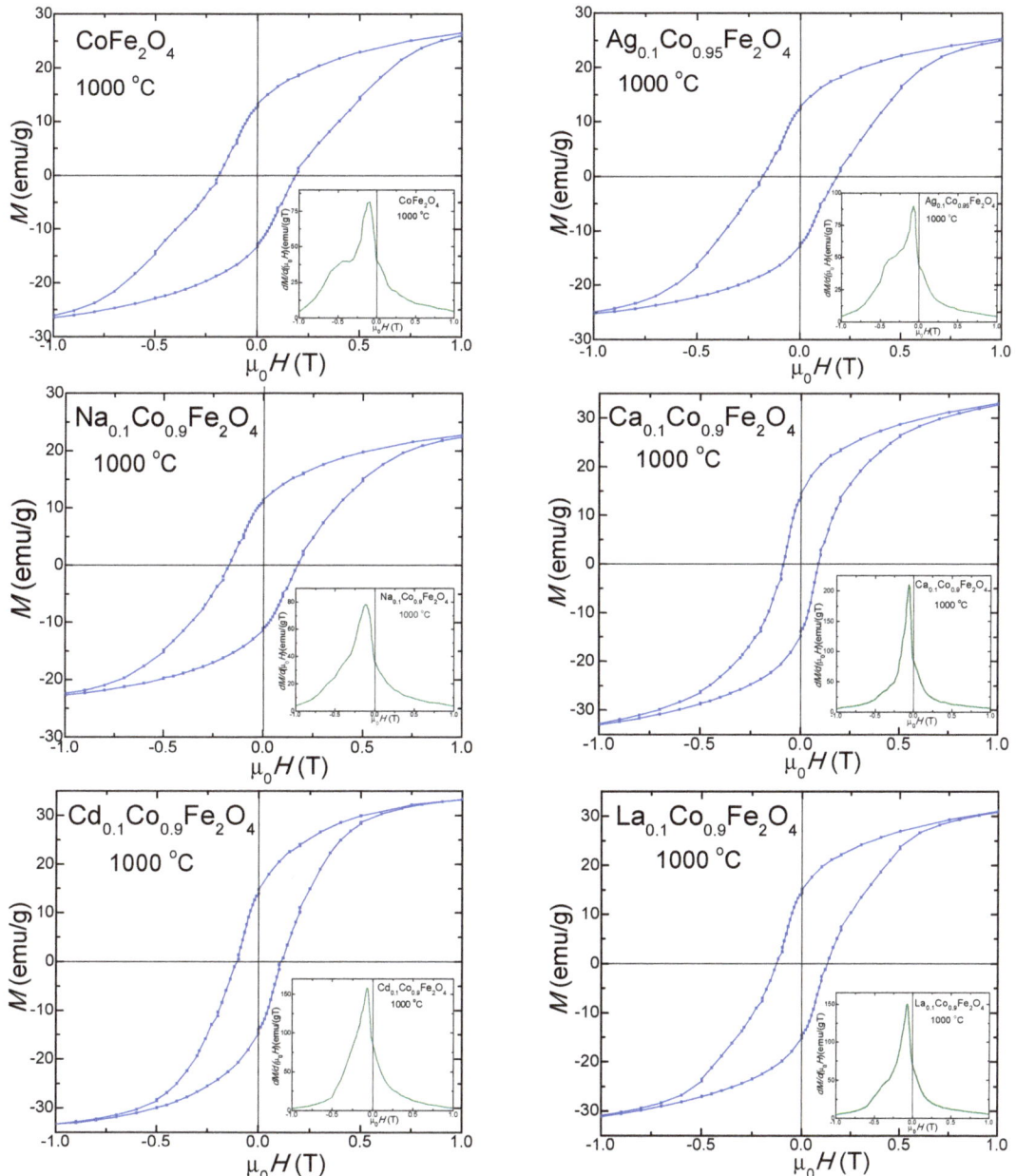

Figure 10. Magnetic hysteresis loops of gels calcined at 1000 °C.

The highest M_s value corresponds to $CoFe_2O_4$ calcined at 700 °C (30 emu/g), with the doping ions leading to lower M_s values. For the sample calcined at 1000 °C, a slight increase in the undoped $CoFe_2O_4$ (31.5 emu/g) is remarked. The doping with monovalent (Ag^+, Na^+) ions results in lower M_s values, while the doping with the divalent (Ca^{2+}, Cd^{2+}) and trivalent (La^{3+}) ions, leads to higher M_s values, with the highest M_s values corresponding to $Cd_{0.1}Co_{0.9}Fe_2O_4$ (39.4 emu/g) for this bunch of gels. The surface effects at the nanoparticle surface include forming of a dead layer, which contains broken chemical bonds; deviations from the bulk cation distribution; randomly oriented magnetic moments; lattice defects and non-saturation effects, resulting in depreciated magnetic properties [21].

$Cd_{0.1}Co_{0.9}Fe_2O_4$ is a good candidate for various technological applications such as communication, data storage, and high-frequency inductors [1,4]. For both calcination temperatures, $Ag_{0.1}Co_{0.95}Fe_2O_4$ exhibits the lowest M_S value. Previous studies also reported that the doping of diamagnetic Ag^+ into the $CoFe_2O_4$ spinel structure substantially decreases the M_S of $CoFe_2O_4$; a possible explanation for this is the high number of uncoordinated magnetic spins that are not able to align in the direction of the external magnetic field. Generally, the Ag doping enhances the nanoparticles' antibacterial activities, suggesting that $Ag_{0.1}Co_{0.95}Fe_2O_4$ may be a potential candidate for antibacterial applications [4]. Moreover, considering the excellent electron conductivity of Ag, it is expected that Ag doping increases the catalytic activity of $CoFe_2O_4$ [25].

As La^{3+} is a non-magnetic ion, it does not participate in the exchange interactions with its nearest neighbor ion; thus, the superexchange interactions between the A and B sites' cations are depreciated [55]. Above the single-domain critical size, the competition between the magneto-static energy and the domain-wall energy favors forming domain walls and splitting the single-domain particle into multi-domain particles [22]. Mariosi et al. reported that the M_S of the undoped $CoFe_2O_4$ (44.6 emu/g) decreased to 29.0 emu/g for the first increase in the La^{3+} concentration (sample $CoLa_{0.025}Fe_{1.975}O_4$); the possible mechanisms for the magnetic behavior of these nanoparticles are still widely discussed [6]. Moreover, a disorder in the crystal's surface results in a lack of collinearity of magnetic moments; this effect is generally attributed to a single magnetic domain configuration [6]. When non-magnetic La^{3+} ions substitute Fe^{3+} ions, the content of Fe^{3+} ions at ferrite lattice sites is reduced, resulting in a decrease in the total magnetic moment and a weakening of the Fe^{3+}–Fe^{3+} interactions and, consequently, a lower M_S value [9].

The remanent magnetization (M_R) for the undoped $CoFe_2O_4$ calcined at 700 °C is 3.5 emu/g. The doping with monovalent cations (Ag^+, Na^+) increases the M_R to 7.7–7.9 emu/g, while doping with the divalent (Ca^{2+}, Cd^{2+}) and trivalent (La^{3+}) cations decreases the M_R to 1.8–3.0 emu/g. In the samples calcined at 1000 °C, the M_R of the doped ferrites increases, compared to the M_R of $CoFe_2O_4$ (13.4 emu/g), except for the M_R of $Na_{0.1}Co_{0.95}Fe_2O_4$, which slightly decreases (11.3 emu/g).

The slight decrease in the Hc values in the doped samples could result from the magnetocrystalline anisotropy, microstrain, size distribution and the decrease in the magnetic domain size [27]. For the gels calcined at 700 °C, the H_C of the undoped $CoFe_2O_4$ is 600 Oe, while that of the doped ferrites is lower, which is most probably due to changes in the crystallite size, anisotropy and formation of agglomerates that increase the average particle size above the critical single-domain, which results in a multi-domain structure and the reduction in pinning effects on the domain wall mobility at the grain boundary [22,55]. Oppositely, for the gels calcined at 1000 °C, the H_c of the doped $CoFe_2O_4$ with monovalent cations (Ag^+ si Na^+) are comparable, while for those doped with the di- and trivalent cations, the H_c is lower than that of the undoped $CoFe_2O_4$ (1750 Oe). The lower H_c values of the obtained gels indicate a spin distortion on the surface, owing to the magnetocrystalline anisotropy [24]. The presence of SiO_2 generates stress on the surface of the ferrite particles, which hinders the rotation of the dead layer's magnetic moments and contributes to the reduction in the H_c [21,22]. The increase in the surface potential barrier caused by crystalline lattice defects, such as the deviation of atoms from the normal positions in the surface layers, also determines the increase in the H_C [29].

The n_B of the gels calcined at 700 °C decreases from 0.935 ($CoFe_2O_4$) to 0.857–0.815 by doping. Additionally, in the samples calcined at 1000 °C, except for doping with Cd^{2+} and La^{3+}, the n_B of the doped samples is lower than that of the undoped $CoFe_2O_4$ (0.977).

To calculate the magnetic anisotropy constant (K), we assumed that the spinel ferrite particles have a spherical shape. The value of K depends on the crystalline symmetry of the lattice, the crystalline anisotropy, and the particle size and shape [21]. The highest K

was obtained for the undoped $CoFe_2O_4$ ($1.13 \cdot 10^{-3}$ erg/cm^3 in gels calcined at 700 °C, and $3.46 \cdot 10^{-3}$ erg/cm^3 in gels calcined at 1000 °C). The value of K increases with the increasing calcination temperature and decreases by doping. A possible explanation for this decrease could be the pinning of some surface spins in the magnetically disordered surface layer, which needs a higher magnetic field for magnetic saturation [22]. In addition, the magnetic disorder may originate in randomly oriented grains of different sizes and disordered vacancies [22]. The individual K of particles acts as an energy barrier and delays the switch of the magnetization direction to the easy axis [22]. Crystalline anisotropy is strongly affected by the volume strain in the crystal, which is determined by the substitution of Fe^{3+} ions by the different sized (La^{3+}) ions [55].

To summarize, the embedding of the undoped and doped $CoFe_2O_4$ in the non-magnetic SiO_2 matrix promotes both the formation of single-phase spinel and minimization of the spin disorder and surface roughness, thus enhancing the magnetic properties of the ferrites. Combining the best magnetic properties and morphological configuration of the undoped and doped $CoFe_2O_4$ can be of interest for several applications, such as high-density storage and biomedicine. Moreover, since SiO_2 is non-toxic, biologically inert, and widely accepted material by the living body, and even reduces the inflammatory risk, embedding the undoped and doped $CoFe_2O_4$ could enhance their biocompatibility [21]. Although the properties of the obtained doped $CoFe_2O_4$ could be further improved by optimizing the amount of dopant ions, the calcination temperature, or the SiO_2-to-ferrite ratio, our study brings valuable baseline data on the properties of doped $CoFe_2O_4$-SiO_2 nanocomposites.

3. Materials and Methods

3.1. Reagents

All chemicals and reagents were used as received, without additional purification. Tetrahydrate calcium nitrate ($Ca(NO_3)_2 \cdot 4H_2O$, 99%), silver nitrate ($AgNO_3$, 99%), hexahydrate lanthanum nitrate ($La(NO_3)_3 \cdot 6H_2O$, 98%) and tetrahydrate cadmium nitrate ($Cd(NO_3)_2 \cdot 4H_2O$, 99%) were purchased from Carlo Erba (Milan, Italy), while nonahydrate ferric nitrate ($Fe(NO_3)_3 \cdot 9H_2O$, 98%), sodium nitrate ($NaNO_3$, 99%), hexahydrate cobalt nitrate ($Co(NO_3)_2 \cdot 6H_2O$, 98%), 1,4-BD 99%, TEOS (99%) and ethanol 96% were purchased from Merck (Darmstadt, Germany).

3.2. Synthesis

$CoFe_2O_4$, $Ag_{0.1}Co_{0.95}Fe_2O_4$, $Na_{0.1}Co_{0.95}Fe_2O_4$, $Ca_{0.1}Co_{0.9}Fe_2O_4$, $Cd_{0.1}Co_{0.9}Fe_2O_4$ and $La_{0.1}CoFe_{1.9}O_4$ embedded in SiO_2 gels containing 50 wt.% ferrite and 50 wt.% SiO_2 were prepared through a sol-gel route using different X/Co/Fe (X= Ag, Na, Ca, Cd, La) molar ratios, i.e., 0/1/2 ($CoFe_2O_4$), 1/9.5/20 ($Ag_{0.1}Co_{0.95}Fe_2O_4$), 1/9.5/20 ($Na_{0.1}Co_{0.95}Fe_2O_4$), 1/9/20 ($Ca_{0.1}Co_{0.9}Fe_2O_4$), 1/9/20 ($Cd_{0.1}Co_{0.9}Fe_2O_4$) and 1/10/19 ($La_{0.1}CoFe_{1.9}O_4$). The sols were prepared by stirring the metal nitrates with 1,4-BD, TEOS and ethanol using a NO_3^-/1,4-BD/TEOS molar ratio of 1/1/1. The resulting sols were stirred continuously for 30 min and kept at room temperature until complete gelation; the formed gel enclosed a homogenous mixture of metal nitrates and 1,4-BD. As high-purity gels with large crystallites are obtained by using a thermal pretreatment [5], the obtained gels were ground, heated at 40 °C (2 h) and 200 °C (5 h) and calcined at 400 °C (5 h), 700 °C (5 h) and 1000 °C (5 h) using a Nabertherm LT9 (Lilienthal, Germany) muffle furnace.

3.3. Characterization

The reaction progress was investigated via thermogravimetry (TG) and differential thermal analysis (DTA) in air, up to 1000 °C, at 10 °C·min^{-1} using alumina standards and a simultaneous SDT Q600 (TA Instruments, New Castle, DE, USA) thermal analyzer. The Fourier transform infrared (FT-IR) spectra of samples were recorded on KBr pellets containing 1% sample using a BX II FT-IR (Perkin Elmer, Waltham, MA, USA) spectrometer. The crystalline phases were investigated via X-ray diffraction using a D8 Advance (Bruker, Karlsruhe, Germany) diffractometer at ambient temperature with CuKα radiation

(λ = 1.5418 Å), operating at 40 kV and 35 mA. The composition of gels calcined at 400, 7000 and 1000 was confirmed via Optima 5300 DV (Perkin Elmer, Norwalk, CT, USA) ICP-OES after microwave digestion using a Speedwave Xpert (Berghof, Germany) system. Atomic force microscopy (AFM) was performed using a JSPM 4210 (JEOL, Tokyo, Japan) microscope in tapping mode using an NSC 15 (Mikromasch, Sofia, Bulgaria) silicon cantilever with a nominal resonant frequency of 325 kHz and a nominal force constant of 40 N/m. Three different 1 µm x 1 µm areas of the thin films obtained by transferring nanoparticles onto glass slides via adsorption from aqueous suspension were scanned for each sample. Image processing and topography were performed using a WinSPM 2.0 software (JEOL, (Tokyo, Japan). Cantilever characteristics were considered in the particle size determination. The particles' morphology was visualized using an HD-2700 (Hitachi, Tokyo, Japan) transmission electron microscope (TEM). The magnetic measurements were performed using a 7400 vibrating sample magnetometer (VSM) (Lake Shore, Carson, CA, USA). The hysteresis loops were recorded at room temperature, up to an applied field of 2 T, while the magnetization (M) was measured in a high magnetic field of up to 5 T.

4. Conclusions

The influence of doping with monovalent (Ag^+, Na^+), divalent (Ca^{2+}, Cd^{2+}), and trivalent (La^{3+}) ions on the structural, morphological, surface, and magnetic properties of $CoFe_2O_4$ was investigated. The kinetic formation of the doped and undoped $CoFe_2O_4$ showed that the activation energy of $CoFe_2O_4$ (1.236 kJ/mol) increased to 1.487–1.747 kJ/mol by Ag^+, Ca^{2+}, Cd^{2+}, and La^{3+} doping, and decreased to 1.102 kJ/mol by Na^+ doping, while the rate constant increased with the calcination temperature and depended on the doping ion. Poorly crystalline ferrites at 400 and 700 °C, and a well-crystallized single-phase ferrite in the undoped $CoFe_2O_4$ at 1000 °C, were observed. By doping, besides the well-crystallized ferrite, crystalline silica phases (cristobalite and quartz) were also formed. Although all the obtained gels have a cubic spinel structure, doping with different ions changed in the structural parameters determined via XRD. The AFM revealed that a low calcination temperature generated mainly spherical particles with a polycrystalline structure containing ferrite crystallites mixed with amorphous material. The increase in the calcination temperature led to a larger crystallite size, forming particles by a single-phase ferrite core covered by traces of secondary phases. The TEM measurements also indicate that thermal treatment is the main cause of the large size of the obtained nanoparticles; these results indicate a coalescence of nanoparticles, increasing the mean size. At 700 and 1000 °C, a single magnetic phase is generally observed, except in the case of doping with monovalent (Ag^+, Na^+) ions at 1000 °C, when the formation of two magnetic phases is favored. Moreover, the magnetic parameters of the gels calcined at 1000 °C were higher than those at 700 °C. The doping with monovalent ions decreased the M_S and increased the H_C, while the doping with multivalent ions increased the M_S and decreased the H_C. The K value decreased with doping, with the undoped $CoFe_2O_4$ displaying the highest anisotropy constant. The obtained results confirm that doping plays an important role in the tuning of the physical properties of promising $CoFe_2O_4$, which may be of great importance in the exploration of new applications in high-density information storage, drug delivery and tissue imaging.

Author Contributions: Conceptualization, T.D.; methodology, T.D., E.A.L., I.P., R.A.M. and O.C.; formal analysis, T.D., E.A.L., I.G.D., I.P., R.A.M. and O.C.; investigation, T.D.; resources, T.D., E.A.L., I.G.D., I.P., R.A.M. and O.C.; data curation, T.D.; writing—original draft preparation, T.D., O.C., I.P., I.G.D. and E.A.L.; writing—review and editing, T.D., O.C. and E.A.L.; visualization, T.D.; supervision, T.D. All authors have read and agreed to the published version of the manuscript.

Funding: The APC was funded by the Technical University of Cluj-Napoca.

Institutional Review Board Statement: Not applicable.

Informed Consent Statement: Not applicable.

Data Availability Statement: The data presented in this study are available on request from the corresponding author.

Acknowledgments: This study was supported by the Ministry of Research, Innovation and Digitization through Program 1—Development of the National Research and Development System and Subprogram 1.2—Institutional Performance—Projects that Finance the RDI Excellence, contract No. 18PFE/30.12.2021 (E.A.L. and O.C.). The authors acknowledge the Research Centre in Physical Chemistry "CECHIF" for their assistance with AFM.

Conflicts of Interest: The authors declare no conflict of interest.

References

1. Kaur, H.; Singh, A.; Kumar, V.; Ahlawat, D.S. Structural, thermal and magnetic investigations of cobalt ferrite doped with Zn^{2+} and Cd^{2+} synthesized by auto combustion method. *J. Magn. Magn. Mater.* **2019**, *474*, 505–511. [CrossRef]
2. Ahmad, R.; Gul, I.H.; Zarrar, M.; Anwar, H.; Niazi, M.B.K.; Khan, A. Improved electrical properties of cadmium substituted cobalt ferrites nano-particles for microwave application. *J. Magn. Magn. Mater.* **2016**, *405*, 28–35. [CrossRef]
3. Ghorbani, H.; Eshraghi, M.; Dodaran, A.A.S. Structural and magnetic properties of cobalt ferrite nanoparticles doped with cadmium. *Phys. B Condens. Matter.* **2022**, *634*, 413816. [CrossRef]
4. Sharifianjazia, F.; Moradi, M.; Parvin, N.; Nemati, A.; Rad, A.J.; Sheysi, N.; Abouchenari, A.; Mohammadi, A.; Karbasi, S.; Ahmadi, Z.; et al. Magnetic $CoFe_2O_4$ nanoparticles doped with metal ions: A review. *Ceram. Int.* **2020**, *46*, 18391–18412. [CrossRef]
5. Pinheiro, A.V.B.; da Silva, R.B.; Morales, M.A.; Silva Filho, E.D.; Soares, J.M. Exchange bias and superspin glass behavior in nanostructured $CoFe_2O_4$–Ag composites. *J. Magn. Magn. Mater.* **2020**, *497*, 165940. [CrossRef]
6. Mariosi, F.R.; Venturini, J.; da Cas Viegas, A.; Bergmann, C.P. Lanthanum-doped spinel cobalt ferrite ($CoFe_2O_4$) nanoparticles for environmental applications. *Ceram. Int.* **2020**, *462*, 772–2779. [CrossRef]
7. Waki, T.; Takao, K.; Tabata, Y.; Nakamura, H. Single crystal synthesis and magnetic properties of Co^{2+}-substituted and non-substituted magnetoplumbite-type Na–La ferrite. *J. Solid State Chem.* **2020**, *282*, 121071. [CrossRef]
8. Mansour, S.F.; Hemeda, O.M.; El-Dek, S.I.; Salem, B.I. Influence of La doping and synthesis method on the properties of $CoFe_2O_4$ nanocrystals. *J. Magn. Magn. Mater.* **2016**, *420*, 7–18. [CrossRef]
9. Shang, T.; Lu, Q.; Chao, L.; Qin, Y.; Yun, Y.; Yun, G. Effects of ordered mesoporous structure and La-doping on the microwave absorbing properties of $CoFe_2O_4$. *Appl. Surf. Sci.* **2018**, *434*, 232–242. [CrossRef]
10. Thakur, P.; Gahlawat, N.; Punia, P.; Kharbanda, S.; Ravelo, B.; Thakur, A. Cobalt nanoferrites: A review on synthesis, characterization, and applications. *J. Supercond. Nov. Magn.* **2022**, *35*, 2639–2669. [CrossRef]
11. Dippong, T.; Levei, E.A.; Cadar, O. Recent advances in synthesis and applications of MFe_2O_4 (M = Co, Cu, Mn, Ni, Zn) Nanoparticles. *Nanomaterials* **2021**, *11*, 1560. [CrossRef] [PubMed]
12. Ahmad, S.I. Nano cobalt ferrites: Doping, Structural, Low-temperature, and room temperature magnetic and dielectric properties–A comprehensive review. *J. Magn. Magn. Mater.* **2022**, *562*, 169840. [CrossRef]
13. Hua, J.; Cheng, Z.; Chen, Z.; Dong, H.; Li, P.; Wang, J. Tuning the microstructural and magnetic properties of $CoFe_2O_4/SiO_2$ nanocomposites by Cu^{2+} doping. *RSC Adv.* **2021**, *11*, 26336–26343. [CrossRef]
14. Gupta, M.; Das, A.; Das, D.; Mohapatra, S.; Datta, A. Chemical synthesis of rare earth (La, Gd) doped cobalt ferrite and a comparative analysis of their magnetic properties. *J. Nanosci. Nanotechnol.* **2020**, *20*, 5239–5245. [CrossRef] [PubMed]
15. Singh, S.; Singhal, S. Transition metal doped cobalt ferrite nanoparticles: Efficient photocatalyst for photodegradation of textile dye. *Mater. Today Proc.* **2019**, *14*, 453–460. [CrossRef]
16. Abbas, N.; Rubab, N.; Sadiq, N.; Manzoor, S.; Khan, M.I.; Fernandez Garcia, J.; Barbosa Aragao, I.; Tariq, M.; Akhtar, Z.; Yasmin, G. Aluminum-doped cobalt ferrite as an efficient photocatalyst for the abatement of methylene blue. *Water* **2020**, *12*, 2285. [CrossRef]
17. Aslam, A.; Razzaq, A.; Naz, S.; Amin, N.; Arshad, M.I.; Nabi, M.A.U.; Nawaz, A.; Mahmood, K.; Bibi, A.; Iqbal, F.; et al. Impact of lanthanum-doping on the physical and electrical properties of cobalt ferrites. *J. Supercond. Nov. Magn.* **2021**, *34*, 1855–1864. [CrossRef]
18. Kashid, P.; Suresh, H.K.; Mathad, S.N.; Shedam, R.; Shedam, M. A review on synthesis, properties and applications on cobalt ferrite. *Int. J. Adv. Sci. Eng.* **2022**, *9*, 2567–2583. [CrossRef]
19. Sakhawat, K.; Ullah, F.; Ahmed, I.; Iqbal, T.; Maqbool, F.; Khan, M.; Tabassum, H.; Abrar, M. Simple synthesis of Ag-doped ferrites nanoparticles for its application as bactericidal activity. *Appl. Nanosci.* **2021**, *11*, 2801–2809. [CrossRef]
20. El Nahrawy, A.A.; Soliman, A.A.; Sakr, E.M.M.; El Attar, H.A. Sodium–cobalt ferrite nanostructure study: Sol–gel synthesis, characterization, and magnetic properties. *J. Ovonic Res.* **2018**, *14*, 193–200.
21. Dippong, T.; Deac, I.G.; Cadar, O.; Levei, E.A. Effect of silica embedding on the structure, morphology and magnetic behavior of $(Zn_{0.6}Mn_{0.4}Fe_2O_4)_\delta/(SiO_2)_{(100-\delta)}$ Nanoparticles. *Nanomaterials* **2021**, *11*, 2232. [CrossRef] [PubMed]
22. Dippong, T.; Levei, E.A.; Deac, I.G.; Petean, I.; Cadar, O. Dependence of structural, morphological and magnetic properties of manganese ferrite on Ni-Mn substitution. *Int. J. Mol. Sci.* **2022**, *23*, 3097. [CrossRef]
23. Dippong, T.; Levei, E.A.; Cadar, O.; Deac, I.G.; Lazar, M.; Borodi, G.; Petean, I. Effect of amorphous SiO_2 matrix on structural and magnetic properties of $Cu_{0.6}Co_{0.4}Fe_2O_4/SiO_2$ nanocomposites. *J. Alloys Compd.* **2020**, *849*, 156695. [CrossRef]

24. Ajeesha, T.L.; Manikandan, A.; Anantharaman, A.; Jansi, S.; Durka, M.; Almessiere, M.A.; Slimani, Y.; Baykal, A.; Asiri, A.M.; Kasmery, H.A.; et al. Structural investigation of Cu doped calcium ferrite ($Ca_{1-x}Cu_xFe_2O_4$; x = 0, 0.2, 0.4, 0.6, 0.8, 1) nanomaterials prepared by co-precipitation method. *J. Mater. Res. Technol.* **2022**, *18*, 705–719. [CrossRef]
25. Sadegh, F.; Tavakol, H. Synthesis of $Ag/CoFe_2O_4$ magnetic aerogel for catalytic reduction of nitroaromatics. *Res. Chem.* **2022**, *4*, 100592. [CrossRef]
26. He, S.; Yang, C.; Niu, M.; Wei, D.; Chu, S.; Zhong, M.; Wang, J.; Su, X.; Wang, L. Coordination adsorption of Ag(I) on cobalt-ferrous oxalates and their derived $Ag/CoFe_2O_4$ for catalytic hydrogenation reactions. *Colloids Surf. A Physicochem. Eng. Asp.* **2019**, *583*, 124007. [CrossRef]
27. Satheeshkumar, M.K.; Kumar, E.R.; Srinivas, C.; Suriyanarayanan, N.; Deepty, M.; Prajapat, C.L.; Rao, T.V.C.; Sastry, D.L. Study of structural, morphological and magnetic properties of Ag substituted cobalt ferrite nanoparticles prepared by honey assisted combustion method and evaluation of their antibacterial activity. *J. Magn. Magn. Mater.* **2019**, *469*, 691–697. [CrossRef]
28. Guo, L.; Okinaka, N.; Zhang, L.; Watanabe, S. Molten salt-assisted shape modification of $CaFe_2O_4$ nanorods for highly efficient photocatalytic degradation of methylene blue. *Opt. Mater.* **2021**, *119*, 111295. [CrossRef]
29. Dippong, T.; Levei, E.A.; Cadar, O. Investigation of structural, morphological and magnetic properties of MFe_2O_4 (M = Co, Ni, Zn, Cu, Mn) obtained by thermal decomposition. *Int. J. Mol. Sci.* **2022**, *23*, 8483. [CrossRef] [PubMed]
30. Dippong, T.; Cadar, O.; Levei, E.A.; Deac, I.G.; Goga, F.; Borodi, G.; Barbu Tudoran, L. Influence of polyol structure and molecular weight on the shape and properties of $Ni_{0.5}Co_{0.5}Fe_2O_4$ nanoparticles obtained by sol-gel synthesis. *Ceram. Int.* **2019**, *45*, 7458–7467. [CrossRef]
31. Malyshev, A.V.; Vlasov, V.A.; Nikolaev, E.V.; Surzhikov, A.P.; Lysenko, E.N. Microstructure and thermal analysis of lithium ferrite pre-milled in a high-energy ball mill. *J. Therm. Anal. Calorim.* **2018**, *134*, 127–133.
32. Becherescu, D.; Cristea, V.; Marx, F.; Menessy, I.; Winter, F. *Metode Fizice in Chimia Silicatilor*; Editura Stiintifica si Enciclopedica: Bucuresti, Romania, 1977.
33. Kumar, H.; Singh, J.P.; Srivastava, R.C.; Negi, P.; Agrawal, H.M.; Asokan, K. FTIR and electrical study of dysprosium doped cobalt ferrite nanoparticles. *J. Nanosci.* **2014**, *2014*, 862415. [CrossRef]
34. Xu, Z.; Tang, Z.; Chen, F.; Bo, X.; Wu, H.; Li, Z.; Jiang, S. Study of lateral assembly of magnetic particles in magnetorheological fluids under magnetic fields. *J. Magn. Magn. Mater.* **2023**, *556*, 170293. [CrossRef]
35. Singh, R.; Sharma, V. Investigations on sintering mechanism of nano tungsten carbide powder based on molecular dynamics simulation and experimental validation. *Adv. Powder Technol.* **2022**, *33*, 103724. [CrossRef]
36. Guo, L.; Zhang, W.Y.; Zhao, X.J.; Xiao, L.R.; Cai, Z.Y.; Pu, R.; Xu, X.Q. Enhancing dispersion of ultra-fine WC powders in aqueous media. *Colloids Surf. A Physicochem. Eng. Asp.* **2019**, *567*, 63–68. [CrossRef]
37. Damm, C.; Peukert, W. Chapter 3-Particle dispersions in liquid media. In *Handbooks in Separation Science*; Contado, C., Ed.; Particle Separation Techniques; Elsevier: Amsterdam, The Netherlands, 2022; pp. 27–62.
38. Shao, Y.; Tian, M.; Zhen, Z.; Cui, J.; Xiao, M.; Qi, B.; Wang, T.; Hou, X. Adsorption of Ag/Au nanoparticles by ordered macro-microporous ZIF-67, and their synergistic catalysis application. *J. Clean. Prod.* **2022**, *346*, 131032. [CrossRef]
39. Zhao, C.; Fu, H.; Yang, X.; Xiong, S.; Han, D.; An, X. Adsorption and photocatalytic performance of Au nanoparticles decorated porous Cu_2O nanospheres under simulated solar light irradiation. *Appl. Surf. Sci.* **2021**, *545*, 149014. [CrossRef]
40. Johan, A.; Setiabudidaya, D.; Arsyad, F.S.; Ramlan; Adi, W.A. Strong and weak ferromagnetic of cobalt ferrite: Structural, magnetic properties and reflection loss characteristic. *Mater. Chem. Phys.* **2023**, *295*, 127086. [CrossRef]
41. Lakshmi, K.; Bharadwaj, S.; Chanda, S.; Reddy, C.V.K.; Pola, S.; Kumar, K.V.S. Iron ion non-stoichiometry and its effect on structural, magnetic and dielectric properties of cobalt ferrites prepared using oxalate precursor method. *Mater. Chem. Phys.* **2023**, *295*, 127172. [CrossRef]
42. Prabagar, C.J.; Anand, S.; Janifer, M.A.; Pauline, S.; Agastian, P.; Theoder, S. Effect of metal substitution (Zn, Cu and Ag) in cobalt ferrite nanocrystallites for antibacterial activities. *Mater. Today Proc.* **2021**, *16*, 1999–2006. [CrossRef]
43. Mahajan, P.; Sharma, A.; Kaur, B.; Goyal, N.; Gautam, S. Green synthesized (*Ocimum sanctum* and *Allium sativum*) Ag-doped cobalt ferrite nanoparticles for antibacterial application. *Vacuum* **2019**, *16*, 1389–1397. [CrossRef]
44. Soltani, N.; Salavati, H.; Moghadasi, A. The role of Na-montmorillonite/cobalt ferrite nanoparticles in the corrosion of epoxy coated AA 3105 aluminum alloy. *Surf. Interfaces* **2019**, *15*, 89–99. [CrossRef]
45. Kumar, P.; Kar, M. Correlation between lattice strain and magnetic behavior in non-magnetic Ca substituted nano-crystalline cobalt ferrite. *Ceram. Int.* **2016**, *42*, 6640–6647. [CrossRef]
46. Jauhar, S.; Kaur, J.; Goyala, A.; Singhal, S. Tuning the properties of cobalt ferrite: A road towards diverse applications. *RSC Adv.* **2016**, *6*, 97694–97719. [CrossRef]
47. Shakil, M.; Inayat, U.; Arshad, M.I.; Nabi, G.; Khalid, N.R.; Tariq, N.H.; Shah, A.; Iqbal, M.Z. Influence of zinc and cadmium co-doping on optical and magnetic properties of cobalt ferrites. *Ceram. Int.* **2020**, *46*, 7767–7773. [CrossRef]
48. Kamran, M.; Anisur-Rehman, M. Influence of La^{3+} substitutions on structural, dielectric and electrical properties of spinel cobalt ferrite. *Ceram. Int.* **2023**, *49*, 7017–7029. [CrossRef]
49. Kumar, L.; Kumar, P.; Kar, M. Non-linear behavior of coercivity to the maximum applied magnetic field in La substituted nanocrystalline cobalt ferrite. *Phys. B Condens. Matter* **2014**, *448*, 38–42. [CrossRef]
50. Channagoudra, G.; Nunez, J.P.J.; Hadimani, R.L.; Dayal, V. Study of cation distribution in La^{3+} and Eu^{3+} substituted cobalt ferrite and its effect on magnetic properties. *J. Magn. Magn. Mater.* **2022**, *559*, 169550. [CrossRef]

51. Jiang, C.; Leung, C.W.; Pong, P.W.T. Self-assembled thin films of Fe_3O_4-Ag composite nanoparticles for spintronic applications. *Appl. Surf. Sci.* **2017**, *419*, 692–696. [CrossRef]
52. Zang, D.; Huo, Z.; Yang, S.; Li, Q.; Dai, G.; Zeng, M.; Ruhlmann, L.; Wei, Y. Layer by layer self-assembled hybrid thin films of Porphyrin/Polyoxometalates@Pt nanoparticles for photo & electrochemical application. *Mater. Today Commun.* **2022**, *31*, 103811.
53. Li, X.; Sun, Y.; Zong, Y.; Wei, Y.; Liu, X.; Li, X.; Peng, Y.; Zheng, X. Size-effect induced cation redistribution on the magnetic properties of well-dispersed $CoFe_2O_4$ nanocrystals. *J. Alloys Compd.* **2020**, *841*, 155710. [CrossRef]
54. Nadeem, K.; Traussnig, T.; Letofsky-Papst, I.; Krenn, H.; Brossmann, U. Sol-gel synthesis and characterization of single-phase Ni ferrite nanoparticles dispersed in SiO_2 matrix. *J. Alloys Compd.* **2010**, *493*, 385–390. [CrossRef]
55. Haque, S.U.; Saikia, K.K.; Murugesan, G.; Kalainathan, S. A study on dielectric and magnetic properties of lanthanum substituted cobalt ferrite. *J. Alloys Compd.* **2017**, *701*, 612–618. [CrossRef]

Disclaimer/Publisher's Note: The statements, opinions and data contained in all publications are solely those of the individual author(s) and contributor(s) and not of MDPI and/or the editor(s). MDPI and/or the editor(s) disclaim responsibility for any injury to people or property resulting from any ideas, methods, instructions or products referred to in the content.

Review
Gold-Based Nanostructures for Antibacterial Application

Chinmaya Mutalik [1,†], Muhammad Saukani [2,3,†], Muhamad Khafid [4], Dyah Ika Krisnawati [5], Widodo [6], Rofik Darmayanti [5], Betristasia Puspitasari [5], Tsai-Mu Cheng [7,8,9,*] and Tsung-Rong Kuo [1,10,*]

1. Graduate Institute of Nanomedicine and Medical Engineering, College of Biomedical Engineering, Taipei Medical University, Taipei 11031, Taiwan; d845108002@tmu.edu.tw
2. International Ph.D. Program in Biomedical Engineering, College of Biomedical Engineering, Taipei Medical University, Taipei 11031, Taiwan; d845110002@tmu.edu.tw
3. Department of Mechanical Engineering, Faculty of Engineering, Universitas Islam Kalimantan MAB, Banjarmasin 70124, Kalimantan Selatan, Indonesia
4. Department of Nursing, Faculty of Nursing and Midwifery, Universitas Nahdlatul Ulama Surabaya, Surabaya 60237, East Java, Indonesia; khafid@unusa.ac.id
5. Dharma Husada Nursing Academy, Kediri 64117, East Java, Indonesia; dyahkrisna77@gmail.com (D.I.K.); rofik.darmayanti@gmail.com (R.D.); betristasya@gmail.com (B.P.)
6. College of Information System, Universitas Nusantara PGRI, Kediri 64112, East Java, Indonesia; widodoido7@gmail.com
7. Graduate Institute for Translational Medicine, College of Medical Science and Technology, Taipei Medical University, Taipei 11031, Taiwan
8. Taipei Heart Institute, Taipei Medical University, Taipei 11031, Taiwan
9. Cardiovascular Research Center, Taipei Medical University Hospital, Taipei Medical University, Taipei 11031, Taiwan
10. Stanford Byers Center for Biodesign, Stanford University, Stanford, CA 94305, USA
* Correspondence: tmcheng@tmu.edu.tw (T.-M.C.); trkuo@tmu.edu.tw (T.-R.K.)
† These authors contributed equally to this work.

Abstract: Bacterial infections have become a fatal threat because of the abuse of antibiotics in the world. Various gold (Au)-based nanostructures have been extensively explored as antibacterial agents to combat bacterial infections based on their remarkable chemical and physical characteristics. Many Au-based nanostructures have been designed and their antibacterial activities and mechanisms have been further examined and demonstrated. In this review, we collected and summarized current developments of antibacterial agents of Au-based nanostructures, including Au nanoparticles (AuNPs), Au nanoclusters (AuNCs), Au nanorods (AuNRs), Au nanobipyramids (AuNBPs), and Au nanostars (AuNSs) according to their shapes, sizes, and surface modifications. The rational designs and antibacterial mechanisms of these Au-based nanostructures are further discussed. With the developments of Au-based nanostructures as novel antibacterial agents, we also provide perspectives, challenges, and opportunities for future practical clinical applications.

Keywords: bacterial infection; gold nanoparticles; gold nanoclusters; gold nanorods; gold nanobipyramids; gold nanostars; antibacterial mechanism

1. Introduction

Emerging nanostructures have achieved various applications in catalysis [1], biofuels [2], energy storage, and electronics [3–7]. Discussing specifically nanomedicines, various nanostructures have been applied for microfluidic devices [8], biosensors [9,10], drug delivery systems [11], medical imaging, disease diagnosis [12–15], and antibacterial agents [16]. With these advancements in nanomedicine, nanostructures have been intensively demonstrated to be antibacterial agents to fight contagious infections induced by bacteria [17,18]. Antibiotic abuse adds to the specific hazards posed by bacterial infections by developing antibiotic resistance, resulting in treatment failures, boosting the spread of resistant bacteria, restricting treatment alternatives, and influencing numerous medical processes. To preserve antibiotic effectiveness and limit these hazards to public health, it is critical to use antibiotics carefully and in accordance with approved prescribing guidelines [19–21]. Antibiotics are the main drugs to treat bacterial infections, but the overuse of antibiotics has increased the opportunity for bacterial mutations which have resulted in resistant bacteria. A systematic study estimated that the 1.27 million deaths directly resulting from bacterial antimicrobial resistance were higher than the 864,000 deaths from the human immunodeficient virus (HIV)/acquired immunodeficiency syndrome (AIDS) or 643,000 from deaths malaria in 2019 [22]. Bacterial antimicrobial resistance is a globally urgent problem that requires immediate action from health communities, including basic research and clinical medicine [23–26]. Therefore, the development of superior antibacterial nanostructures is a promising approach against bacterial antimicrobial resistance.

Semiconductor, metal, and polymer nanostructures have been extensively investigated as antibacterial agents because of their unique chemical and physical characteristics [27–34]. Among these nanostructures, metal-based nanostructures such as gold (Au), silver, platinum, and copper have been explored as antibacterial agents [35–38]. Considering their inherent toxicity, chemically functionalized silver and titanium dioxide nanoparticles have been used to demonstrate antimicrobial activity. In an alternative strategy, photo-thermal properties of metallic nanoparticles have been theoretically researched and experimentally tested against several temperature-sensitive (mesophilic) bacteria using a plasmonic-based heating therapy [39].

Recently, gold nanostructures have emerged as promising candidates in the field of biomedicine due to their unique physical and chemical properties, including excellent stability, facile modification, and high surface area-to-volume ratio [40–44]. Most importantly, the unique optical properties of gold nanostructures, such as surface plasmon resonance, also make these nanostructures highly attractive for various theranostic techniques, including photothermal therapy, photoacoustic imaging, and surface-enhanced Raman spectroscopy. Additionally, gold nanostructures exhibit excellent electrical conductivity and catalytic activity, enabling their use in biosensing devices and electrochemical sensors. The physical and chemical properties of gold nanostructures make them versatile tools in biomedicine.

Great achievements have proven that Au-based nanostructures with different morphologies, including Au nanoparticles (NPs), Au nanoclusters (NCs), and anisotropic Au nanostructures such as Au nanorods (NRs), Au nanobipyramids (NBPs), and Au nanostars (NSs) can serve as potential antibacterial agents owing to their outstanding structural and optical properties [45,46]. In this literature review, we collected and summarized available data on AuNPs, AuNCs, and anisotropic Au nanostructures for antibacterial applications. Furthermore, we emphasized the antibacterial mechanisms of these Au-based nanostructures to reveal their excellent antibacterial activities. Critical challenges and future perspectives of Au-based antibacterial nanostructures for fundamental investigations and clinical applications are also provided and discussed.

2. Antibacterial Nanostructures of AuNPs

Because of their distinctive optical and electrical characteristics, as well as their potential in numerous biomedical applications, including drug administration, imaging, and therapy, AuNPs have garnered a lot of attention in recent years [47–49]. The antibacterial activities of AuNPs are some of the most researched uses of these particles. We talk about current developments in the synthesis of AuNPs and their claimed mechanisms of antibacterial action in this review. Recent studies showed that AuNPs can be synthesized using various methods, including physical methods such as laser ablation and sonication, and chemical methods such as the reduction of gold ions using reducing agents. Among these methods, the most widely used method is the chemical reduction method, which involves the reduction of gold ions using reducing agents such as sodium citrate, ascorbic acid, and thiols. AuNPs can be coupled to a variety of functionalized molecules, including ligands, drugs, peptides, proteins, and so forth [50].

Recent Advancements in the Synthesis of AuNPs and Their Antibacterial Applications

In recent work, cefotaxime (CTX)-loaded AuNPs (C-AuNPs) were prepared, and their antibacterial effectiveness against diverse bacterial strains was assessed. CTX is loaded onto the surface of AuNPs during their manufacture and acts as a reducing and capping agent. It was shown that C-AuNPs could be produced using a simple one-pot synthesis process (Figure 1a). By eliminating the use of external chemicals or biomolecules as reducing or capping agents, this technique prevented the development of leftover contaminants that might affect the antibacterial results. The progressive transition of the reaction solution's hue from light yellow to ruby red after incubation with the CTX antibiotic served as evidence that C-AuNPs had successfully been synthesized. The surface plasmon resonance (SPR) that took place in C-AuNPs is responsible for this color change. Minimum inhibitory concentration (MIC_{50}) values for CTX and C-AuNPs represent the concentrations that block 50% of populations of tested bacterial strains. According to Figure 1b–e, the quantified MIC_{50} values for *Staphylococcus aureus* (*S. aureus*) were 1.34 µg/mL (for CTX) and 0.68 µg/mL (for C-AuNPs), respectively. Values for *Escherichia coli* (*E. coli*), *Klebsiella oxytoca*, and *Pseudomonas aeruginosa* (*P. aeruginosa*) were 1.48, 3.03, and 1.92 µg/mL for CTX and 0.73, 1.03, and 1.87 µg/mL for C-AuNPs. In terms of the mechanism of action, C-AuNPs bind to the bacterial cell membrane and disrupt the transport of ions across the membrane, leading to a decrease in the production of adenosine triphosphate (ATP), the cell's energy currency. The enhanced antibacterial efficacy of C-AuNPs over free CTX and AuNPs may be due to AuNPs' substantial concentration of CTX, which is easily absorbed by bacteria and evades bacterial enzyme breakdown. Additionally, AuNPs themselves have strong antibacterial potential due to their capacity to damage bacterial DNA by direct contact and by preventing its unwinding during transcription. According to the study's findings, a sufficient amount (83.94%) of CTX was coupled to AuNPs, which effectively delivered CTX sodium to bacterial cells. The higher concentration of CTX molecules per unit volume of the system may have contributed to the overall antibacterial activity of C-AuNPs. Increased porosity in cell walls may have allowed CTX loaded onto AuNPs to enter gram-positive bacterial cells with ease. It was also noted that AuNPs may interact with proteins and lipopolysaccharides (LPSs) on the outer membranes of gram-negative bacterial strains. This might make it easier for AuNPs to infiltrate and distribute CTX to gram-negative bacteria [51–55].

According to a recent report, in vitro ciprofloxacin (CIP)-mediated sonodynamic antimicrobial chemotherapy (SACT) is facilitated by AuNPs. As depicted in Figure 2, 0.2 g/L AuNPs and 0.1 g/L of a CIP aqueous solution were gently stirred for 10 min of reaction time at room temperature. At 4 °C, the solution mix was centrifuged for 5 min at a speed of 12,000 rpm. The desired outcome, precipitates, could be dissolved in distilled water for subsequent experiments. Using plate colony-counting techniques, the bactericidal efficiency measurement threshold for SACT was determined. The length of ultrasonic (US) exposure, solution temperature, and CIP:AuNP concentration were all found to have significant impacts on SACT. *E. coli* and *S. aureus* were severely injured by the US, losing their usual microbial shape and revealing their contents. The aforementioned experimental findings thus originally supported the hypothesis that AuNPs could improve the bacteriostasis of CIP-mediated SACT. Additionally, it was demonstrated by intracellular reactive oxygen species (ROS) detection experiments that this acceleration may be related to ROS produced by US mechanics [56]. Research in this area usually suggests that one of the key processes explaining how US inhibits bacteria is the physical harm brought about by the cavitation effect. Scanning electron microscopy (SEM) under US irradiation allowed researchers to monitor the collapse of the membrane integrity and the efflux of cell contents from *E. coli* and *S. aureus*. The fact that CIP:AuNPs and US worked together to synergistically suppress the bacterium while the level of cell damage increased supports the idea that AuNPs might improve the cavitation impact of US by expanding nuclear sites and lowering the cavitation threshold. In addition to the physical harm caused by cavitation, the chemical impact caused by ROS also significantly contributes to the suppression of bacteria by US. When cavitation vesicles break, free radicals are created. These radicals mix with oxygen to generate ROS, which can harm cells by oxidizing proteins, lipids, DNA, and monosaccharides (Figure 2) [57–63].

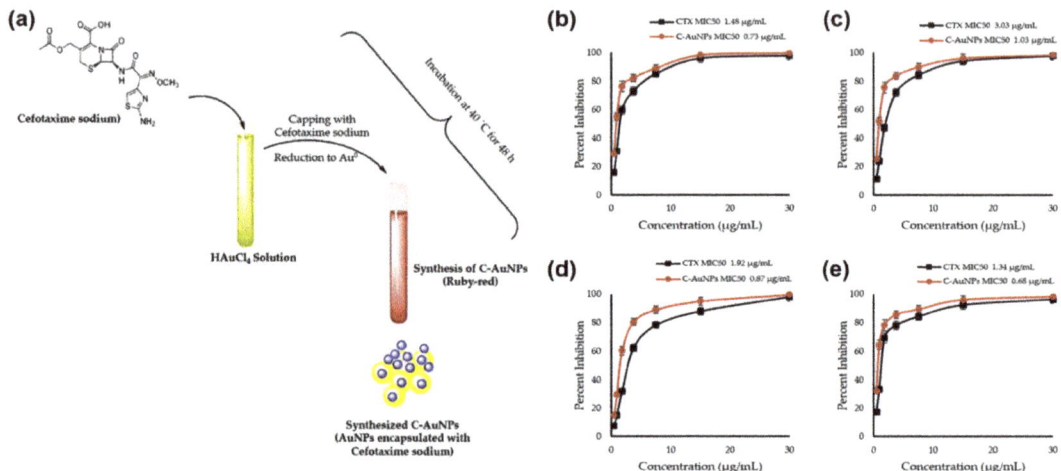

Figure 1. (**a**) Schematic representation of cefotaxime (CTX)-loaded gold nanoparticle (C-AuNP) synthesis and CTX and C-AuNP minimal inhibitory concentrations for (**b**) *Escherichia coli*, (**c**) *Pseudomonas aeruginosa*, (**d**) *Klebsiella oxytoca*, and (**e**) *Staphylococcus aureus*. Reproduced from ref. [51].

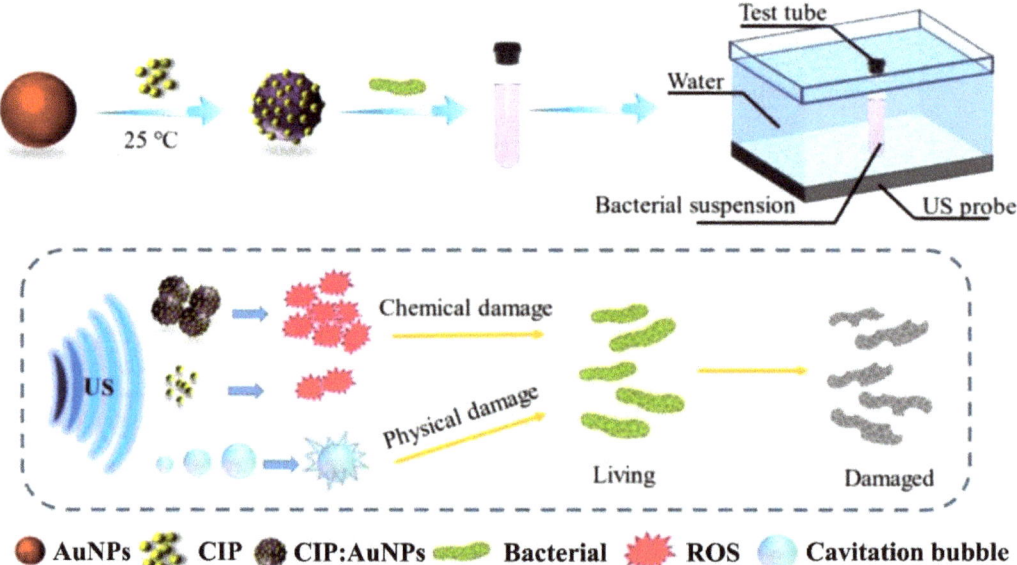

Figure 2. Ciprofloxacin: gold nanoparticle (CIP:AuNP) synthesis schematic and representation of the antibacterial pathway using sonodynamic antimicrobial chemotherapy (SACT). Reproduced from ref. [57].

Recently Dong et al. reported on adding epigallocatechin gallate (EGCG) to hydrogels modified with AuNPs (E-Au@H) to achieve combined effects of bactericidal, bactericidal photosensitization, and periodontal tissue regeneration as shown in Figure 3 [64]. In the synthesis process, a 5 mL beaker was first filled with 0.4 mL solution of HAuCl$_4$ (17 mg/mL), and then the beaker was submerged in an ice-water bath. Under rapid stirring conditions (800 rpm), 2 mL of the freshly made, ice-cold NaBH$_4$ solution (1.75 mg/mL) was added. Following the reaction, the product was separated by centrifugation (15 min at 15,000 rpm) and rinsed with deionized water. The cleaned product was then mixed again in 2 mL of deionized water with 1 mL of a 1 mg/mL EGCG solution, and the mixture was agitated (at 500 rpm) for 10 h at room temperature to allow for optimal adsorption. The homogenous hydrogel solution was then mixed with 0.05 g of the manufactured E-Au powder for 1 h of sonication and left to sit for 6 h before the E-Au@H product was retrieved. The mouth microbiome is interconnected with the outside environment, which can cause complicated periodontal irritation with microbial pathogens accumulating in plaque biofilms, thus making periodontal therapy very challenging. An agar plate test was used to evaluate the antibacterial efficacy of E-Au@H against gram-positive and gram-negative bacteria in order to determine whether composite materials were practical for treating periodontitis. While the near-infrared (NIR) spectrum may completely manage the release of tea polyphenols, boost the antibacterial action, promote angiogenesis, and improve osteogenesis, E-Au@H swiftly heated up to 50.7 °C in less than 5 min when exposed to NIR light. The NIR-irradiated nanocomposite inhibited *S. aureus*, *E. coli*, and *S. aureus* biofilms by 94%, 92%, and 74%, respectively, according to in vitro studies. They also increased alkaline phosphatase activity 5-fold after 7 days and the rate of extracellular matrix mineralization by 3-fold after 21 days. E-Au@H with NIR laser illumination was shown to reduce dental plaque biofilms by 87% in an animal model [64–68].

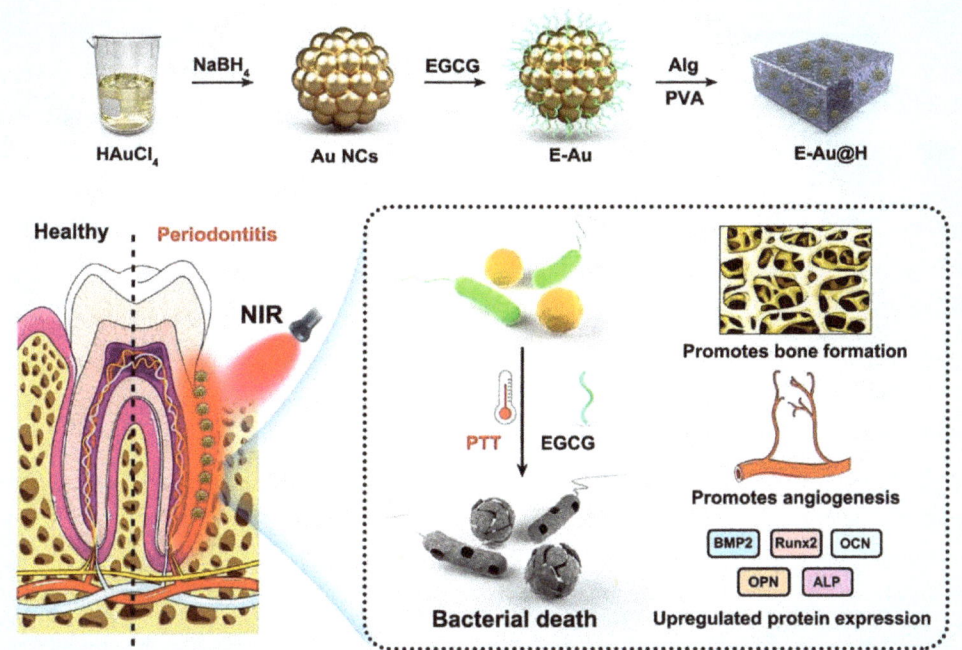

Figure 3. Epigallocatechin gallate (EGCG) added to hydrogels modified with gold nanoparticles (AuNPs; E-Au@H) synthesis schematic and representation of the antibacterial pathway. Reproduced from ref. [64].

A recent paper described a straightforward procedure for creating customized gold-decorated magnetic NPs with a high capacity to alter conventional formulations of antibiotics like sulfamethoxazole (SMX) and a subsequent investigation of the drug's adsorption-desorption (release) cycle. Several techniques, including zeta potential testing and field emission scanning electron microscopy, were used to analyze these produced NPs. AuNPs decorated with magnetic NPs on SMX bind to the cell membranes of bacteria and impair ion transport across the membrane, resulting in a decrease in ATP synthesis, the cell's energy currency. This results in the destruction of bacterial cell walls, and there is a possibility of ROS generation, leading to microbial elimination. In this study, there was a unique representation of an antibacterial activity and mechanism that was associated with the study of material kinetics and dynamics. Temkin and pseudo-first-order Lagergren models were respectively utilized to study the drug sorbate's adsorption isotherms and kinetics; a zero-order model was used to study the kinetics of the drug's release from this carrier. MIC values of pure SMX and SMX-conjugated magnetic NPs and AuNPs in antibacterial tests were calculated to be 14 and 2.5 µg/mL against *E. coli* and 24 and 1.25 µg/mL against *S. aureus*, respectively, displaying the highest order of antibacterial activity (Figure 4) [69–74].

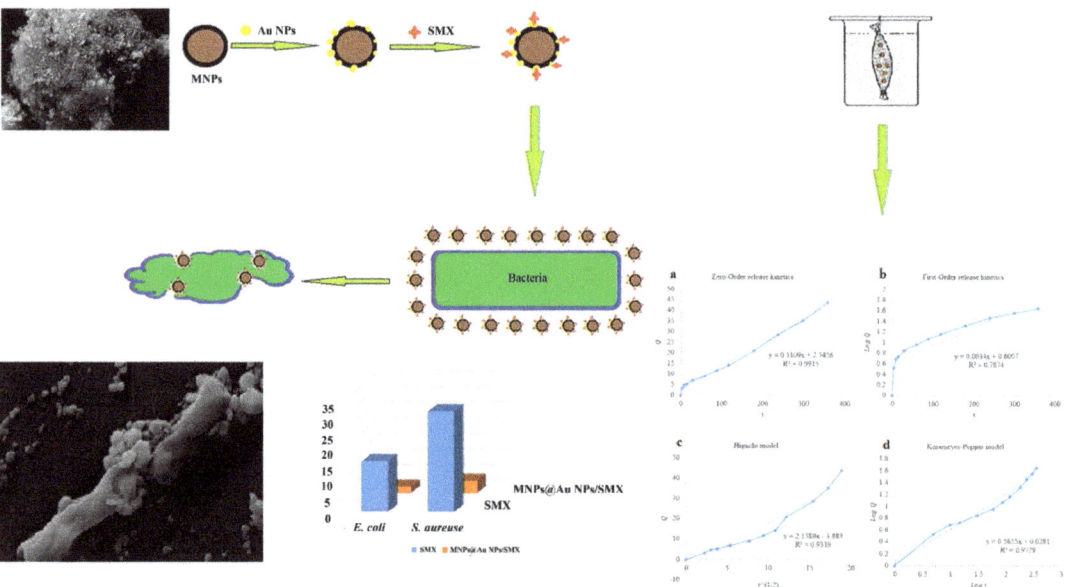

Figure 4. Sulfomethoxazole (SMX)-conjugated magnetic and gold nanoparticle (AuNP) synthesis and antibacterial studies. Reproduced with permission from ref. [69]. Copyright 2022 Elsevier.

Another mechanism of AuNPs is affected by the particle dispersibility, size, and shape, as well as the presence of other molecules such as oxygen, water, and organic substances. Following the above-mentioned trial, one recent study reported that aggregation was a major element impacting NPs' functions using thioproline (T) and Boc-protected thioproline (B) in combination with AuNPs [74]. The antibacterial activity of thioproline-modified AuNPs attained its minimum potential by thioproline-mediated particle aggregation. The maximum antimicrobial activities of AuNPs were attained by fine-tuning the balance between thioproline exposure and shielding. A strategy for balancing NP monodispersity with antibacterial activity was described. Maximum antibacterial properties were obtained by fine-tuning the equilibrium between the exposition and sheltering of active chemicals on AuNPs (Figure 5). The AuNPs synthesized in this study were altered with organic substance concentrations to describe the mechanistic antibacterial behavior of functionally modified AuNPs. The study was conducted using multidrug-resistant (MDR) and non-MDR *E. coli* and *Klebsiella pneumonia*, and many more gram-negative bacteria were found to be affected by T_1B_1 (1:1)-AuNPs, which showed the highest antibacterial activity in a subsequent study. T_1B_1-AuNPs promoted cell membrane permeability and, hence, killed bacteria. Although the negative charge of T_1B_1-AuNPs may prevent them from interacting with negatively charged bacteria, the carboxylic acid groups of T ligands might compete for hydrogen bonding interactions with LPSs/peptidoglycans on bacterial surfaces. This action may disrupt hydrogen bonds within the cell wall, causing instability and disintegration of the cell walls, which is a plausible mechanism for T_1B_1-AuNPs' antibacterial activity [75–79].

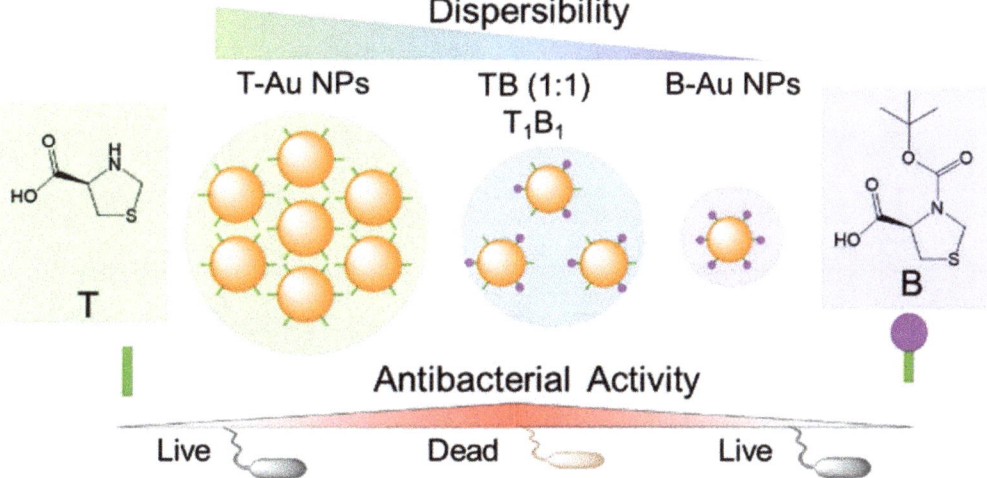

Figure 5. The synthesis approach of thioproline (T) (Boc-protected (B))-gold nanoparticles (AuNPs) to modify their antibacterial activity as schematically depicted. Reproduced with permission from ref. [75]. Copyright 2022 Royal Society of Chemistry.

3. Antibacterial Nanostructures of AuNCs

AuNCs consist of several to hundreds of gold atoms with a size of <3 nm which are protected by organic ligands [80]. AuNCs, or ultrasmall AuNPs, are promising nanostructures that have unique molecule-like properties for several applications, including theranostic, antibacterial, sensing, and catalysis [81–83]. Au-based NPs are known as materials with good biocompatibility with mammalian cells even if the size is reduced to NC size [84–86]. AuNPs with particle sizes of >3 nm are inert for bacteria, whereas AuNCs have shown potent antibacterial activity [87]. AuNCs have shown effective antibacterial activity due to their positive charge, while the bacteria have a negative charge, which can help the NCs penetrate cell membranes. In addition, the ultrasmall size of AuNCs makes it easier for them to internalize into bacteria, thereby facilitating the process of bacterial inhibition. The antibacterial mechanism of AuNCs against bacteria is complex and not fully understood. Generally, the antibacterial mechanism of AuNCs includes attachment, internalization, and destruction of bacterial membranes through ROS reactions [84]. Ultrasmall AuNCs are able to readily pass through cell wall pores and become internalized in bacteria, where they induce the production of ROS, which oxidize bacterial membranes and disrupt bacterial metabolic processes [88]. The antibacterial mechanism was investigated at the cellular level through the destruction of membrane integrity, disruption of the antioxidant defense system, metabolic disturbances, and DNA damage (Figure 6), and also at the molecular level through a transcriptome analysis (RNA sequencing) [89]. Furthermore, AuNCs have exhibited antibacterial mechanisms of photodynamic therapy (PDT) and photothermal therapy (PTT). Therefore, AuNCs have strong antibacterial activity against a wide range of bacterial strains, including both gram-positive and gram-negative bacteria, MDR bacteria, and even bacterial biofilms [90–94].

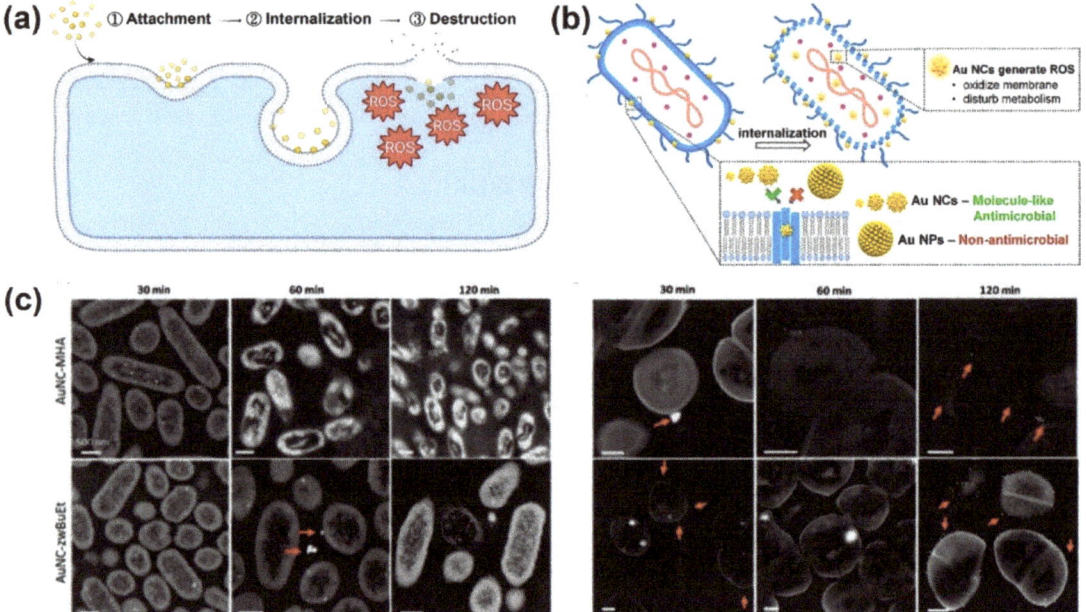

Figure 6. (**a**) Stages of antibacterial mechanism of gold nanoclusters (AuNCs). Reproduced with permission from ref. [84]. Copyright 2022 American Chemical Society. (**b**) Illustration of bacterial eradication by AuNCs. Reproduced with permission from ref. [88]. Copyright Elsevier 2021. (**c**) Interaction analysis of bacteria and AuNCs at different time points. Reproduced with permission from ref. [91]. Copyright 2022 American Chemical Society.

3.1. Antibacterial AuNCs for Gram-Positive and Gram-Negative Bacteria

Gram-positive and gram-negative bacteria are two major categories of bacteria based on their cell wall structure. Gram-positive bacteria have a thick peptidoglycan cell wall, whereas gram-negative bacteria have a thinner peptidoglycan layer and an outer membrane containing LPSs. Gram-negative bacteria are generally considered more resistant to antibiotics and other antimicrobial agents than are gram-positive bacteria [93]. Both kinds of bacteria can effectively be killed by AuNCs. Ligands are an important factor that can improve the antibacterial capability of AuNCs. $Au_{25}MHA_{18}$ (Au_{25} protected by the 6-mercaptohexanoic acid (MHA) ligand) effectively killed ~95% of *S. aureus* and ~96% of *E. coli* after 2 h of treatment with increases in ROS levels that reached 2~3-fold for *S. epidermidis, Bacillus subtilis, E. coli,* and *P. aeruginosa* [87,88,91]. AuNCs conjugated with cysteine (Cys-AuNCs) also increased ROS levels 2.1-fold in *E. coli,* whereas (11-mercaptoundecyl)-N,N,N-trimethylammonium bromide (MUTAB)-conjugated AuNCs increased ROS levels 5-fold [88]. In addition to effectively killing gram-positive and gram-negative bacteria, AuNCs eradicated fungi (*Candida albicans*) [94] with the N-heterocyclic carbene (NHC) ligand, with a reduction in the colony-forming unit (CFU) rate of more than 99% at a concentration of 20 μg/mL (Figure 7) [86].

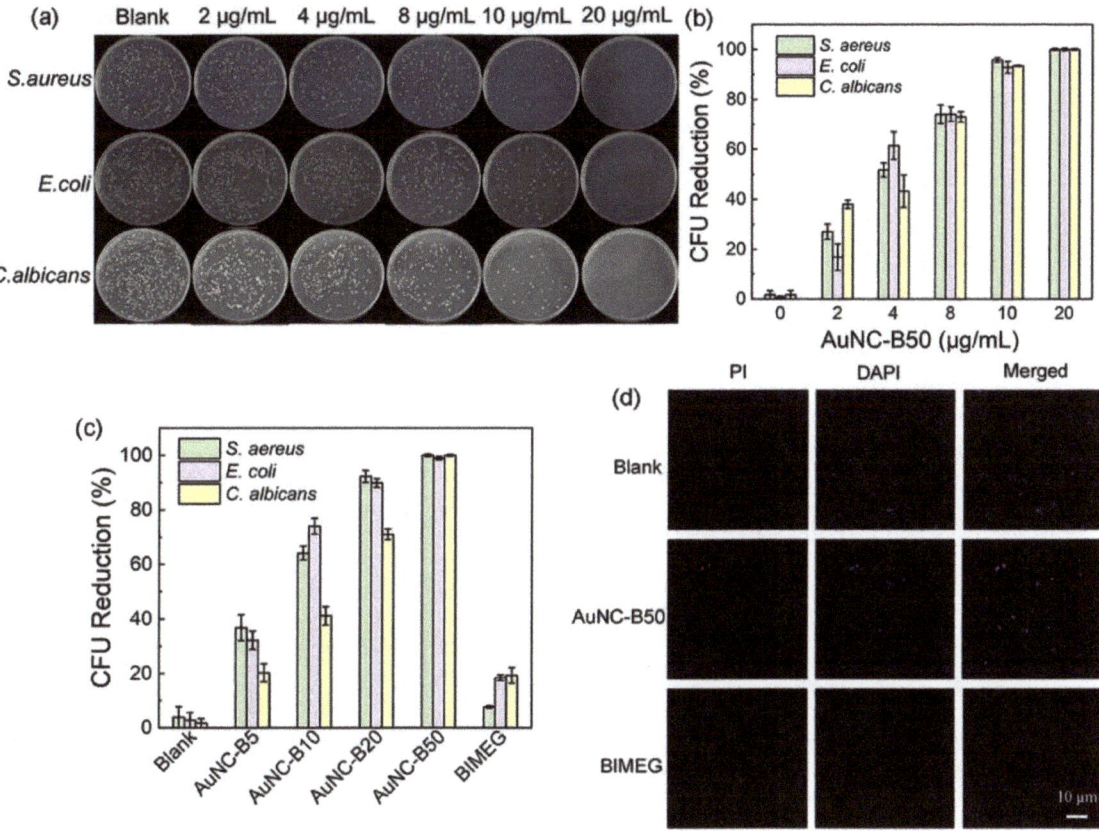

Figure 7. Antibacterial performance of gold nanoclusters (AuNCs) with the N-heterocyclic carbene ligand. (**a**) Photographs of agar plates and (**b**) colony forming unit (CFU) reduction rates of *S. aureus*, *E. coli*, and *C. albicans* incubated with AuNCB50 at various concentrations in the dark for 10 min. (**c**) CFU reduction rates of *S. aureus*, *E. coli*, and *C. albicans* incubated with AuNCs at different ligand ratios. (**d**) Live/dead staining of *S. aureus* before and after incubation with AuNC-B50 (20 μg/mL) for 30 min. Reproduced with permission from ref. [86]. Copyright 2022 Elsevier.

3.2. Antibacterial AuNCs for MDR Bacteria

MDR bacteria are strains of bacteria that are resistant to multiple types of antibiotics. These bacteria have developed mechanisms to protect themselves from the effects of antibiotics, making it difficult to treat the infections they cause. One of the ways that AuNCs can fight MDR bacteria is through the disruption of bacterial cell membranes. The small size and high surface-area-to-volume ratio of AuNCs allow them to penetrate cell membranes and disrupt their integrity, leading to bacterial cell death. AuNCs functionalized with quaternary ammonium (QA) salt (QA-AuNCs) were designed to combat methicillin-resistant *S. aureus* (MRSA). In vitro and in vivo studies showed their excellent performance as a safe and efficient strategy for clinical treatment of MRSA [89]. In addition, the gram-positive bacterium, *Clostridium difficile*, was inhibited by MHA-AuNCs. MHA-AuNCs (100 μM) generated a 5-fold increase in ROS levels, which drastically killed *C. difficile* by destroying membrane integrity without toxic effects on human cells [90]. Not only MDR gram-positive strains can be eradicated when interacting with AuNCs, but also MDR gram-negative strains. MUTAB-AuNCs were reported to eradicate vancomycin-resistant enterococci from MDR isolates [94]. AuNC-Mal/TU effectively inhibited MDR *P. aeruginosa* after 15 min of incubation. The antibacterial mechanisms of AuNC-Mal/TU were proven through multiple modes of actions in bacteria, such as inhibiting the TrxR enzyme, depleting ATP, and interfering with copper regulation [90]. Enhanced ROS generation by AuNCs and the overcoming of poor drug penetration can be induced by light irradiation. Under NIR 405-nm light irradiation, AuNCs combined with lysozyme and curcumin (Lys-AuNCs-Cur) destroyed the integrity of the outer membranes of MRSA bacteria with a killing rate of 99.92% [91]. Zhuo et al. demonstrated QA-AuNCs with a positive surface charge and an average size of ~2 nm which efficiently bound to bacteria. With NIR irradiation, QA-AuNCs can be easily internalized by bacteria and then accelerate ROS generation and toxic hyperthermia to result in bacterial eradication with triple mechanisms, including direct killing, PDT, and PTT (Figure 8). With irradiation for 10 min, QA-AuNCs at a low concentration (50 µg/mL) eradicated MRSA in biofilms without developing resistance, and decreased inflammation [92].

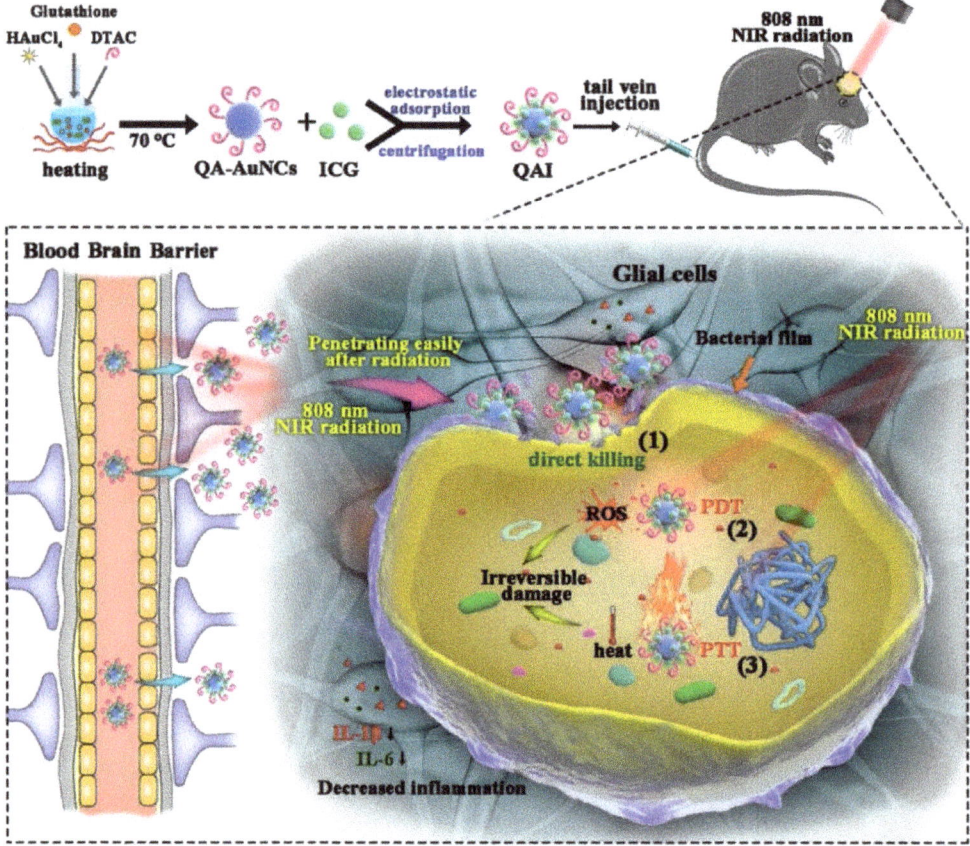

Figure 8. Quaternary ammonium (QA)-gold nanoclusters (AuNCs) conjugated with indocyanine green (ICG) revealed triple weapons, including direct killing, photodynamic therapy (PDT), and photothermal therapy (PTT) for methicillin-resistant *Staphylococcus aureus* (MRSA) eradication under near-infrared (NIR) irradiation. Reproduced with permission from ref. [92]. Copyright 2022 Elsevier.

3.3. Bacterial Biofilms

In the medical field, bacterial biofilms can form on a variety of surfaces, including medical devices such as catheters, artificial joints, and dental implants. These biofilms can cause infections that are difficult to treat because the bacteria are protected by extracellular polymeric substances, making them less susceptible to antibiotics and the host's immune response. The use of conventional antibiotics to treat bacterial biofilms can lead to the development of MDR. Therefore, ultrasmall AuNCs are a promising material which can combat bacterial biofilms. AuNCs were found to be able to penetrate biofilms and disrupt the structural integrity of the matrix, allowing antibiotics to reach the bacteria more effectively. *Enterococcus faecium, K. pneumoniae, Acinetobacter baumannii, S. aureus, P. aeruginosa*, and *Enterobacter* species are often found in biofilms and are major microorganisms that adhere to surfaces and form a protective matrix around themselves. AuNCs protected by mercaptopropionic acid (MPA-AuNCs) exhibited electronegative properties and revealed a 10-fold higher cellular internalization than AuNCs protected with glutathione (GSH-AuNCs) [79]. Surface chemical modifications were conducted by Srinivasulu et al. to utilize MPA-AuNCs as an antibiofilm nanostructure [95]. MPA-AuNCs further conjugated with protoporphyrin IX (PpIX) and chitosan to form PpIX-Chito-Au18 nanocomposites were used to destroy biofilms of *S. aureus* and *P. aeruginosa*. Under white light irradiation, the PpIX-Chito-Au18 nanocomposites demonstrated better biofilm penetration and elimination capacity of biofilms, including *S. aureus* and *P. aeruginosa*, compared to that of nanocomposites without white light irradiation. *Fusobacterium nucleatum* is known to be a keystone species in the formation of oral biofilms, particularly in periodontal diseases such as gingivitis and periodontitis. Ultrasmall MHA-AuNCs showed superior antibacterial effects against oral biofilms of *F. nucleatum* due to their effective penetration and destruction. In vivo studies confirmed that periodontal inflammation and bone loss may be reduced after topical application of MHA-AuNCs, because MHA-AuNCs destroyed the biofilms caused by the *F. nucleatum* bacterium. Microbiome investigations also showed that MHA-AuNCs could fix the damage in the mouth and gut microbiota caused by infection with *F. nucleatum* (Figure 9) [96].

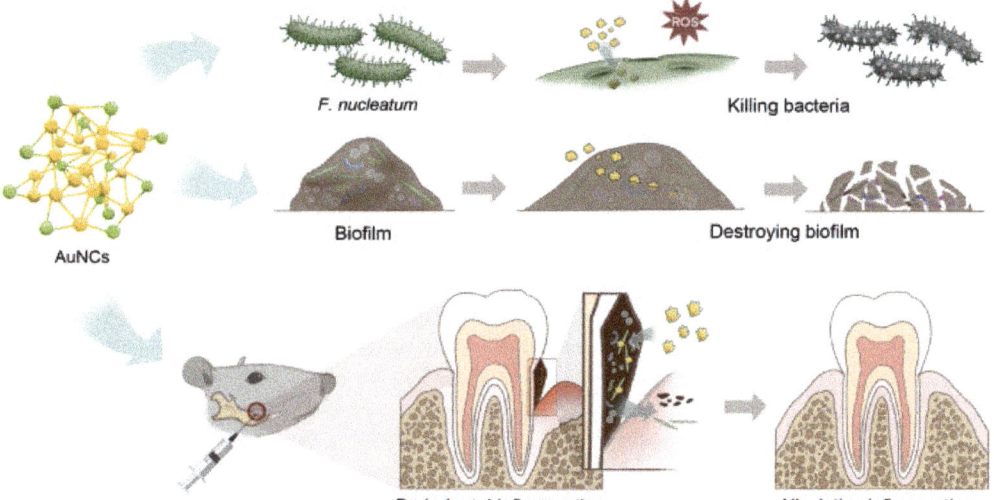

Figure 9. Excellent ability of gold nanoclusters (AuNCs) to fix problems due to oral biofilms. Reproduced from ref. [96].

4. Anisotropic Au-Based Nanostructures as Antibacterial Agents

Anisotropic Au-based nanostructures have size- and shape-dependent chemical and physical features [97–99]. Currently, various Au-based nanostructures have been designed and prepared, including NRs, NBPs, NSs, nanowires, triangles, cubes, octahedrons, and plates. Studies have reported uses of various Au-based nanostructures with exceptional antibacterial activities. Herein, we focused on Au-based nanostructures for highly frequent utilization as antibacterial agents, including AuNRs, AuNBPs, and AuNSs. The antibacterial mechanisms of antibacterial nanostructures of AuNRs, AuNBPs, and AuNSs are also emphasized to disclose their future perspectives.

4.1. Nanostructures of AuNRs for Antibacterial Applications

Anisotropic AuNRs have achieved several antibacterial applications because of their unique surface plasmon resonance (SPR) [100]. AuNRs have two SPR absorptions, including transverse and longitudinal surface plasma absorptions. Based on their SPR, AuNRs can interact with incident light and transform light energy into heat for PTT. A recent study reported the antibacterial activities of AuNRs636 (with a longitudinal plasmon peak at 636 nm), AuNRs772 (with a longitudinal plasmon peak at 772 m), AuNPs, and AgNPs under incandescent light illumination [101]. With white light illumination, AuNR636 and AuNRs772 exhibited significant antibacterial activities, whereas AuNPs had no significant antibacterial activity against *E. coli*, *S. aureus*, *Salmonella enterica* serovar typhimurium, or MRSA. Furthermore, compared to AuNRs772, AuNRs636 presented higher antibacterial activity due to higher dangling bonds. Most importantly, under white light illumination, both AuNRs772 and AuNRs636 were proven to express PTT and PDT because of their damage to bacterial cell membranes, which decreases the cell membrane potential of bacteria and increases DNA degradation. Plasmonic photothermal therapy (PPTT) is a non-invasive and drug-free treatment method that utilizes the unique properties of noble metal nanoparticles to convert bio-transparent electromagnetic radiation into heat. When subjected to resonant laser irradiation, gold nanorods (GNRs) become highly efficient nano-converters, effectively generating heat for PPTT applications. In this study, the goal was to evaluate the antimicrobial impact of easily synthesizable, purified, and water-dispersible GNRs on *E. coli* bacteria. It was crucial to control the concentration of GNRs used in the process to avoid cytotoxic effects on cells while still producing sufficient heat, under near-infrared illumination, to raise the temperature to approximately 50 °C within approximately 5 min. Viability experiments demonstrated that the proposed system achieved a killing efficiency capable of reducing the population of Escherichia coli by approximately 2 log CFU (colony-forming unit) [102]. In our previous work, visible-light-activated metallic molybdenum disulfide nanosheets (1T-MoS_2 NSs) and AuNRs with plasmonic absorption at a wavelength of 808 nm were combined to form nanocomposites by electrostatic adsorption. 1T-MoS_2 NSs were synthesized through a solvothermal method. Aqueous solutions of AuNRs at three different concentrations, namely 100 μg/mL, 50 μg/mL, and 33.3 μg/mL, were prepared in CTAB aqueous solutions (125 μM). For the preparation of AuNR-decorated 1T-MoS_2 NSs, three samples of the 1T-MoS_2 NS solution (500 μL, 100 μg/mL) were combined with 500 μL of AuNR solutions with varying concentrations: 100 μg/mL (MoS_2@AuNRs), 50 μg/mL (MoS_2@1/2AuNRs), and 33.3 μg/mL (MoS2@1/3AuNRs). The photothermal properties were examined using an 808 nm NIR laser for MoS_2@1/2AuNRs. The total amount of reactive oxygen species (ROS) generated was determined by measuring the fluorescence intensity produced by mixing 1 mL of the working solution with 200 μL of each sample. The antimicrobial phototherapy effect of MoS_2@AuNRs on *E. coli* was assessed using the agar plate counting method. After exposing the bacterial samples to NIR laser irradiation for 2 min, the MoS_2@1/3AuNRs, MoS_2@1/2AuNRs, and MoS_2@AuNRs achieved bacterial reductions of 84.4%, 97.5%, and 99.0%, respectively. Additionally, when subjected to visible light irradiation for 1 min, the antibacterial rates of MoS_2@1/3AuNRs, MoS_2@1/2AuNRs, and MoS_2@AuNRs were found to be 83.8%, 93.3%, and 98.5%, respectively [103]. Under 808-nm NIR laser irradiation for 10 min, the temperature of the MoS_2@AuNR nanocomposites increased from 25 to 66.7 °C based on a photothermal effect. Moreover, MoS_2@AuNR nanocomposites revealed a capability to generate ROS under visible light irradiation due to a photodynamic effect. With the combination of PTT and PDT, the MoS_2@AuNR nanocomposites exhibited higher antibacterial activity compared to that of only PTT or PDT (Figure 10). To sum up, light-activated MoS₂@AuNR nanocomposites showed a brilliant synergistic effect of PTT and PDT to provide an alternative approach to fight bacterial infections.

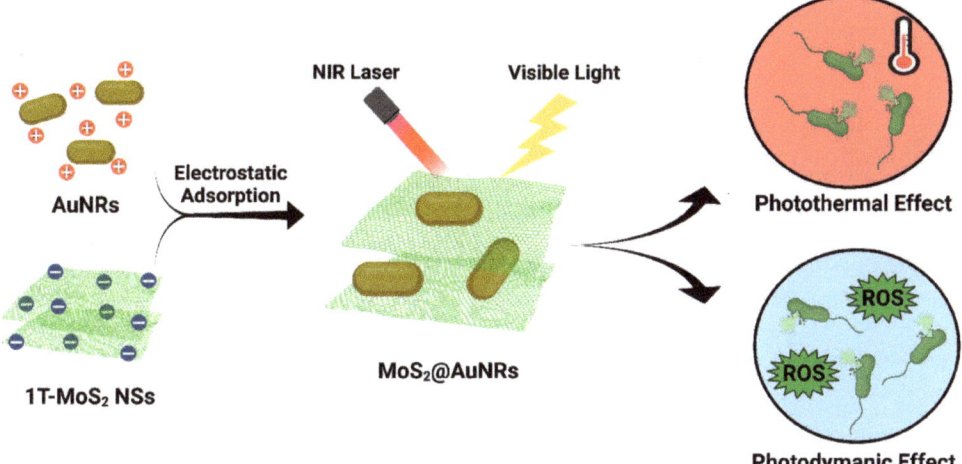

Figure 10. Illustration of the preparation of molybdenum-disulfide-conjugated gold nanorod (MoS₂@AuNR) nanocomposites and their antibacterial application based on photothermal therapy (PTT) and photodynamic therapy (PDT). Reproduced from ref. [103].

4.2. Nanostructures of AuNBPs for Antibacterial Applications

Extensive explorations of plasmonic AuNBPs for the photothermal killing of bacteria have significantly impacted the development of antibacterial agents in recent years [104]. In our recent work, the photothermal effects of AuNBPs and AuNRs with similar longitudinal surface plasma bands at ~808 nm were compared under NIR laser irradiation [105]. With 808-nm laser irradiation, AuNBPs with the (111) plane exhibited a better photothermal effect than that of AuNRs with the (200) plane. According to density function theory simulations, the water adsorption energy of Au(111) is higher than that of Au(100). Therefore, based on simulations, water molecules can more easily desorb from AuNBP surfaces for photothermal heating compared to AuNRs (Figure 11a). Furthermore, for PTT, AuNBPs exhibited higher antibacterial activity than did AuNRs, according to the growth rates of *E. coli*. In combinations of experimental PTT and DFT simulations, AuNBPs were found to be a potential antibacterial agent for noninvasive PTT. Moreover, AuNBPs were functionalized with chiral glutamic acid (D/L-Glu-AuNBPs) for PTT against bacteria and biofilms [106]. Based on chemical and physical interactions with bacteria, D/L-Glu-AuNBPs were enhanced to target and interact with bacterial cell walls. With sharp tips and a nanoscale size, AuNBPs can easily penetrate bacteria and biofilms and further can induce damage to bacterial cell walls and cause leakage of bacterial components. D/L-Glu-AuNBPs also showed enhancement of DNA and nucleic acid leakage under NIR laser irradiation. In vitro and in vivo antibacterial and antibiofilm investigations demonstrated remarkable performances of D/L-Glu-AuNBPs in killing bacteria and eradicating biofilms based on synergistic effects, including chemotherapy, physiotherapy, and PTT. Overall, AuNBPs and their nanocomposites provide a novel strategy to fight bacteria and biofilms.

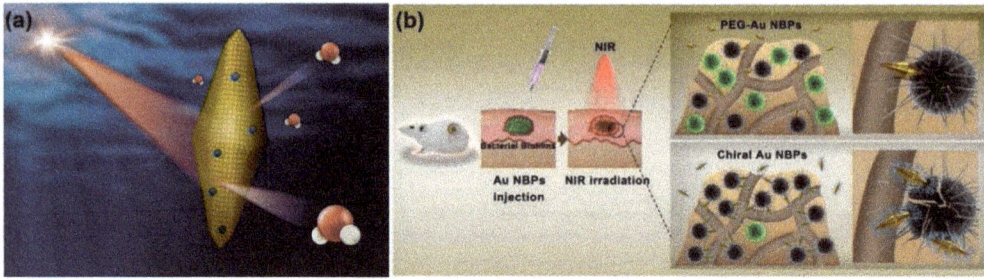

Figure 11. (**a**) Schematic illustration of gold nanobipyramids (AuNBPs) with the (111) plane for water desorption under near-infrared (NIR) laser irradiation. Reproduced with permission from ref. [105]. Copyright 2021 Elsevier. (**b**) Evaluation of in vivo antibacterial activity and antibacterial mechanisms of AuNBPs functionalized with chiral glutamic acid (D/L-Glu-AuNBPs). Reproduced with permission from ref. [106]. Copyright 2020 Elsevier.

4.3. Nanostructures of AuNSs for Antibacterial Applications

Au-based nanostructures with special star structures, AuNSs, were proven to have excellent antibacterial effects. AuNSs have a spiked structure that can help disrupt outer bacterial cell membranes, leading to the death of bacteria. A recent study showed that AuNPs (250 μg/mL) cultured with *S. aureus* revealed no antibacterial effect [107]. However, at the same concentration (250 μg/mL), AuNSs exhibited an obvious antibacterial effect against *S. aureus*. The antibacterial mechanism of AuNSs can be attributed to the high specific surface area (surface area/volume) and star-shaped spikes of AuNSs, which induce high local stress for bacterial cell membranes, resulting in the rupture of bacterial cell membranes. Most importantly, AuNSs with unique SPR can provide superior NIR photothermal conversion performance [108]. A recent achievement was to prepare vancomycin-coated AuNSs (AuNSs@Van) for selective targeting of MRSA and elimination of MRSA under NIR laser illumination (Figure 12) [109]. For in vivo studies, AuNSs@Van showed excellent biocompatibility and outstanding antibacterial activity against bacterial infections based on physical effects in a mouse model. Furthermore, AuNSs@Van were proven to be a potential antibacterial agent for PTT in fighting gram-positive bacteria. For future health care, AuNS-based antibacterial agents have revealed promising abilities as nanoantibiotics. Moreover, in Table 1, we have summarized gold-based nanostructures and their antibacterial mechanisms in this review.

Table 1. Summary of gold-based nanostructures and their antibacterial mechanisms.

No	Nanostructures	Antibacterial Mechanism	Bacteria	Reference
1	Cefotaxime (CTX)-loaded AuNP(C-AuNP)	The C-AuNPs prevented the peptidoglycan cell wall synthesis by inhibiting transpeptidation in the bacterial cell wall.	*S. aureus*, *E. coli*, *Klebsiella oxytoca*, *P. aeruginosa*	[51]
2	Ciprofloxacin- AuNCs(CIP:AuNP)	ROS generation and damage cell membrane	*S. aureus* and *E. coli*	[57]
3	Epigallocatechin gallate (EGCG) to hydrogels modified with AuNPs(E-Au@H)	PTT	*S. aureus*, *E. coli*, and *S. aureus* biofilms	[64]
4	Gold-decorated magnetic nanoparticles of Fe0 conjugated with sulfamethoxazole (MNPs@AuNPs/SMX)	Cell wall damage and ROS generation	*S. aureus* and *E. coli*,	[69]
5	Thioproline (T) and Boc-protected thioproline (B) in combination with AuNPs(TB-AuNPs)	Cell wall damage	*E. coli*, MDR *E. coli* and *Klebsiella pneumonia*	[75]

Table 1. *Cont.*

No	Nanostructures	Antibacterial Mechanism	Bacteria	Reference
6	GSH-AuNCs	ROS generation	*Acetobacter aceti*	[84]
7	Phosphine-capped AuNCs	DNA damage and ROS generation	MDR *P. aeruginosa*	[85]
8	NHC-protected AuNCs	ROS generation	*S. aureus*, *E. coli*, and *C. albicans*	[86]
9	MHA-AuNCs	ROS generation, membrane damage, DNA damage	*S. aureus*, *E. coli*, *S. epidermidis*, *B. subtilis*, *C. difficile*, *F. nucleatum* and *P. aeruginosa*	[87,88,91, 96,110]
10	DTAC-AuNCs/ICG	ROS generation, PTT	MRSA	[92]
11	Lys-AuNCs-Cur	ROS generation	*S. aureus*, *E. coli*, and MRSA	[93]
12	AuNC-ZwBuEt	ROS generation	*P. aeruginosa* and *S. aureus*	[91]
13	MUTAB-AuNCs	ROS generation	*E. faecalis*, VRE, *S. pneumoniae* and *P. aeruginosa*	[94]
14.	QA-AuNCs	Membrane damage, ROS generation, and disturbance	MRSA	[111]
15	Polyurethane consists of gold nanorods and polyethylene glycol (PU-Au-PEG)	PTT	*P. aeruginosa*, *S. aureus*, MRSA	[100]
16	AuNRs636, AuNRs772, AuNPs, and AgNPs	PDT (ROS generation) and PTT	*E. coli*, *S. aureus*, *Salmonella enterica*, and MRSA	[101]
17	MoS2@AuNRs	PDT (ROS generation) and PTT	*E. coli*	[103]
18	AuNRs-(200) and AuNBPs-(111)	PTT	*E. coli*	[105]
19	D/L-Glu-Au NBPs	Chemotherapy, physiotherapy and PTT	*S. epidermidis*	[106]
20	Gold nanostar (AuNSs)	Cell membrane damage and PTT	*S. aureus*	[107]
21	Vancomycin-modified gold nanostars (AuNSs@Van)	Cell wall damage and PTT	methicillin-resistant *Staphylococcus aureus*	[109]

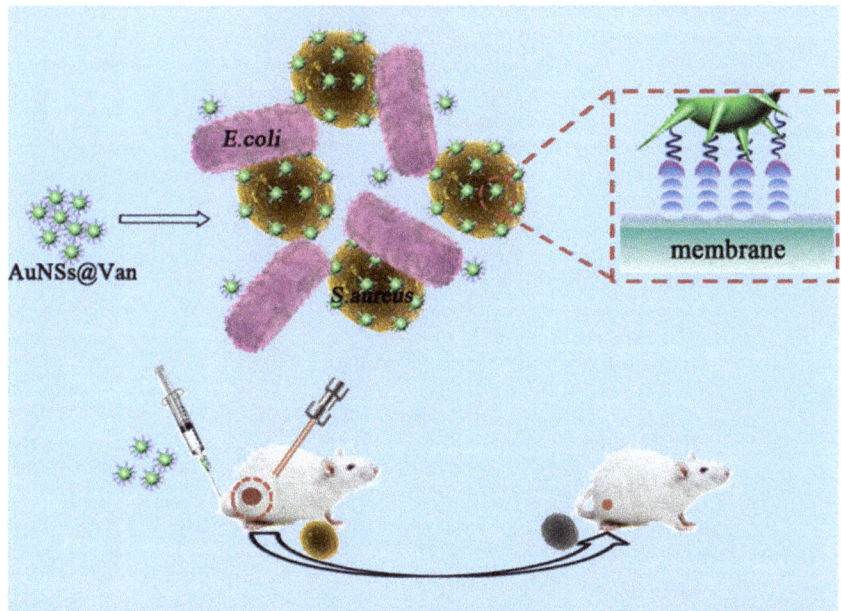

Figure 12. Schematic Illustration of vancomycin-coated gold nanostars (AuNSs@Van) for targeting and killing methicillin-resistant *Staphylococcus aureus* (MRSA) under near-infrared (NIR) laser irradiation. Reproduced with permission from ref. [109]. Copyright 2019 American Chemical Society.

5. Conclusions

In this review, recent achievements of Au-based nanostructures for antibacterial applications were collected and summarized according to their size and shape. Au-based nanostructures, including AuNPs, AuNCs, and AuNBPs, have been intensively utilized as antibacterial agents according to their antibacterial mechanisms, such as increasing ROS generation, PTT, PDT, and physical interactions. These Au-based nanostructures have revealed great potential for antibacterial applications in food conservation, wastewater depuration, and bacterial infections in humans. Combined with two or more antibacterial mechanisms, Au-based nanostructures have shown brilliant bactericidal effects. Although various Au-based nanostructures have been demonstrated as antibacterial agents, their performances of bacterial inactivation still need to be increased. For clinical treatment, antibacterial nanostructures have to reach 99.99% inhibition of bacterial growth. For future perspectives, to increase antibacterial activity, the rational design of Au-based nanostructures could focus on combining various antibacterial mechanisms based on their synergistic effects. Furthermore, with surface modifications, Au-based nanostructures can be conjugated with bactericidal motifs such as ligands and antibiotics. Eventually, for real clinical applications, the biocompatibility and elimination of Au-based antibacterial nanostructures in the human body should be investigated and improved. Gold nanostructures have gained significant attention in recent years due to their unique properties and potential applications in various fields, including medicine. As antibacterial agents, gold nanostructures offer several advantages, such as their small size, high surface-area-to-volume ratio, and ability to be easily functionalized with antibacterial agents. Although gold is generally considered biocompatible, the toxicity of gold nanostructures can vary depending on their size, shape, surface charge, and surface functionalization. It is essential to carefully evaluate the potential toxic effects of these nanostructures on both bacteria and human cells to ensure their safety and minimize any harmful side effects. Other challenges include understanding the precise mechanisms by which gold nanostructures exert their antibacterial activity, which is crucial. While it is believed that their small size and large surface area contribute to enhanced antimicrobial properties, the specific interactions between gold nanostructures and bacteria, such as membrane disruption or oxidative stress induction, need to be elucidated to optimize their antibacterial efficacy. On the positive side is their use to treat bacterial biofilms, which pose a significant challenge in healthcare settings, as they exhibit increased resistance to conventional antibiotics. Gold nanostructures have shown potential in disrupting biofilms and preventing their formation. Further research is needed to develop effective strategies for eradicating biofilms using gold nanostructures. Gold nanostructures have unique optical properties that can be harnessed for diagnostic purposes, such as biosensing or imaging. Additionally, they can serve as drug delivery vehicles, facilitating the targeted and controlled release of therapeutic agents at the site of infection. Human clinical studies of Au-based antibacterial nanostructures are a critical procedure for future practical applications in medicine. With great efforts from the scientific community, we believe that Au-based antibacterial nanostructures will show their potential as antibacterial agents in the coming days.

Author Contributions: Conceptualization, C.M., M.S., M.K., D.I.K., W., R.D., B.P., T.-M.C. and T.-R.K.; writing—original draft preparation, T.-M.C. and T.-R.K.; writing—review and editing, T.-M.C. and T.-R.K.; supervision, T.-R.K.; funding acquisition, T.-R.K. All authors have read and agreed to the published version of the manuscript.

Funding: We thank the financial support from the National Science and Technology Council (Taiwan, MOST 111-2113-M-038-003) and Taipei Medical University.

Institutional Review Board Statement: Not applicable.

Informed Consent Statement: Not applicable.

Data Availability Statement: No new data were created or analyzed in this study. Data sharing is not applicable to this article.

Conflicts of Interest: The authors declare no conflict of interest.

References

1. Zhu, Y.P.; Kuo, T.R.; Li, Y.H.; Qi, M.Y.; Chen, G.; Wang, J.L.; Xu, Y.J.; Chen, H.M. Emerging dynamic structure of electrocatalysts unveiled by in situ X-ray diffraction/absorption spectroscopy. *Energy Environ. Sci.* **2021**, *14*, 1928–1958. [CrossRef]
2. Okoro, G.; Husain, S.; Saukani, M.; Mutalik, C.; Yougbaré, S.; Hsiao, Y.-C.; Kuo, T.-R. Emerging trends in nanomaterials for photosynthetic biohybrid systems. *ACS Mater. Lett.* **2023**, *5*, 95–115. [CrossRef]
3. Chen, T.-Y.; Kuo, T.-R.; Yougbaré, S.; Lin, L.-Y.; Xiao, C.-Y. Novel direct growth of ZIF-67 derived Co_3O_4 and N-doped carbon composites on carbon cloth as supercapacitor electrodes. *J. Colloid Interface Sci.* **2022**, *608*, 493–503. [CrossRef] [PubMed]
4. Huang, C.-Y.; Kuo, T.-R.; Yougbaré, S.; Lin, L.-Y. Design of $LiFePO_4$ and porous carbon composites with excellent high-rate charging performance for lithium-ion secondary battery. *J. Colloid Interface Sci.* **2022**, *607*, 1457–1465. [CrossRef] [PubMed]
5. Kuo, T.-R.; Liao, H.-J.; Chen, Y.-T.; Wei, C.-Y.; Chang, C.-C.; Chen, Y.-C.; Chang, Y.-H.; Lin, J.-C.; Lee, Y.-C.; Wen, C.-Y. Extended visible to near-infrared harvesting of earth-abundant FeS_2-TiO_2 heterostructures for highly active photocatalytic hydrogen evolution. *Green Chem.* **2018**, *20*, 1640–1647. [CrossRef]
6. Chang, Y.H.; Lin, J.C.; Chen, Y.C.; Kuo, T.R.; Wang, D.Y. Facile synthesis of two-dimensional ruddlesden-popper perovskite quantum dots with fine-tunable optical properties. *Nanoscale Res. Lett.* **2018**, *13*, 247. [CrossRef]
7. Kuo, T.R.; Chen, W.T.; Liao, H.J.; Yang, Y.H.; Yen, H.C.; Liao, T.W.; Wen, C.Y.; Lee, Y.C.; Chen, C.C.; Wang, D.Y. Improving hydrogen evolution activity of earth-abundant cobalt-doped iron pyrite catalysts by surface modification with phosphide. *Small* **2017**, *13*, 1603356. [CrossRef]
8. Lo, S.-C.; Li, S.-S.; Yang, W.-F.; Wu, K.-C.; Wei, P.-K.; Sheen, H.-J.; Fan, Y.-J. A Co-printed nanoslit surface plasmon resonance structure in microfluidic device for LMP-1 detection. *Biosensors* **2022**, *12*, 653. [CrossRef] [PubMed]
9. Dutta, P.; Su, T.-Y.; Fu, A.-Y.; Chang, M.-C.; Guo, Y.-J.; Tsai, I.J.; Wei, P.-K.; Chang, Y.-S.; Lin, C.-Y.; Fan, Y.-J. Combining portable solar-powered centrifuge to nanoplasmonic sensing chip with smartphone reader for rheumatoid arthritis detection. *Chem. Eng. J.* **2022**, *434*, 133864. [CrossRef]
10. Hsieh, H.-Y.; Chang, R.; Huang, Y.-Y.; Juan, P.-H.; Tahara, H.; Lee, K.-Y.; Vo, D.N.K.; Tsai, M.-H.; Wei, P.-K.; Sheen, H.-J.; et al. Continuous polymerase chain reaction microfluidics integrated with a gold-capped nanoslit sensing chip for epstein-barr virus detection. *Biosens. Bioelectron.* **2022**, *195*, 113672. [CrossRef]
11. Lu, T.-Y.; Chiang, C.-Y.; Fan, Y.-J.; Jheng, P.-R.; Quinones, E.D.; Liu, K.-T.; Kuo, S.-H.; Hsieh, H.Y.; Tseng, C.-L.; Yu, J. Dual-targeting glycol chitosan/heparin-decorated polypyrrole nanoparticle for augmented photothermal thrombolytic therapy. *ACS Appl. Mater. Interfaces* **2021**, *13*, 10287–10300. [CrossRef]
12. Chuang, E.-Y.; Huang, M.-W.; Ho, T.-L.; Wang, P.-C.; Hsiao, Y.-C. Ir-inspired visual display/response device fabricated using photothermal liquid crystals for medical and display applications. *Chem. Eng. J.* **2022**, *429*, 132213. [CrossRef]
13. Lu, K.-Y.; Jheng, P.-R.; Lu, L.-S.; Rethi, L.; Mi, F.-L.; Chuang, E.-Y. Enhanced anticancer effect of ROS-boosted photothermal therapy by using fucoidan-coated polypyrrole nanoparticles. *Int. J. Biol. Macromol.* **2021**, *166*, 98–107. [CrossRef]
14. Chin, Y.-C.; Yang, L.-X.; Hsu, F.-T.; Hsu, C.-W.; Chang, T.-W.; Chen, H.-Y.; Chen, L.Y.-C.; Chia, Z.C.; Hung, C.-H.; Su, W.-C.; et al. Iron oxide@chlorophyll clustered nanoparticles eliminate bladder cancer by photodynamic immunotherapy-initiated ferroptosis and immunostimulation. *J. Nanobiotechnol.* **2022**, *20*, 373. [CrossRef] [PubMed]
15. Chang, W.-H.; Yang, Z.-Y.; Chong, T.-W.; Liu, Y.-Y.; Pan, H.-W.; Lin, C.-H. Quantifying cell confluency by plasmonic nanodot arrays to achieve cultivating consistency. *ACS Sens.* **2019**, *4*, 1816–1824. [CrossRef] [PubMed]
16. Mei, L.; Zhu, S.; Yin, W.; Chen, C.; Nie, G.; Gu, Z.; Zhao, Y. Two-dimensional nanomaterials beyond graphene for antibacterial applications: Current progress and future perspectives. *Theranostics* **2020**, *10*, 757. [CrossRef]
17. Chang, T.-W.; Ko, H.; Huang, W.-S.; Chiu, Y.-C.; Yang, L.-X.; Chia, Z.-C.; Chin, Y.-C.; Chen, Y.-J.; Tsai, Y.-T.; Hsu, C.-W. Tannic acid-induced interfacial ligand-to-metal charge transfer and the phase transformation of Fe_3O_4 nanoparticles for the photothermal bacteria destruction. *Chem. Eng. J.* **2022**, *428*, 131237. [CrossRef]
18. Hsu, I.-L.; Yeh, F.H.; Chin, Y.-C.; Cheung, C.I.; Chia, Z.C.; Yang, L.-X.; Chen, Y.-J.; Cheng, T.-Y.; Wu, S.-P.; Tsai, P.-J. Multiplex antibacterial processes and risk in resistant phenotype by high oxidation-state nanoparticles: New killing process and mechanism investigations. *Chem. Eng. J.* **2021**, *409*, 128266. [CrossRef]
19. Yougbare, S.; Mutalik, C.; Okoro, G.; Lin, I.H.; Krisnawati, D.I.; Jazidie, A.; Nuh, M.; Chang, C.C.; Kuo, T.R. Emerging trends in nanomaterials for antibacterial applications. *Int. J. Nanomed.* **2021**, *16*, 5831–5867. [CrossRef] [PubMed]
20. Yougbare, S.; Mutalik, C.; Krisnawati, D.I.; Kristanto, H.; Jazidie, A.; Nuh, M.; Cheng, T.M.; Kuo, T.R. Nanomaterials for the photothermal killing of bacteria. *Nanomaterials* **2020**, *10*, 1123. [CrossRef] [PubMed]
21. Mutalik, C.; Wang, D.Y.; Krisnawati, D.I.; Jazidie, A.; Yougbare, S.; Kuo, T.R. Light-activated heterostructured nanomaterials for antibacterial applications. *Nanomaterials* **2020**, *10*, 643. [CrossRef]
22. Murray, C.J.L.; Ikuta, K.S.; Sharara, F.; Swetschinski, L.; Robles Aguilar, G.; Gray, A.; Han, C.; Bisignano, C.; Rao, P.; Wool, E.; et al. Global burden of bacterial antimicrobial resistance in 2019: A systematic analysis. *Lancet* **2022**, *399*, 629–655. [CrossRef] [PubMed]
23. Mutalik, C.; Hsiao, Y.-C.; Chang, Y.-H.; Krisnawati, D.I.; Alimansur, M.; Jazidie, A.; Nuh, M.; Chang, C.-C.; Wang, D.-Y.; Kuo, T.-R. High uv-vis-NIR light-induced antibacterial activity by heterostructured TiO_2-FeS_2 nanocomposites. *Int. J. Nanomed.* **2020**, *15*, 8911. [CrossRef]

24. Chang, T.-K.; Cheng, T.-M.; Chu, H.-L.; Tan, S.-H.; Kuo, J.-C.; Hsu, P.-H.; Su, C.-Y.; Chen, H.-M.; Lee, C.-M.; Kuo, T.-R. Metabolic mechanism investigation of antibacterial active cysteine-conjugated gold nanoclusters in *Escherichia coli*. *ACS Sustain. Chem. Eng.* **2019**, *7*, 15479–15486. [CrossRef]
25. Yougbare, S.; Chang, T.-K.; Tan, S.-H.; Kuo, J.-C.; Hsu, P.-H.; Su, C.-Y.; Kuo, T.-R. Antimicrobial gold nanoclusters: Recent developments and future perspectives. *Int. J. Mol. Sci.* **2019**, *20*, 2924. [CrossRef]
26. Kaur, N.; Aditya, R.N.; Singh, A.; Kuo, T.R. Biomedical applications for gold nanoclusters: Recent developments and future perspectives. *Nanoscale Res. Lett.* **2018**, *13*, 302. [CrossRef]
27. Mutalik, C.; Lin, I.-H.; Krisnawati, D.I.; Khaerunnisa, S.; Khafid, M.; Widodo; Hsiao, Y.-C.; Kuo, T.-R. Antibacterial pathways in transition metal-based nanocomposites: A mechanistic overview. *Int. J. Nanomed.* **2022**, *17*, 6821–6842. [CrossRef] [PubMed]
28. Mutalik, C.; Krisnawati, D.I.; Patil, S.B.; Khafid, M.; Atmojo, D.S.; Santoso, P.; Lu, S.C.; Wang, D.Y.; Kuo, S.R. Phase-dependent MoS_2 nanoflowers for light-driven antibacterial application. *ACS Sustain. Chem. Eng.* **2021**, *9*, 7904–7912. [CrossRef]
29. Mutalik, C.; Okoro, G.; Chou, H.-L.; Lin, I.H.; Yougbaré, S.; Chang, C.-C.; Kuo, T.-R. Phase-dependent $1T/2H-MoS_2$ nanosheets for effective photothermal killing of bacteria. *ACS Sustain. Chem. Eng.* **2022**, *10*, 8949–8957. [CrossRef]
30. Lu, H.T.; Huang, G.Y.; Chang, W.J.; Lu, T.W.; Huang, T.W.; Ho, M.H.; Mi, F.L. Modification of chitosan nanofibers with CuS and fucoidan for antibacterial and bone tissue engineering applications. *Carbohydr. Polym.* **2022**, *281*, 119035. [CrossRef]
31. Dima, S.; Huang, H.T.; Watanabe, I.; Pan, Y.H.; Lee, Y.Y.; Chang, W.J.; Teng, N.C. Sequential application of calcium phosphate and epsilon-polylysine show antibacterial and dentin tubule occluding effects in vitro. *Int. J. Mol. Sci.* **2021**, *22*, 11. [CrossRef]
32. Shen, H.Y.; Liu, Z.H.; Hong, J.S.; Wu, M.S.; Shiue, S.J.; Lin, H.Y. Controlled-release of free bacteriophage nanoparticles from 3d-plotted hydrogel fibrous structure as potential antibacterial wound dressing. *J. Control. Release* **2021**, *331*, 154–163. [CrossRef] [PubMed]
33. Dima, S.; Lee, Y.Y.; Watanabe, I.; Chang, W.J.; Pan, Y.H.; Teng, N.C. Antibacterial effect of the natural polymer epsilon-polylysine against oral pathogens associated with periodontitis and caries. *Polymers* **2020**, *12*, 9. [CrossRef]
34. Huang, H.M.; Chen, F.L.; Lin, P.Y.; Hsiao, Y.C. Dielectric thermal smart glass based on tunable helical polymer-based superstructure for biosensor with antibacterial property. *Polymers* **2021**, *13*, 9. [CrossRef]
35. Tan, S.-H.; Yougbaré, S.; Tao, H.-Y.; Chang, C.-C.; Kuo, T.-R. Plasmonic gold nanoisland film for bacterial theranostics. *Nanomaterials* **2021**, *11*, 3139. [CrossRef]
36. Mutalik, C.; Okoro, G.; Krisnawati, D.I.; Jazidie, A.; Rahmawati, E.Q.; Rahayu, D.; Hsu, W.T.; Kuo, T.R. Copper sulfide with morphology-dependent photodynamic and photothermal antibacterial activities. *J. Colloid Interface Sci.* **2022**, *607*, 1825–1835. [CrossRef]
37. Liao, M.-Y.; Huang, T.-C.; Chin, Y.-C.; Cheng, T.-Y.; Lin, G.-M. Surfactant-free green synthesis of Au@chlorophyll nanorods for NIR PDT-elicited CDT in bladder cancer therapy. *ACS Appl. Bio Mater.* **2022**, *5*, 2819–2833. [CrossRef] [PubMed]
38. Lee, W.F.; Wang, L.Y.; Renn, T.Y.; Yang, J.C.; Fang, L.S.; Lee, Y.H.; Peng, P.W. Characterization and antibacterial properties of polyetherketoneketone coated with a silver nanoparticle-in-epoxy lining. *Polymers* **2022**, *14*, 2906. [CrossRef] [PubMed]
39. Pezzi, L.; Pane, A.; Annesi, F.; Losso, M.A.; Guglielmelli, A.; Umeton, C.; De Sio, L. Antimicrobial effects of chemically functionalized and/or photo-heated nanoparticles. *Materials* **2019**, *12*, 1078. [CrossRef]
40. Monaco, I.; Armanetti, P.; Locatelli, E.; Flori, A.; Maturi, M.; Del Turco, S.; Menichetti, L.; Comes Franchini, M. Smart assembly of mn-ferrites/silica core–shell with fluorescein and gold nanorods: Robust and stable nanomicelles for in vivo triple modality imaging. *J. Mater. Chem. B* **2018**, *6*, 2993–2999. [CrossRef]
41. Xu, W.; Lin, Q.; Yin, Y.; Xu, D.; Huang, X.; Xu, B.; Wang, G. A review on cancer therapy based on the photothermal effect of gold nanorod. *Curr. Pharm. Des.* **2019**, *25*, 4836–4847. [CrossRef] [PubMed]
42. Maturi, M.; Locatelli, E.; Sambri, L.; Tortorella, S.; Šturm, S.; Kostevšek, N.; Comes Franchini, M. Synthesis of ultrasmall single-crystal gold–silver alloy nanotriangles and their application in photothermal therapy. *Nanomaterials* **2021**, *11*, 912. [CrossRef] [PubMed]
43. Li, W.; Chen, X. Gold nanoparticles for photoacoustic imaging. *Nanomedicine* **2015**, *10*, 299–320. [CrossRef] [PubMed]
44. Alchera, E.; Monieri, M.; Maturi, M.; Locatelli, I.; Locatelli, E.; Tortorella, S.; Sacchi, A.; Corti, A.; Nebuloni, M.; Lucianò, R.; et al. Early diagnosis of bladder cancer by photoacoustic imaging of tumor-targeted gold nanorods. *Photoacoustics* **2022**, *28*, 100400. [CrossRef]
45. Chen, Y.-J.; Chang, W.-H.; Lin, C.-H. Selective growth of patterned monolayer gold nanoparticles on su-8 through photoreduction for plasmonic applications. *ACS Appl. Nano Mater.* **2021**, *4*, 229–235. [CrossRef]
46. Mobed, A.; Hasanzadeh, M.; Seidi, F. Anti-bacterial activity of gold nanocomposites as a new nanomaterial weapon to combat photogenic agents: Recent advances and challenges. *RSC Adv.* **2021**, *11*, 34688–34698. [CrossRef]
47. Chen, H.C.; Cheng, C.Y.; Lin, H.C.; Chen, H.H.; Chen, C.H.; Yang, C.P.; Yang, K.H.; Lin, C.M.; Lin, T.Y.; Shih, C.M.; et al. Multifunctions of excited gold nanoparticles decorated artificial kidney with efficient hemodialysis and therapeutic potential. *ACS Appl. Mater. Interfaces* **2016**, *8*, 19691–19700. [CrossRef]
48. Guo, R.H.; Shu, C.C.; Chuang, K.J.; Hong, G.B. Rapid colorimetric detection of phthalates using DNA-modified gold nanoparticles. *Mater. Lett.* **2021**, *293*, 4. [CrossRef]
49. Hong, G.B.; Hsu, J.P.; Chuang, K.J.; Ma, C.M. Colorimetric detection of 1-naphthol and glyphosate using modified gold nanoparticles. *Sustainability* **2022**, *14*, 16. [CrossRef]

50. Sengani, M.; Grumezescu, A.M.; Rajeswari, V.D. Recent trends and methodologies in gold nanoparticle synthesis-a prospective review on drug delivery aspect. *OpenNano* **2017**, *2*, 37–46. [CrossRef]
51. Al Hagbani, T.; Rizvi, S.M.D.; Hussain, T.; Mehmood, K.; Rafi, Z.; Moin, A.; Abu Lila, A.S.; Alshammari, F.; Khafagy, E.-S.; Rahamathulla, M.; et al. Cefotaxime mediated synthesis of gold nanoparticles: Characterization and antibacterial activity. *Polymers* **2022**, *14*, 771. [CrossRef] [PubMed]
52. Yeh, Y.-C.; Creran, B.; Rotello, V.M. Gold nanoparticles: Preparation, properties, and applications in bionanotechnology. *Nanoscale* **2012**, *4*, 1871–1880. [CrossRef] [PubMed]
53. Hu, X.; Zhang, Y.; Ding, T.; Liu, J.; Zhao, H. Multifunctional gold nanoparticles: A novel nanomaterial for various medical applications and biological activities. *Front. Bioeng. Biotechnol.* **2020**, *8*, 990. [CrossRef] [PubMed]
54. Anwar, Y.; Ullah, I.; Ul-Islam, M.; Alghamdi, K.M.; Khalil, A.; Kamal, T. Adopting a green method for the synthesis of gold nanoparticles on cotton cloth for antimicrobial and environmental applications. *Arab. J. Chem.* **2021**, *14*, 103327. [CrossRef]
55. Shaikh, S.; Nazam, N.; Rizvi, S.M.D.; Ahmad, K.; Baig, M.H.; Lee, E.J.; Choi, I. Mechanistic insights into the antimicrobial actions of metallic nanoparticles and their implications for multidrug resistance. *Int. J. Mol. Sci.* **2019**, *20*, 2468. [CrossRef]
56. Yu, Y.; Tan, L.; Li, Z.; Liu, X.; Zheng, Y.; Feng, X.; Liang, Y.; Cui, Z.; Zhu, S.; Wu, S. Single-atom catalysis for efficient sonodynamic therapy of methicillin-resistant *Staphylococcus aureus*-infected osteomyelitis. *ACS Nano* **2021**, *15*, 10628–10639. [CrossRef]
57. Wang, M.; Wang, X.; Liu, B.; Lang, C.; Wang, W.; Liu, Y.; Wang, X. Synthesis of ciprofloxacin-capped gold nanoparticles conjugates with enhanced sonodynamic antimicrobial activity in vitro. *J. Pharm. Sci.* **2023**, *112*, 336–343. [CrossRef]
58. Yang, M.; Xie, S.; Adhikari, V.P.; Dong, Y.; Du, Y.; Li, D. The synergistic fungicidal effect of low-frequency and low-intensity ultrasound with amphotericin b-loaded nanoparticles on c. Albicans in vitro. *Int. J. Pharm.* **2018**, *542*, 232–241. [CrossRef]
59. Costley, D.; Nesbitt, H.; Ternan, N.; Dooley, J.; Huang, Y.-Y.; Hamblin, M.R.; McHale, A.P.; Callan, J.F. Sonodynamic inactivation of gram-positive and gram-negative bacteria using a rose bengal–antimicrobial peptide conjugate. *Int. J. Antimicrob. Agents* **2017**, *49*, 31–36. [CrossRef]
60. Serpe, L.; Giuntini, F. Sonodynamic antimicrobial chemotherapy: First steps towards a sound approach for microbe inactivation. *J. Photochem. Photobiol. B Biol.* **2015**, *150*, 44–49. [CrossRef]
61. Canavese, G.; Ancona, A.; Racca, L.; Canta, M.; Dumontel, B.; Barbaresco, F.; Limongi, T.; Cauda, V. Nanoparticle-assisted ultrasound: A special focus on sonodynamic therapy against cancer. *Chem. Eng. J.* **2018**, *340*, 155–172. [CrossRef] [PubMed]
62. Serpe, L.; Foglietta, F.; Canaparo, R. Nanosonotechnology: The next challenge in cancer sonodynamic therapy. *Nanotechnol. Rev.* **2012**, *1*, 173–182.
63. Pender, D.S.; Vangala, L.M.; Badwaik, V.D.; Thompson, H.; Paripelly, R.; Dakshinamurthy, R. A new class of gold nanoantibiotics-direct coating of ampicillin on gold nanoparticles. *Pharm. Nanotechnol.* **2013**, *1*, 126–135. [CrossRef]
64. Dong, Z.; Lin, Y.; Xu, S.; Chang, L.; Zhao, X.; Mei, X.; Gao, X. NIR-triggered tea polyphenol-modified gold nanoparticles-loaded hydrogel treats periodontitis by inhibiting bacteria and inducing bone regeneration. *Mater. Des.* **2023**, *225*, 111487. [CrossRef]
65. Ma, Y.-C.; Zhu, Y.-H.; Tang, X.-F.; Hang, L.-F.; Jiang, W.; Li, M.; Khan, M.I.; You, Y.-Z.; Wang, Y.-C. Au nanoparticles with enzyme-mimicking activity-ornamented zif-8 for highly efficient photodynamic therapy. *Biomater. Sci.* **2019**, *7*, 2740–2748. [CrossRef] [PubMed]
66. Deng, X.; Liang, S.; Cai, X.; Huang, S.; Cheng, Z.; Shi, Y.; Pang, M.; Ma, P.A.; Lin, J. Yolk–shell structured Au nanostar@ metal–organic framework for synergistic chemo-photothermal therapy in the second near-infrared window. *Nano Lett.* **2019**, *19*, 6772–6780. [CrossRef]
67. Chambrone, L.; Wang, H.L.; Romanos, G.E. Antimicrobial photodynamic therapy for the treatment of periodontitis and peri-implantitis: An american academy of periodontology best evidence review. *J. Periodontol.* **2018**, *89*, 783–803.
68. Tong, X.; Qi, X.; Mao, R.; Pan, W.; Zhang, M.; Wu, X.; Chen, G.; Shen, J.; Deng, H.; Hu, R. Construction of functional curdlan hydrogels with bio-inspired polydopamine for synergistic periodontal antibacterial therapeutics. *Carbohydr. Polym.* **2020**, *245*, 116585. [CrossRef]
69. Shi, N.; Wang, H.; Cui, C.; Afshar, E.A.; Mehrabi, F.; Taher, M.A.; Shojaei, M.; Hamidi, A.S.; Dong, Y. Survey of antibacterial activity and release kinetics of gold-decorated magnetic nanoparticles of fe0 conjugated with sulfamethoxazole against *Escherichia coli* and *Staphylococcus aureus*. *Chemosphere* **2022**, *305*, 135179. [CrossRef]
70. Mehrabi, F.; Shamspur, T.; Sheibani, H.; Mostafavi, A.; Mohamadi, M.; Hakimi, H.; Bahramabadi, R.; Salari, E. Silver-coated magnetic nanoparticles as an efficient delivery system for the antibiotics trimethoprim and sulfamethoxazole against *E. coli* and *S. aureus*: Release kinetics and antimicrobial activity. *BioMetals* **2021**, *34*, 1237–1246. [CrossRef]
71. Mehrabi, F.; Shamspur, T.; Mostafavi, A.; Hakimi, H.; Mohamadi, M. Inclusion of sulfamethoxazole in a novel cufe2o4 nanoparticles/mesoporous silica-based nanocomposite: Release kinetics and antibacterial activity. *Appl. Organomet. Chem.* **2021**, *35*, e6035. [CrossRef]
72. Kim, S.-H.; Kim, M.-J.; Choa, Y.-H. Fabrication and estimation of Au-coated Fe3O4 nanocomposite powders for the separation and purification of biomolecules. *Mater. Sci. Eng. A* **2007**, *449*, 386–388. [CrossRef]
73. Buttersack, C. Modeling of type IV and V sigmoidal adsorption isotherms. *Phys. Chem. Chem. Phys.* **2019**, *21*, 5614–5626. [CrossRef] [PubMed]
74. Butler, M.S.; Paterson, D.L. Antibiotics in the clinical pipeline in October 2019. *J. Antibiot.* **2020**, *73*, 329–364. [CrossRef]
75. Wang, L.; Zheng, W.; Li, S.; Hou, Q.; Jiang, X. Modulating the antibacterial activity of gold nanoparticles by balancing their monodispersity and aggregation. *Chem. Commun.* **2022**, *58*, 7690–7693. [CrossRef]

76. Pillai, P.P.; Kowalczyk, B.; Kandere-Grzybowska, K.; Borkowska, M.; Grzybowski, B.A. Engineering gram selectivity of mixed-charge gold nanoparticles by tuning the balance of surface charges. *Angew. Chem. Int. Ed.* **2016**, *55*, 8610–8614. [CrossRef]
77. Raetz, C.R.; Whitfield, C. Lipopolysaccharide endotoxins. *Annu. Rev. Biochem.* **2002**, *71*, 635–700. [CrossRef]
78. Koerner, H.; MacCuspie, R.I.; Park, K.; Vaia, R.A. In situ uv/vis, saxs, and tem study of single-phase gold nanoparticle growth. *Chem. Mater.* **2012**, *24*, 981–995. [CrossRef]
79. Wang, L.; Li, S.; Yin, J.; Yang, J.; Li, Q.; Zheng, W.; Liu, S.; Jiang, X. The density of surface coating can contribute to different antibacterial activities of gold nanoparticles. *Nano Lett.* **2020**, *20*, 5036–5042. [CrossRef]
80. Tay, C.Y.; Yu, Y.; Setyawati, M.I.; Xie, J.; Leong, D.T. Presentation matters: Identity of gold nanocluster capping agent governs intracellular uptake and cell metabolism. *Nano Res.* **2014**, *7*, 805–815. [CrossRef]
81. Li, S.T.; Nagarajan, A.V.; Alfonso, D.R.; Sun, M.K.; Kauffman, D.R.; Mpourmpakis, G.; Jin, R.C. Boosting CO_2 electrochemical reduction with atomically precise surface modification on gold nanoclusters. *Angew. Chem. Int. Ed.* **2021**, *60*, 6351–6356. [CrossRef]
82. Seong, H.; Efremov, V.; Park, G.; Kim, H.; Yoo, J.S.; Lee, D. Atomically precise gold nanoclusters as model catalysts for identifying active sites for electroreduction of CO2. *Angew. Chem. Int. Ed.* **2021**, *60*, 14563–14570. [CrossRef] [PubMed]
83. Zhang, B.H.; Chen, J.S.; Cao, Y.T.; Chai, O.J.H.; Xie, J.P. Ligand design in ligand-protected gold nanoclusters. *Small* **2021**, *17*, 20. [CrossRef]
84. Kuo, J.-C.; Tan, S.-H.; Hsiao, Y.-C.; Mutalik, C.; Chen, H.-M.; Yougbaré, S.; Kuo, T.-R. Unveiling the antibacterial mechanism of gold nanoclusters via in situ transmission electron microscopy. *ACS Sustain. Chem. Eng.* **2021**, *10*, 464–471. [CrossRef]
85. Ndugire, W.; Raviranga, N.H.; Lao, J.; Ramström, O.; Yan, M. Gold nanoclusters as nanoantibiotic auranofin analogues. *Adv. Healthc. Mater.* **2022**, *11*, 2101032. [CrossRef] [PubMed]
86. Wu, Q.; Peng, R.; Gong, F.; Luo, Y.; Zhang, H.; Cui, Q. Aqueous synthesis of n-heterocyclic carbene-protected gold nanoclusters with intrinsic antibacterial activity. *Colloids Surf. A Physicochem. Eng. Asp.* **2022**, *645*, 128934. [CrossRef]
87. Zheng, K.; Setyawati, M.I.; Leong, D.T.; Xie, J. Antimicrobial gold nanoclusters. *ACS Nano* **2017**, *11*, 6904–6910. [CrossRef]
88. Zheng, K.; Setyawati, M.I.; Leong, D.T.; Xie, J. Overcoming bacterial physical defenses with molecule-like ultrasmall antimicrobial gold nanoclusters. *Bioact. Mater.* **2021**, *6*, 941–950. [CrossRef]
89. Wang, Y.; Malkmes, M.J.; Jiang, C.; Wang, P.; Zhu, L.; Zhang, H.; Zhang, Y.; Huang, H.; Jiang, L. Antibacterial mechanism and transcriptome analysis of ultra-small gold nanoclusters as an alternative of harmful antibiotics against gram-negative bacteria. *J. Hazard. Mater.* **2021**, *416*, 126236. [CrossRef]
90. Ndugire, W.; Truong, D.; Raviranga, L.; Lao, J.; Ramstrom, O.; Yan, M. Turning on the antimicrobial activity of gold nanoclusters against multidrug-resistant bacteria. *Angew. Chem. Int. Ed.* **2022**, *62*, e202214086. [CrossRef]
91. Linklater, D.P.; Le Guével, X.; Bryant, G.; Baulin, V.A.; Pereiro, E.; Perera, P.G.T.; Wandiyanto, J.V.; Juodkazis, S.; Ivanova, E.P. Lethal interactions of atomically precise gold nanoclusters and *Pseudomonas aeruginosa* and *Staphylococcus aureus* bacterial cells. *ACS Appl. Mater. Interfaces* **2022**, *14*, 32634–32645. [CrossRef]
92. Zhuo, Y.; Zhang, Y.; Wang, B.; Cheng, S.; Yuan, R.; Liu, S.; Zhao, M.; Xu, B.; Zhang, Y.; Wang, X. Gold nanocluster & indocyanine green based triple-effective therapy for mrsa infected central nervous system. *Appl. Mater. Today* **2022**, *27*, 101453.
93. Mai-Prochnow, A.; Clauson, M.; Hong, J.; Murphy, A.B. Gram positive and gram negative bacteria differ in their sensitivity to cold plasma. *Sci. Rep.* **2016**, *6*, 38610. [CrossRef] [PubMed]
94. Li, Y.; Zhen, J.; Tian, Q.; Shen, C.; Zhang, L.; Yang, K.; Shang, L. One step synthesis of positively charged gold nanoclusters as effective antimicrobial nanoagents against multidrug-resistant bacteria and biofilms. *J. Colloid Interface Sci.* **2020**, *569*, 235–243. [CrossRef] [PubMed]
95. Srinivasulu, Y.G.; Mozhi, A.; Goswami, N.; Yao, Q.; Xie, J. Gold nanocluster based nanocomposites for combinatorial antibacterial therapy for eradicating biofilm forming pathogens. *Mater. Chem. Front.* **2022**, *6*, 689–706. [CrossRef]
96. Zhang, Y.; Chen, R.; Wang, Y.; Wang, P.; Pu, J.; Xu, X.; Chen, F.; Jiang, L.; Jiang, Q.; Yan, F. Antibiofilm activity of ultra-small gold nanoclusters against fusobacterium nucleatum in dental plaque biofilms. *J. Nanobiotechnol.* **2022**, *20*, 1–17. [CrossRef]
97. Liao, Y.T.; Liu, C.H.; Chin, Y.; Chen, S.Y.; Liu, S.H.; Hsu, Y.C.; Wu, K.C.W. Biocompatible and multifunctional gold nanorods for effective photothermal therapy of oral squamous cell carcinoma. *J. Mater. Chem. B* **2019**, *7*, 4451–4460. [CrossRef]
98. Lin, M.Y.; Hsieh, H.H.; Chen, J.C.; Chen, C.L.; Sheu, N.C.; Huang, W.S.; Ho, S.Y.; Chen, T.W.; Lee, Y.J.; Wu, C.Y. Brachytherapy approach using Lu-177 conjugated gold nanostars and evaluation of biodistribution, tumor retention, dosimetry and therapeutic efficacy in head and neck tumor model. *Pharmaceutics* **2021**, *13*, 13. [CrossRef]
99. Yang, X.; Chen, Y.; Zhang, X.; Xue, P.; Lv, P.; Yang, Y.; Wang, L.; Feng, W. Bioinspired light-fueled water-walking soft robots based on liquid crystal network actuators with polymerizable miniaturized gold nanorods. *Nano Today* **2022**, *43*, 101419. [CrossRef]
100. Zhao, Y.-Q.; Sun, Y.; Zhang, Y.; Ding, X.; Zhao, N.; Yu, B.; Zhao, H.; Duan, S.; Xu, F.-J. Well-defined gold nanorod/polymer hybrid coating with inherent antifouling and photothermal bactericidal properties for treating an infected hernia. *ACS Nano* **2020**, *14*, 2265–2275. [CrossRef]
101. Shao, L.; Majumder, S.; Liu, Z.; Xu, K.; Dai, R.; George, S. Light activation of gold nanorods but not gold nanospheres enhance antibacterial effect through photodynamic and photothermal mechanisms. *J. Photochem. Photobiol. B Biol.* **2022**, *231*, 112450. [CrossRef] [PubMed]
102. Annesi, F.; Pane, A.; Losso, M.A.; Guglielmelli, A.; Lucente, F.; Petronella, F.; Placido, T.; Comparelli, R.; Guzzo, M.G.; Curri, M.L.; et al. Thermo-plasmonic killing of *Escherichia coli* tg1 bacteria. *Materials* **2019**, *12*, 1530. [CrossRef]

103. Yougbaré, S.; Mutalik, C.; Chung, P.-F.; Krisnawati, D.I.; Rinawati, F.; Irawan, H.; Kristanto, H.; Kuo, T.-R. Gold nanorod-decorated metallic MoS_2 nanosheets for synergistic photothermal and photodynamic antibacterial therapy. *Nanomaterials* **2021**, *11*, 3064. [CrossRef] [PubMed]
104. Chow, T.H.; Li, N.; Bai, X.; Zhuo, X.; Shao, L.; Wang, J. Gold nanobipyramids: An emerging and versatile type of plasmonic nanoparticles. *Acc. Chem. Res.* **2019**, *52*, 2136–2146. [CrossRef] [PubMed]
105. Yougbaré, S.; Chou, H.-L.; Yang, C.-H.; Krisnawati, D.I.; Jazidie, A.; Nuh, M.; Kuo, T.-R. Facet-dependent gold nanocrystals for effective photothermal killing of bacteria. *J. Hazard. Mater.* **2021**, *407*, 124617. [CrossRef]
106. Zhang, M.; Zhang, H.; Feng, J.; Zhou, Y.; Wang, B. Synergistic chemotherapy, physiotherapy and photothermal therapy against bacterial and biofilms infections through construction of chiral glutamic acid functionalized gold nanobipyramids. *Chem. Eng. J.* **2020**, *393*, 124778. [CrossRef]
107. Penders, J.; Stolzoff, M.; Hickey, D.J.; Andersson, M.; Webster, T.J. Shape-dependent antibacterial effects of non-cytotoxic gold nanoparticles. *Int. J. Nanomed.* **2017**, *12*, 2457. [CrossRef]
108. You, Y.-H.; Lin, Y.-F.; Nirosha, B.; Chang, H.-T.; Huang, Y.-F. Polydopamine-coated gold nanostar for combined antitumor and antiangiogenic therapy in multidrug-resistant breast cancer. *Nanotheranostics* **2019**, *3*, 266. [CrossRef]
109. Wang, H.; Song, Z.; Li, S.; Wu, Y.; Han, H. One stone with two birds: Functional gold nanostar for targeted combination therapy of drug-resistant *Staphylococcus aureus* infection. *ACS Appl. Mater. Interfaces* **2019**, *11*, 32659–32669. [CrossRef]
110. Yang, H.; Cai, R.; Zhang, Y.; Chen, Y.; Gu, B. Gold nanoclusters as an antibacterial alternative against clostridium difficile. *Int. J. Nanomed.* **2020**, *15*, 6401. [CrossRef]
111. Xie, Y.; Liu, Y.; Yang, J.; Liu, Y.; Hu, F.; Zhu, K.; Jiang, X. Gold nanoclusters for targeting methicillin-resistant *Staphylococcus aureus* in vivo. *Angew. Chem. Int. Ed.* **2018**, *57*, 3958–3962. [CrossRef]

Disclaimer/Publisher's Note: The statements, opinions and data contained in all publications are solely those of the individual author(s) and contributor(s) and not of MDPI and/or the editor(s). MDPI and/or the editor(s) disclaim responsibility for any injury to people or property resulting from any ideas, methods, instructions or products referred to in the content.

Article

Influence of the Structure on Magnetic Properties of Calcium-Phosphate Systems Doped with Iron and Vanadium Ions

Dania Racolta [1], Constantin Andronache [1], Maria Balasoiu [2,3,4], Leonard Mihaly-Cozmuta [1], Vadim Sikolenko [2,5,6,*], Oleg Orelovich [2], Andrey Rogachev [2,7], Gheorghe Borodi [8] and Gheorghe Iepure [1]

[1] Faculty of Sciences, North University Center Baia Mare, Technical University of Cluj Napoca, 430122 Baia Mare, Romania
[2] Joint Institute for Nuclear Research, Dubna 141980, Russia
[3] Horia Hulubei National Institute of Physics and Nuclear Engineering, 077125 Magurele, Romania
[4] R&D CSMBA, Faculty of Physics, West University of Timisoara, 300223 Timișoara, Romania
[5] Karlsruhe Institute of Technology, 76131 Karlsruhe, Germany
[6] REC "Functional Materials", Immanuel Kant Baltic Federal University, Kaliningrad 236016, Russia
[7] Moscow Institute of Physics and Technology, Dolgoprudniy 141701, Russia
[8] National Institute for Research and Development of Isotopic and Molecular Technologies, 400293 Cluj-Napoca, Romania
* Correspondence: vadim.sikolenko@jinr.ru

Abstract: The aim of this study was to prepare and characterize the glasses made of $x(Fe_2O_3 \cdot V_2O_5) \cdot (100 - x)[P_2O_5 \cdot CaO]$ with x ranging of 0–50%. The contribution of Fe_2O_3 and V_2O_5 amount on the structure of $P_2O_5 \cdot CaO$ matrix was investigated. The vitreous materials were characterized by XRD (X-ray diffraction analysis), EPR (Electron Paramagnetic Resonance) spectroscopy, and magnetic susceptibility measurements. A hyperfine structure typical for isolated V^{4+} ions was noticed to all spectra containing low amount of V_2O_5. The XRD spectra show the amorphous nature of samples, apart x = 50%. An overlap of the EPR spectrum of a broad line without the hyperfine structure characteristic of clustered ions was observed with increasing V_2O_5 content. The results of magnetic susceptibility measurements explain the antiferromagnetic or ferromagnetic interactions expressed between the iron and vanadium ions in the investigated glass.

Keywords: calcium phosphate glasses; XRD; EPR; magnetic susceptibility

1. Introduction

Studies of phosphate glasses have attracted great interest from science, engineering, and technological fields due to their valuable physical properties, which are different from silicate and borate glasses [1,2]. These properties include a lower melting temperature, high transparency in the UV domain, reduced viscosity, low glass transition temperatures, and elevated thermal expansion coefficients [1,2]. Phosphorus-based glassy materials have a disordered and partially disordered structure and by doping with different transition metal ions acquire some special electrical, optical and magnetic properties [3–7] compared to crystalline materials. In this way, they become useful in many fields such as electronics, optics, sealing materials, bio-glass fabrication, and also in some cases in microbiological and biomedical applications [8–11]. By adding various elements and transition metals such as iron, vanadium, and copper oxides to the phosphate glass network, the dissolution rate of the new compounds decreases while their chemical durability increases [12].

The addition of elements such as calcium, sodium, and lithium also make the compounds suitable as biomaterials [13–17]. Calcium phosphate materials are currently known as materials used by the human body to build bone or to automatically produce material for bone repair and regeneration. Some of them are osteoconductive and others are osteoinductive [18].

The forming oxide in phosphate glasses is P2O5. It has a different structure compared to other glass formers due to the existence of a terminal oxygen on each network cation. This exhibits a covalent P=O double bond and influences the additional valence electron. In fact, the P_2O_5 glass structure consists of a network in which three of the oxygens are bridged (P-O-P) and one is non-bridged (P=O). The basic elements of the network are PO_4 tetrahedra linked by covalent oxygen bridges to form different phosphate anions [19]. The number of oxygen bonds present in the phosphate tetrahedra are used to define the structure of the phosphate glasses.

The properties of phosphate glasses can be modified by adding alkali and alkaline-earth oxides, such as CaO or Li_2CO_3, to the glass network, thereby achieving a partial structural modification [20,21]. In this way, non-bridging oxygens are created at the expense of bridging oxygens, leading to depolymerization of the phosphate network.

Ferric or ferrous oxides (e.g., Fe_2O_3 or FeO) confer interesting effects on the structure and properties of the phosphate-based glasses. Among the two iron ions, only Fe^{3+} shows EPR absorptions at room temperature, although both Fe^{3+} and Fe^{2+} display paramagnetic properties [22]. Although Vanadium ions incorporated in the phosphate glasses present two oxidation states, V^{4+} and V^{5+}, only vanadium in its +4 oxidation state is paramagnetic. In the oxide matrix of the glass, the V^{4+} ions tend to form their specific VO^{2+} complexes. As a consequence of hopping an unpaired $3d^1$ electron from the V^{4+} site to the V^{5+} site, electrical conduction occurs. The possibility of the vanadium ion to change its oxidation state during the melting and quenching process of glass preparation was reported in the literature [23–26]. The data reveal structural changes of the units due to the formation of bonds between non-bridging oxygen and iron atoms in different coordination, while the increase in the content of vanadium ions in the samples leads to the decrease of the activation energy [23–26].

The present paper reports structural and magnetic investigations on new calcium phosphate glasses systems. The synthesis, structural, and magnetic properties of these glass systems, based on CaO and P_2O_5 as network formers doped with Fe and V ions, were investigated by XRD, EPR, and magnetic susceptibility measurements in a large concentration range with x varying between in the range $0 \leq x \leq 50\%$.

XRD analyzes have been employed to characterize the short-range order and electronic structures of the samples, as well as earlier for other types of glassy materials [27,28].

The EPR spectroscopy and magnetic susceptibility measurements are used to obtain complementary data regarding the influence of $Fe_2O_3 \cdot V_2O_5$ content on the local symmetry, and interactions between iron and vanadium ions in the $P_2O_5 \cdot CaO$ glass matrix.

Investigations by the EPR technique have been reported on similar phosphate glasses to provide the most direct and accurate descriptions of the ground states and neighborhood effects on the energy levels of the paramagnetic centers, and to make it possible to determine the crystal field parameters [29,30].

The concentration of the 3D transition metal ions combined with the ratio of valence states and the structure of the glassy matrix influence the magnetic properties of sample. Through magnetic susceptibility measurements, the valence states of transition metal ions and the type of interactions involving them can be measured [31]. An antiferromagnetic coupling of Fe_2O_3-P_2O_5-CaO and Fe_2O_3-V_2O_5-P_2O_5-CaO can be considered responsible for the super-exchange interaction of the iron ions in the oxide glasses. The magnetic properties and the antiferromagnetic coupling between iron ions in different phosphate, borate, aluminosilicate, and oxide glasses have been reported previously [32–35].

The sample preparation conditions, the structure of the glass matrix, and the Fe^{3+}/Fe^{2+} ratio determine the concentration range in which antiferromagnetic interactions occur.

2. Results and Discussion
2.1. XRD Data

The X-ray powder diffraction pattern for $x(Fe_2O_3 \cdot V_2O_5) \cdot (100 - x)[P_2O_5 \cdot CaO]$ glass systems are presented in Figure 1a,b. It is observed that apart from sample x = 50 mol%, all the other samples do not show characteristic diffraction peaks for crystalline phases, these being in an amorphous state. The diffraction lines for sample x = 50mol% (Figure 1b) are attributed to the crystalline phase V_3O_5 (PDF: 72-0524), which crystallizes in the monoclinic system having the space group Cc and the following lattice parameters: a = 9.98 Å, b = 5.03 Å, c = 9.84 Å, and β = 138.8⁰. For all other samples with 0 ≤ x < 50 mol%, it can be noted that they each have a diffraction halo characteristic for the amorphous phase, which reflects the local order. The location of this broadening halo is difficult to determine precisely, but it can be stated that the position of its maximum moves to bigger angles when the sample changes from x = 0, 1, 3, 5, 10, 20, 35 mol%.

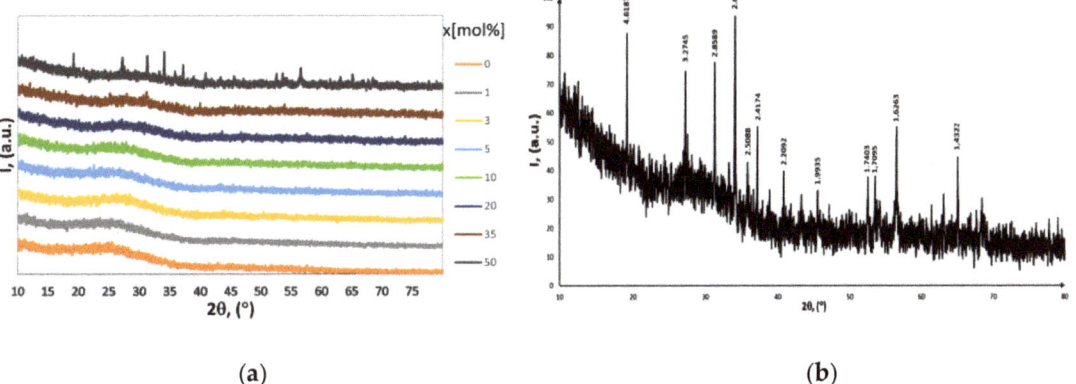

(a) (b)

Figure 1. (a) The XRD patterns of $x(Fe_2O_3 \cdot V_2O_5) \cdot (100 - x)[P_2O_5 \cdot CaO]$ glasses systems (0 ≤ x ≤ 50 mol%), (b) The XRD pattern of $x(Fe_2O_3 \cdot V_2O_5) \cdot (100 - x)[P_2O_5 \cdot CaO]$ for x = 50%mol, normalized to maximum.

The average distance between the atoms in the first coordination sphere R can be evaluated from the position of the halo diffraction maximum using the relation R = (5λ)/(8 sinθ). From this relationship, it was found that when Fe and V ions are introduced into the matrix containing Ca and Fe ions, the average distance between the ions increases from approximately 4 Å to 4.5 Å.

It can be seen that the diffraction halos are a little more prominent, which means that the local order also decreases with x increasing.

2.2. EPR Data

The glass systems $x(Fe_2O_3 \cdot V_2O_5) \cdot (100 - x)[P_2O_5 \cdot CaO]$ were investigated by applying the EPR technique for x values in the range 0–50 mol%. The EPR spectrum of the prepared samples is presented in Figure 2.

As illustrated in Figure 2, there is a strong dependence between the absorption spectral structure and transition metal content parameters.

At low concentrations of $Fe_2O_3 \cdot V_2O_5$, 0.5 ≤ x ≤ 10 mol %, the resulting spectra can be considered as an overlay of two EPR signals: (i) one with a well-resolved hyperfine structure typical of isolated V^{4+} ions; (ii) the other, with a broad line without hyperfine structure, typical for associated ions (Fe^{3+} and/or V^{4+}).

For 20 ≤ x ≤ 50 mol %, the hyperfine structure and line resolution is significantly reduced, leaving only a broad line, as a result of the increase in the number of ions associated with the $Fe_2O_3 \cdot V_2O_5$ content.

The g~4.3 line disappears for x ≥ 20% of transitional oxides, indicating that isolated Fe^{3+} ions form either Fe^{3+}-O-Fe^{3+} bonds or Fe^{3+}-O-V^{4+} interaction pairs.

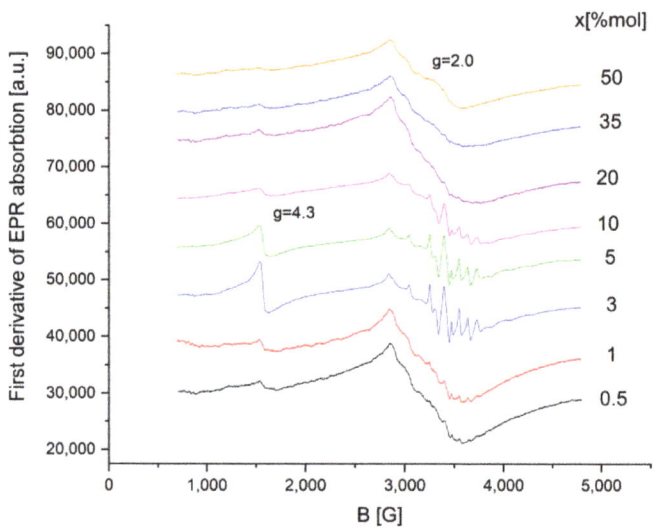

Figure 2. The first derivative of EPR absorption for $x(Fe_2O_3 \cdot V_2O_5) \cdot (100 - x)[P_2O_5 \cdot CaO]$ glasses systems.

The appearance of the g~2 line may be caused by the dipole-dipole interactions between Fe^{3+} ions attributed to the formation of Fe^{3+}-O-Fe^{3+} bonds. These interactions lead to the formation of iron ions clusters.

The g~2 line for x ≥ mol% content is also unresolved due to the presence of $Fe^{3+} \rightarrow V^{4+}$ electron transitions, but $V^{4+} \rightarrow V^{5+}$ transitions cannot be neglected.

The resonance line evolution with the increasing of iron and vanadium ions content in the samples can be determined using the approximate relation J = I(ΔH)2, where I represents the amplitude of the resonance line, and ΔH is the line-width.

The intensity of the resonance line, J, indicates the number of active species in resonant absorption, ΔH reflects the competition between different broadening mechanisms: dipole-dipole interactions, increasing disorder in the matrix structure, and interactions between ions with different valence states. The resonance lines centered at g~4.3 and g~2.0 in the spectra are typically for Fe^{3+} and V^{4+} ions present in the oxide glasses, their prevalence depending on x concentrations (Figure 3a,b).

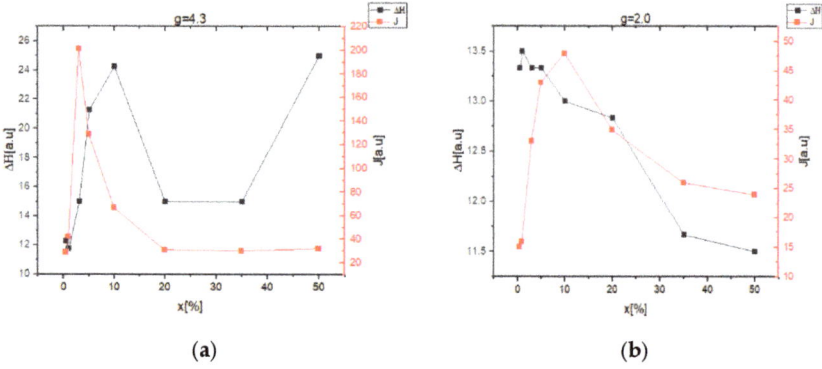

Figure 3. Correlation between intensity (J) of resonance line and the line-width (ΔH) for $x(Fe_2O_3 \cdot V_2O_5) \cdot (100 - x)[P_2O_5 \cdot CaO]$ glass systems: (**a**) g~4.3, (**b**) g~2.0.

The typical line-width (ΔH) increases with increasing of ($Fe_2O_3 \cdot V_2O_5$) content for $x \leq 10$ mol% (Figure 3a) and similarly for $x \leq 1$% (Figure 3b), suggesting that the dipole-dipole interactions prevails among V^{4+} ions even for clustered ions.

The decrease in the line-width with the increase in ($Fe_2O_3 \cdot V_2O_5$) content to more than 10 mol% (Figure 3a) and $x > 1$% (Figure 3b) shows that, in this composition range, the super-exchange interaction become dominant between resonance centers.

The increase in the concentration of Fe and V ions in the system can explain the J = f(x) dependence. The increase in the intensity of the resonance line (J) in the low concentration range 0–5%, (Figure 3a) and 0–10% (Figure 3b) indicates that the number of absorption centers is increasing. The decrease of J in the concentration range 5–50% (Figure 3a) and 10–50% (Figure 3b) shows the formation of iron and vanadium clusters.

2.3. Magnetic Susceptibility Data

The correlation between temperature and the reciprocal magnetic susceptibility for the glass samples with x varying in the range 0 < x < 50 mol% is presented in Figure 4.

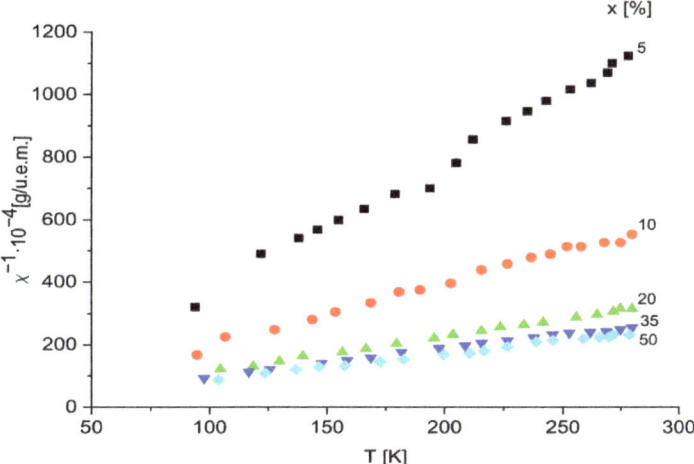

Figure 4. Relation between temperature and the reciprocal magnetic susceptibility for $x(Fe_2O_3 \cdot V_2O_5) \cdot (100 - x)[P_2O_5 \cdot CaO]$ glasses with 0 < x < 50 mol %.

The temperature correlation with the reciprocal magnetic susceptibility involves a Curie–Weiss-type behavior ($\chi = C/T - \theta$) with negative paramagnetic Curie temperature, and suggests that magnetic transition ions are isolated and/or participating in dipolar interactions in this concentration range. Therefore, we can consider that these ions are randomly distributed in the vitreous matrix, being located at distances that do not allow for the magnetic overload interaction through oxygen ions.

This shows that in all the studied concentration ranges, iron, and vanadium ions participate in different proportions depending on x, in super-exchange magnetic interactions, of antiferromagnetic type, behaving magnetically similar to other oxide glasses [36–38]. However, the ferric oxide concentration above which magnetic super-exchange interactions occur is lower.

In this way, one can conclude that the concentration range of transition metal ions in which these interactions occur depends on the nature of the glass matrix. Using the representation $1/\chi = f(T)$, we calculated the molar Curie constant, C_M, for each sample. C_M values increase with the increase in the content of transition metals.

The presence of magnetic ions: V^{4+} (μ_{eff} = 1.8 μB), Fe^{3+} (μ_{eff} = 5.9 μB), Fe^{2+} (μ_{eff} = 5.1 μB) is confirmed by experimental Curie constants. Next, using the relation $\mu_{eff} = 2.827[C_m/2x]1/2$, the magnetic moments of the samples under study were calculated.

The obtained results are for x = 10 mol%, μ_{eff} = 5.3 µB (where µB is magneton Bohr), while for the samples with higher concentration, μ_{eff} decreases to μ_{eff} = 3.67 µB. In Table 1, the obtained values of CM and μ_{eff} are given for samples with x ranging from 0–50 mol%.

Table 1. CM, μ_{eff} for $x(Fe_2O_3 \cdot V_2O_5) \cdot (100 - x)[P_2O_5 \cdot CaO]$ with x ranging from 0–50 mol%.

x [%]	CM [u.e.m./mol]	μ_{eff} [µB]
5	0.73	5.40
10	1.41	5.3
20	2.54	5.03
35	3.10	4.20
50	3.38	3.67

3. Materials and Methods

Glasses of the $x(Fe_2O_3 \cdot V_2O_5) \cdot (100 - x)[P_2O_5 \cdot CaO]$ system were prepared. For the systems studied, reagent grade purity substances were used for analysis. Samples were obtained by a proportional weighing of components, mixed, and melted in sintered corundum crucibles at 1523 K for a 5 min timeframe, and sudden cooling (Figure 5). Eight samples of $x(Fe_2O_3 \cdot V_2O_5) \cdot (100 - x)[CaO \cdot Li2O]$ with x = 0, 1, 3, 5, 10, 20, 35 mol% were obtained.

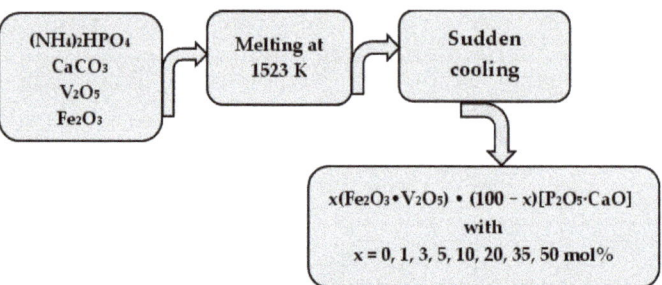

Figure 5. Scheme of the preparation procedure of the compounds $x(Fe_2O_3 \cdot V_2O_5) \cdot (100 - x)[P_2O_5 \cdot CaO]$ with x = 0, 1, 3, 5, 10, 20, 35, 50 mol%.

The structural analysis did not reveal any crystalline formations in the samples up to 50 mol% Fe_2O_3.

Diffraction data were obtained using a Bruker D8 Advance diffractometer using a Cu X-ray tube, the diffractometer being equipped with a germanium (1 1 1) monochromator in the incident beam and a LINXEYE position detector.

The EPR studies were conducted with a Portable Adani PS8400 spectrometer, at room temperature, in the X-ray frequency band.

For EPR investigations the samples were milled and inserted into sample tubes of the same caliber to ensure the same filling factor of the resonant cavity of the spectrometer for all samples. The mass of all samples was 100 mg.

The determination of the magnetic susceptibility (χ) of the samples was achieved by measuring the force with which a magnetic field acts on each of these glass systems, using a Faraday balance, in the temperature range 80–300 K.

For this purpose, the sample connected to an analytical balance, placed in a quartz cup attached to the end of a quartz rod, was introduced into the magnetic field of an electromagnet. The mass of the sample was measured using the analytical balance with and without the field. In order to obtain the real magnetic susceptibility of iron ions in the studied glass compounds, due to the diamagnetism of the P_2O_5, CaO, and Fe_2O_3, corrections were considered.

4. Conclusions

A new system of phosphate glasses doped with transition ions such as $x(Fe_2O_3 \cdot V_2O_5) \cdot (100-x)[P_2O_5 \cdot CaO]$ was obtained and investigated in a large concentration range, i.e., $0 \leq x \leq 50\%$.

In this matrix system, P-O bonds are broken and P-O-Fe bonds are formed when iron ions are added. The gradual decrease in the number of bridging oxygen ions between Fe-O-Fe and Fe-O-V is determined by the increase in the iron ions content of the samples.

In calcium phosphate glasses, vanadium is found as isolated V^{4+} ions in C4V coordination for small transitional metal oxide content.

The XRD spectra of the glass systems $x(Fe_2O_3 \cdot V_2O_5) \cdot (100-x)[P_2O_5 \cdot CaO]$ show that, apart from the x = 50%mol sample, all the other samples do not show diffraction peaks characteristic of crystalline phases, those being in an amorphous state.

EPR measurements show that at low concentrations of $Fe_2O_3 \cdot V_2O_5$, we find the presence of the two species of magnetic ions Fe^{3+} and V^{4+} in isolated positions, respectively, with the resolved hyperfine structure, the dominant interaction being the dipole-dipole one. With the increase in the concentration, the interactions of super-exchange and pair formation become dominant. The binding of isolated Fe^{3+} ions is carried out either through Fe^{3+}-O-Fe^{3+} bridges, or they interact as Fe^{3+}-O-V^{4+} pairs.

The magnetic measurements show that the transition ions in the glasses of the system $x(Fe_2O_3 \cdot V_2O_5) \cdot (100-x)[P_2O_5 \cdot CaO]$ are in the valence states V^{4+}, V^{5+}, Fe^{3+}, and Fe^{2+}. For increased concentrations, the inverse of the magnetic susceptibility is determined by a Curie–Weiss-type law with negative Curie temperature.

In the preparation of the new phosphate glasses, the addition of transition metal, such as iron and vanadium ions, at various concentrations, increases the chemical resistance of the new compound. Compared to other types of phosphate glass, we believe that the new compounds containing calcium oxide as network modifiers could become good candidates for applications in various fields such as, for example, biomedical application, and tissue engineering.

Author Contributions: Conceptualization, D.R., C.A., M.B. and V.S.; methodology, D.R., C.A., M.B., L.M.-C., V.S., G.B., G.I., O.O. and A.R.; investigation, D.R., C.A., M.B., L.M.-C., V.S., G.B., G.I., O.O. and A.R.; validation, D.R., C.A., M.B., L.M.-C., V.S., O.O. and A.R.; writing—original draft preparation D.R., C.A., M.B. and V.S.; writing—review and editing, D.R., C.A., M.B., L.M.-C., V.S., O.O. and A.R.; funding acquisition, D.R., V.S., O.O. and A.R.; All authors have read and agreed to the published version of the manuscript.

Funding: This study was supported by JINR-Romania scientific project, topic no. 02-1-1107-2011/2021 item 6, 04-4-1142-2021/2025 item 37, 04-5-1131-2017/2021 item 97.

Institutional Review Board Statement: Not applicable.

Informed Consent Statement: Not applicable.

Data Availability Statement: Not applicable.

Conflicts of Interest: The authors declare no conflict of interest.

References

1. Gorzkowska, I.; Jozwiak, P.; Garbarczyk, J.E.; Wasiucionek, M.; Julien, C.M. Studies on glass transition of lithium-iron phosphate glasses. *J. Therm. Anal. Calorim.* **2008**, *93*, 759–762. [CrossRef]
2. Aqdim, S.; Ouchetto, M. Elaboration and Structural Investigation of Iron (III) Phosphate Glasses. *Adv. Mater. Phys. Chem.* **2013**, *3*, 332. [CrossRef]
3. Venkateswara, R.G.; Shashikala, H.D. Optical and mechanical properties of calcium phosphate glasses. *Glass. Phys. Chem.* **2014**, *40*, 303–309. [CrossRef]
4. Singh, S.P.; Chakradhar, R.P.S.; Rao, J.L.; Karmakar, B. EPR optical absorption and photoluminescence properties of MnO_2 doped $23B_2O_3$–$5ZnO$–$72Bi_2O_3$ glasses. *Phys. B* **2010**, *405*, 2157. [CrossRef]
5. Chung, W.J.; Choi, J.; Choi, Y.G. Compositional effect on structural and spectroscopic properties of P_2O_5–SnO–MnO ternary glass system. *J. Alloys Compd.* **2010**, *505*, 661. [CrossRef]

6. Takahashi, H.; Karasawa, T.; Sakuma, T.; Garbarczyk, J.E. Electrical conduction in the vitreous and crystallized Li_2O–V_2O_5–P_2O_5 system. *Solid State Ion.* **2010**, *181*, 27–32. [CrossRef]
7. Bingham, P.A.; Hand, R.J.; Forder, S.D.; La Vaysierre, A. Vitrified metal finishing wastes: II. Thermal and structural characterization. *J. Hazard. Mater.* **2005**, *122*, 129. [CrossRef] [PubMed]
8. Brow, R.K. Review: The Structure of Simple Phosphate Glasses. *J. Non-Cryst. Solids* **2000**, *263–264*, 1–28. [CrossRef]
9. Ahmed, I.; Ren, H.; Booth, J. Developing Unique Geometries of Phosphate-Based Glasses and their Prospective Biomedical Applications. *Johns. Matthey Technol. Rev.* **2019**, *63*, 34. [CrossRef]
10. Hossain, K.M.Z.; Patel, U.; Kennedy, A.R.; Macri-Pellizzeri, L.; Sottile, V.; Grant, D.M.; Scammell, B.E.; Ahmed, I. Porous calcium phosphate glass microspheres for orthobiologic applications. *Acta Biomater.* **2018**, *72*, 396–406. [CrossRef]
11. Sheng, L.; Hongting, L.; Fengnian, W.; Ziyuan, C.; Yunlong, Y. Effects of alkaline-earth metal oxides on structure and properties of iron phosphate glasses. *J. Non-Cryst. Solids* **2016**, *434*, 108–114.
12. Bingham, P.A.; Hand, R.; Hannant, O.M.; Forder, S.; Kilcoyne, S. Effects of modifier additions on the thermal properties, chemical durability, oxidation state and structure of iron phosphate glasses. *J. Non-Cryst. Solids* **2009**, *355*, 1526–1538. [CrossRef]
13. Ahmed, I.; Collins, C.A.; Lewis, M.P.; Olsen, I.; Knowles, J.C. Processing, characterization and biocompatibility of iron-phosphate glass fibres for tissue engineering. *Biomaterials* **2004**, *25*, 3223. [CrossRef] [PubMed]
14. Vedeanu, N.S.; Lujerdean, C.; Zăhan, M.; Dezmirean, D.S.; Barbu-Tudoran, L.; Damian, G.; Stefan, R. Synthesis and Structural Characterization of CaO-P_2O_5-CaF:CuO Glasses with Antitumoral Effect on Skin Cancer Cells. *Materials* **2022**, *15*, 1526. [CrossRef] [PubMed]
15. Dutta, D.P.; Roy, M.; Mishra, R.K.; Meena, S.S.; Yadav, A.; Kaushik, C.P.; Tyagi, A.K. Structural investigations on Mo, Cs and Ba ions-loaded iron phosphate glass for nuclear waste storage application. *J. Alloys Compd.* **2021**, *850*, 156715. [CrossRef]
16. Hoppe, A.; Nusret, S.G.; Boccaccini, A.R. A review of the biological response to ionic dissolution products from bioactive glasses and glass-ceramics. *Biomaterials* **2011**, *32*, 2757–2774. [CrossRef]
17. Omidian, S.; Nazarpak, M.H.; Bagher, Z.; Moztardeh, F. The effect of vanadium ferrite doping on the bioactivity of mesoporous bioactive glass-ceramics. *RSC Adv.* **2022**, *12*, 25639–25653. [CrossRef] [PubMed]
18. Hench, L.L. Bioactive Glass: Chronology, Characterization, and Genetic Control of Tissue Regeneration. In *Advances in Calcium Phosphate Biomaterials*; Springer Series in Biomaterials Science and Engineering; Ben-Nissan, B., Ed.; Springer: Berlin/Heidelberg, Germany, 2014; Volume 2, pp. 51–70.
19. Ardelean, I.; Rusu, D.; Andronache, C.; Ciobotă, V. Raman study of $xMeO·(100 - x)[P_2O_5·Li_2O]$ ($MeO \Rightarrow Fe_2O_3$ or V_2O_5) glass systems. *Mater. Lett.* **2007**, *61*, 3301–3304. [CrossRef]
20. Andronache, C. The local structure and magnetic interactions between Fe^{3+} and V^{4+} ions in lithium–phosphate glasses. *Mater. Chem. Phys.* **2012**, *136*, 281–285. [CrossRef]
21. Andronache, C.; Balasoiu, M.; Orelovich, A.; Roghacev, V.; Mihaly-Cozmuta, L.; Balasoiu-Gaina, A.M.; Racolta, D. On the structure of Lithium-Phosphate glasses doped with iron and vanadium ions. *J. Optoelectron. Adv. Mater.* **2019**, *21*, 726–732.
22. Ardelean, I.; Andronache, C.; Cimpean, C.; Pascuta, P. EPR and magnetic investigation of calcium—Phosphate glasses containing iron ions. *J. Optoelectron. Adv. Mater.* **2006**, *8*, 1372.
23. Vedeanu, N.S.; Magdas, D.A. The influence of some transition metal ions in lead-and calcium-phosphate glasses. *J. Alloys Compd.* **2012**, *534*, 93–96. [CrossRef]
24. Andronache, C.; Balasoiu, M.; Racolta, D. Magnetic Interaction between Iron Particles in Lithium Phosphate Systems. *Russ. J. Phys. Chem.* **2017**, *91*, 2686–2689. [CrossRef]
25. Stoch, P.; Stoch, A.; Ciecinska, M.; Krakowiak, I.; Sitarz, M. Structure of phosphate and iron-phosphate glasses by DFT calculations and FTIR/Raman spectroscopy. *J. Non-Cryst. Solids* **2016**, *450*, 48–60. [CrossRef]
26. Shapaan, M.; Shabaan, E.R.; Mostafa, A.G. Study of the hyperfine structure, thermal stability and electric–dielectric properties of vanadium iron phosphate glasses. *Phys. B Condens. Matter* **2009**, *404*, 2058–2064. [CrossRef]
27. Klug, H.P.; Alexander, L.E. *X-ray Diffraction Procedures: For Polycrystalline and Amorphous Materials*, 2nd ed.; Wiley: New York, NY, USA, 1970.
28. Chavan, V.K.; Rupa, B.; Rupa, V.R.; Sreenivasulu, M. Transition metal doped phosphate glass. *Int. J. Innov. Eng. Res. Manag.* **2018**, *5*, 1–7.
29. Andronache, C.; Balasoiu, M.; Racolta, D. Structural properties of different phosphate glasses by EPR analysis. *AIP Conf. Proc.* **2019**, *2071*, 030004.
30. Agarwal, A.; Khasa, S.; Seth, V.P.; Arora, M. Study of EPR, optical properties and dc conductivity of VO^{2+} ion doped $TiO_2R_2OB_2O_3$ (R = Li and K) glasses. *J. Alloys Compd.* **2013**, *568*, 112–117. [CrossRef]
31. Ivanova, O.S.; Ivantsov, R.D.; Edelman, I.S.; Petrakovskaja, E.A.; Velikanov, D.A.; Zubavichus, Y.V.; Zaikovskii, V.I.; Stepanov, S.A. Identification of ε-Fe_2O_3 nano-phase in borate glasses doped with Fe and Gd. *J. Magn. Magn. Mater.* **2016**, *401*, 880–889. [CrossRef]
32. Akamatsu, H.; Oku, S.; Fujita, K.; Murai, S.; Tanaka, K. Magnetic properties of mixed-valence iron phosphate glasses. *Phys. Rev. B* **2009**, *80*, 134408. [CrossRef]
33. Akamatsu, H.; Oku, S.; Fujita, K.; Murai, S.; Tanaka, K. Magnetic properties and photoluminescence of thulium-doped calcium aluminosilicate glasses. *Opt. Mater. Express* **2019**, *9*, 4348–4359.
34. Khattak, G.D.; Mekki, A.; Wenger, L.E. X-ray photoelectron spectroscopy (XPS) and magnetic susceptibility studies of vanadium phosphate glasses. *J. Non-Cryst. Solids* **2000**, *262*, 66–79. [CrossRef]

35. Islam, M.T.; Felfel, R.M.; Abou Neel, E.A.; Grant, D.M.; Ahmed, I.; Hossain, K.M.Z. Bioactive calcium phosphate-based glasses and ceramics and their biomedical applications: A review. *J. Tissue Eng.* **2017**, *8*, 2041731417719170. [CrossRef] [PubMed]
36. Andronache, C.; Racolta, D.; Ardelean, G. Magnetic properties of x(Fe_2O_3)·(100 − x)[P_2O_5·Li_2O] and x(Fe_2O_3)·(100 − x)[P_2O_5·CaO] glass systems. *AIP Conf. Proc.* **2017**, *1916*, 030004.
37. Pascuta, P.; Borodi, G.; Popa, A.; Dan, V.; Culea, E. Influence of iron ions on the structural and magnetic properties of some zinc-phosphate glasses. *Mater. Chem. Phys.* **2010**, *123*, 767–771. [CrossRef]
38. Winterstein, A.; Akamatsu, H.; Möncke, D.; Tanaka, K.; Schmidt, M.A.; Wondraczek, L. Magnetic and magneto-optical quenching in (Mn^{2+}, Sr^{2+}) metaphosphate glasses. *Opt. Mater. Express* **2013**, *3*, 184–193. [CrossRef]

Disclaimer/Publisher's Note: The statements, opinions and data contained in all publications are solely those of the individual author(s) and contributor(s) and not of MDPI and/or the editor(s). MDPI and/or the editor(s) disclaim responsibility for any injury to people or property resulting from any ideas, methods, instructions or products referred to in the content.

Article

Water Decontamination from Cr(VI) by Transparent Silica Xerogel Monolith

Marco Zannotti [1,*], Andrea Rossi [1], Marco Minicucci [2], Stefano Ferraro [1], Laura Petetta [1] and Rita Giovannetti [1,*]

[1] Chemistry Interdisciplinary Project, School of Science and Technology, Chemistry Division, University of Camerino, 62032 Camerino, Italy; andrea.rossi@unicam.it (A.R.); stefano.ferraro@unicam.it (S.F.)
[2] School of Science and Technology, Physics Division, University of Camerino, 62032 Camerino, Italy; marco.minicucci@unicam.it
* Correspondence: marco.zannotti@unicam.it (M.Z.); rita.giovannetti@unicam.it (R.G.)

Abstract: Cr(VI) is highly soluble and mobile in water solution and extremely toxic. In order to obtain a specific material with adsorption properties towards Cr(VI), and that can be used in environmental remediation of water contaminated with Cr(VI), one-step sol-gel technique, at low temperature (50 °C), has been optimized to prepare transparent silica-based xerogel monolith by using tetraethyl orthosilicate as precursor. The obtained xerogel, with disk shape, was fully characterized by Raman, BET, FE-SEM and XRD analysis. The results indicated that the material showed silica amorphous phase and high porosity. The study of the adsorption properties towards different concentrations of Cr(VI), in the form of $HCrO_4^-$ in acidic condition, showed prominent results. The absorption kinetics were evaluated by studying different models, the final result showing that the absorption of Cr(VI) occurred through intra-particle diffusion process, following two steps, and that the absorption equilibrium is regulated by Freundlich isotherm model. The material can be restored by reducing the hazardous Cr(VI) to Cr(III), a less toxic form of chromium, by 1,5-diphenylcarbazide, and with successive treatment in acidic water.

Keywords: chromium; environmental remediation; silica xerogel; absorption kinetics; equilibrium study

1. Introduction

Nowadays, chromium pollution of water and soil is a serious environmental problem; after lead, chromium is the second most common inorganic pollutant discharged in the environment [1].

Chromium is used in different industrial processes such as leather tanning, electroplating, textile dyeing and plastic productions, with the generations of hazardous waste containing relatively high amounts of chromium. In the environment, chromium exists as Cr(III) and Cr(VI), and, with respect to the trivalent form, the hexavalent chromium is 100 times more toxic, carcinogenic and mutagenic; direct exposition to Cr(VI) causes eyes irritation, allergic reactions, asthma, and lung and kidney cancer [2–4].

Cr(VI) is highly soluble and mobile in water solution and can be present in different species, depending on its concentration and the pH of the aqueous environment. Chromate ions (CrO_4^{2-}) exist only at basic pH, the hydrogen chromate ion ($HCrO_4^-$) between acidic pH 1 and neutral pH 7, while the acidic form of Cr(VI)—the chromic acid (H_2CrO_4)—at a very acidic pH of less than 1. The dichromate species ($Cr_2O_7^{2-}$) can be produced in solution when the concentration of Chromium is higher than 1 g L^{-1}. These features with the strong oxidant ability of Cr(VI) compounds make chromium pollution even more dangerous in water solutions [2,5].

Several methods such as reduction to Cr(III) and precipitation, chemical extraction, dialysis, electrochemical separation, coagulation and ion-exchange have been applied to reduce the Cr(VI) concentration and its effect [6–10].

Adsorption method is probably the most effective, with low-cost easy operation and being environmentally friendly; the adsorption process applied to Cr(VI)-contaminated water shows high removal efficiency, low energy cost and chemical investment, and the possibility to reuse the absorbent material [11–13]. Mesoporous silica-based materials for their unique and tunable physicochemical properties such as great chemical, mechanical and thermal stability with a large surface area, are ideal candidates as novel adsorbent materials [2,14]. In particular, silica-based xerogel is a material with high porosity, than can be produced by sol-gel synthesis through hydrolysis and condensation of alkoxysilanes (Si(OR)$_4$) precursor, obtained in the presence of acid or base as catalyst. In this case, in contrast to aerogel, the formed gel is then dried under atmospheric conditions to remove the liquid and to produce a xerogel with the desired shape [15,16]. The molar ratio between alkoxides/water/solvent, pH and temperature, influence the different polymeric structures such as linear, entangled chain, clusters and colloidal particles [17].

Xerogels obtained by conventional drying have high density due to shrinkage during drying normally and are affected by cracking and shrinkage; to overcome this problem, drying control additives are added on the precursor solution [18,19]. The transparency of the material makes it suitable for optical devices and spectroscopic investigations amounts and furthermore the high degree of silanol groups gives high porosity to the materials and available binding groups for chemical absorptions or interaction with inorganic or organic species.

In this work, a silica-based xerogel transparent monolith was produced by a simple one-step sol-gel technique at very low temperatures (50 °C), with tetraethyl orthosilicate (TEOS) as a precursor. The obtained monolithic disk was deeply characterized by Raman, BET, FE-SEM and XRD analysis and used as an adsorbent for Cr(VI) in the form of $HCrO_4^-$. The material was used as prepared, without modification and with no addition of other organic compounds; its complete transparency permitted to monitor the adsorption by easy operation, such as UV-Vis spectroscopy. The absorption of Cr(VI) was monitored during time, at different concentrations, in order to define the adsorption properties and to obtain equilibrium and kinetic results. The Cr(VI) can be converted to Cr(III) by a treatment with 1,5-diphenylcarbazide and subsequently the xerogel restored by washing in acid solution.

2. Results and Discussion

2.1. Silica-Based Xerogel Monolith

The silica-xerogel produced by sol-gel method at low temperature, produced a uniform and transparent disk as reported in Figure 1.

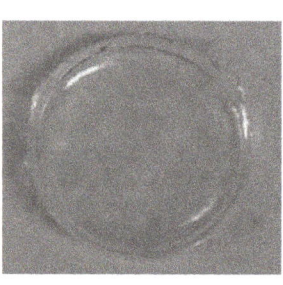 

Figure 1. Silica-based xerogel monolith obtained by sol-gel method.

In this procedure, the addition of DMF, as drying control chemical additive (DCCA), permitted to obtain a crack-free monolith. Despite that the procedure involved HCl, DMF decreased the acidity of the precursor solution by acting as hydrogen bond acceptor, promoting thereby the hydrolysis rate of TEOS. In this case, the formation of hydrogen bonds of DMF with silanol groups facilitated the removal of water molecules preventing the interaction of them with the silanol groups [20]. At the same time, the nucleophilicity of the silanol groups of TEOS is enhanced, with the result of faster condensation. Furthermore, DMF can hydrolyze with the formation of formic acid promoting the subsequent decrease of pH during sol-gel process. It is possible to say that initially there is a faster acid-catalyzed process, and later a based catalyzed process, that lead to a high interconnectivity of small oligomers with a decrease of the gelation time [20]. The addition of DMF, a polar aprotic solvent, by hydrogen bonding to the silanol groups, inhibits the coalescence of the silica particles. The final result is a material without crack, but with larger pore size; the capillary forces in this case will be weaker and thus the stress exerted on the gel during drying will be smaller.

2.2. Silica-Based Xerogel Characterization

The silica-based xerogel was characterized by XRD analysis, as reported in Figure 2. The dried material after two hours at 70 °C showed a broadened peak in the 2θ range of 8–15°; the feature corresponds to the amorphous silica matrix [21]. The results indicated that no crystalline phase was formed during the initial drying of thin films prepared with TEOS at low temperature.

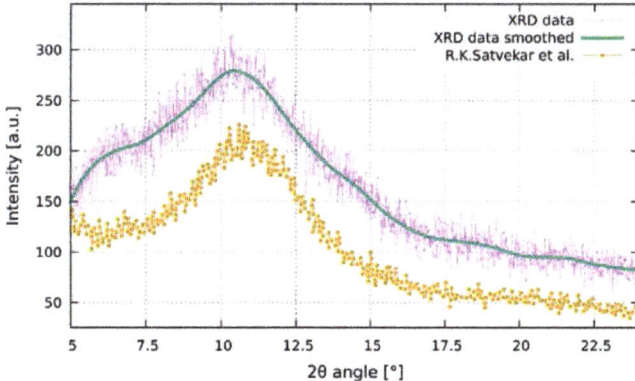

Figure 2. X-ray power diffraction pattern of silica-based xerogel [21].

The Raman spectrum of silica-based xerogel, reported in Figure 3, shows the typical bands of amorphous SiO_2. The broadened peak at 440 cm^{-1}, denoted as R-band, is attributed to the coupled stretching–bending mode of the Si-O-Si bridges. The sharp peak at 495 cm^{-1} is assigned to the symmetric stretching modes of four-membered of SiO_2 rings and usually is called as D_1 peak. The band at 790 cm^{-1} is related to bending modes of the Si-O-Si inter-tetrahedral angle, while the bands at 995 and 1110 cm^{-1} are related to the transversal optic (TO) and longitudinal optic (LO) asymmetric stretching modes of Si-O [22–25].

The morphology of the silica-based xerogel was evaluated by SEM microscopy as reported in Figure 4a where the SEM image shows a monodisperse sample with uniform spherical particles. Figure 4b reports the SEM image of the xerogel after the absorption of Cr(VI); the morphology and dimensions of particles remains the same, to prove that the material is not altered after the absorption of $HCrO_4^-$ on the silica matrix. The EDX spectrum reported in Figure 4c confirms the adsorption of Cr(VI) by the detection of the peak at 5.41 eV, while the map distribution image shows an homogeneous distribution of Cr on the sample.

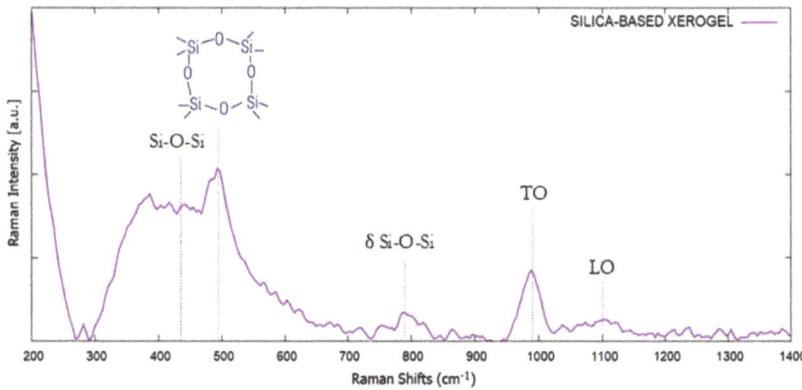

Figure 3. Raman spectrum of silica-based xerogel.

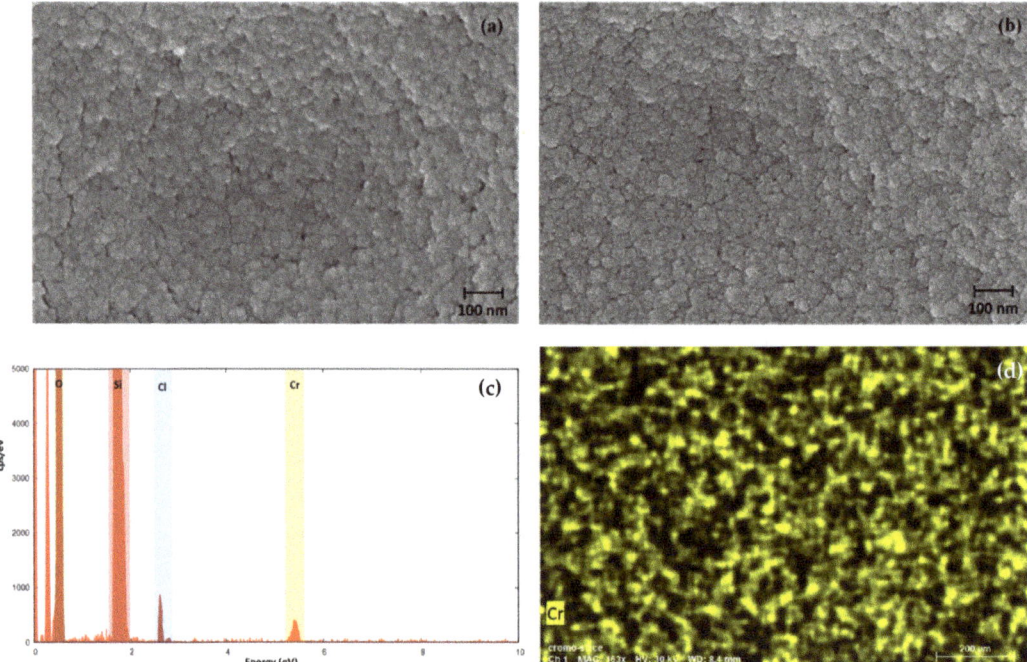

Figure 4. SEM images of silica-based xerogel (**a**) before and (**b**) after the absorption of Cr(VI), (**c**) EDX spectrum of the silica-based xerogel after the adsorption of Cr(VI) and (**d**) map distribution image for Cr.

BET gas adsorption indicates that the silica xerogel is porous with the size of the pores at around 2.38 nm and specific surface are of 547.33 m^2/g. Figure 5 shows the N$_2$ gas adsorption with a hysteresis loop during the desorption, the latter is common for porous materials like inorganic oxides and glasses [26].

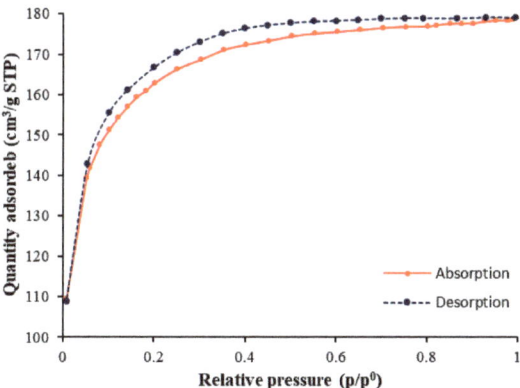

Figure 5. N$_2$ gas adsorption–desorption isotherm for the silica-based xerogel.

2.3. Kinetic Studies

The kinetic in the absorption of solute onto a solid adsorbent surface can be describe by different kinetic models; as example a first order mechanism is showed in the Equation (1) [27]:

$$\frac{dq_t}{dt} = k_1(q_e - q_t) \tag{1}$$

where q_t is the adsorbate (mg/g), in this case HCrO$_4^-$, adsorbed on the xerogel at time t in mg/g, q_e is the equilibrium absorption capacity of the absorbent at the equilibrium (mg/g), and k_1 is the rate constant of the absorption process.

The linear form of the Equation (1) is reported in the Equation (2) as follow:

$$\ln(q_e - q_t) = lnq_e - k_1 t \tag{2}$$

By plotting t vs. $ln(q_e - q_t)$, the value of k_1 can be determined; in this case, the constant is proportional to the starting concentration of the absorbate in solution. The absorption process takes place only on localized sites and no interaction between the absorbed solute are present; the absorption occurs as a monolayer on the surface of the absorbent material and the surface coverage does not affect the energy related to the absorption [28].

A second approach to absorption can be the pseudo second order kinetic, described by Equation (3):

$$\frac{dq_t}{dt} = k_2(q_e - q_t)^2 \tag{3}$$

and by the linear form of Equation (4):

$$\frac{t}{q_t} = \frac{1}{k_2 q_e^2} + \frac{t}{q_e} \tag{4}$$

where k_2 is the pseudo-second order constant.

This model presumes the same conditions for the pseudo-firs order model except that the absorption rate is explained by a rate equation of order 2. The absorption rate depends on the solute amount on the absorbent surface, and the driving force $(q_e - q_t)$ is proportional to the active sites available on the absorbent material. In this case, by plotting t/q_e vs. t/q_t, the intercept on the x-axis represented by $1/k^2 q_e^2$, permits to calculate the pseudo-second order kinetic constant, k_2 [29].

Another kinetic model to explain the absorption process is that of Elovich, which initially was applied to the absorption of the gas molecule. The Elovich model permits the prediction of the mass and surface diffusion, activation and deactivation energy. In this case, the assumption is that the adsorption rate of the solute molecules decreases exponentially with the increasing amount of adsorbed solute. The model is described by the following Equations (5) and (6) [29]:

$$\frac{dq_t}{dt} = \alpha exp^{-\beta q_t} \tag{5}$$

$$q_t = \frac{1}{\beta}\ln[\alpha\beta] + \frac{1}{\beta}\ln t \tag{6}$$

where α is the initial absorption rate (mg/g min), β is the desorption constant. The plot of q_t vs. t permits to calculate the parameter and evaluate the absorption on the heterogeneous surface of the absorbent. The Elovich model predicts that the absorption energy increases linearly with the surface coverage and interactions between the absorbed molecules on the surface of the material are possible [28].

The absorption of a solute can occur in several steps and the overall process can be controlled by different steps such as film or external diffusion, surface diffusion, pore diffusion and adsorption on the pore surface with a combination between them. This absorption model is defined as intra-particle diffusion and described by Equation (7) [30]

$$q_t = k_{id}t^{1/2} + C \tag{7}$$

where k_{id} is the intraparticle diffusion constant (mg/g half minute) and C indicates the thickness of the boundary layer (mg/g). Plotting q_t vs. $t^{0.5}$, the resultant plot is a straight line, proving that the absorption process is regulated by intraparticle diffusion; on the other hand, if the plot exhibits multi-linear plots, two or more steps regulate the absorption process [30].

In order to study the adsorption properties of silica xerogel monolith, different concentrations of Cr(VI) from 36.53 to 269 mg/L were tested in acidic condition; as evidenced in Figure 6a, after the Cr(VI) adsorption, the color of xerogel changed from colorless to yellow. The adsorption processes were spectrophotometrically analyzed, monitoring over time the decrease of the absorbance solution containing Cr(VI) (Figure 6b) with the aim of defining the appropriate kinetic models.

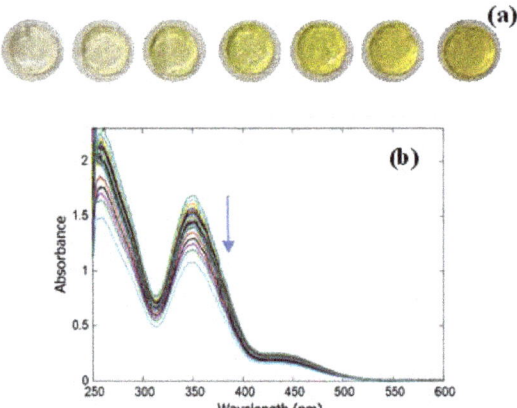

Figure 6. (a) Silica-based xerogel samples after absorption treatment by using different concentrations of Cr(VI) solutions from 36.53 to 269 mg/L, (b) UV-Vis spectra of Cr(VI) solution, with initial concentration of 55.41 mg/L, at different times during the absorption treatment with silica-based xerogel, the blue arrow represents the decrease in absorption during time.

The UV-Vis spectra reported in Figure 6b are typical of water $HCrO_4^-$ solution with the presence of the absorption band at 350 nm and a shoulder at around 445 nm; another intense peak is detected at 260–265 nm that is related to the O-Cr^{6+} charge transfer between oxygen and Cr(VI) of chromate ions in its tetrahedral structure [31,32]. During the adsorption process on the xerogel material, the UV-Vis spectrum of $HCrO_4^-$ decreases, confirming the uptake of Cr(VI) by the silica-based material; the obtained spectra maintain the same profile, demonstrating that the xerogel did not influence the spectrum of the aqueous solution and that therefore the Cr(VI) remains stable as $HCrO_4^-$ ion during the absorption process. The results of the kinetic study and the fitting with the different kinetic equation model are reported in Table 1. As it is possible to observe for the pseudo first order, the correlation factor is lower with respect to the other kinetic models; therefore, it is possible to confirm that, for the three studied concentrations, the absorption process is not described by pseudo-first order kinetic. On the other hand, for the pseudo-second order and the Elovich model, higher correlation factors were obtained, but, in this case, the fitting of the curve and the calculated q_e did not match very well the experimental results.

Table 1. Kinetic constants and correlation factors for Cr(VI) absorption, analyzed by pseudo-first and second order and by Elovich model.

Cr(VI) (mg/L)	q_{exp} (mg/g)	Pseudo-First Order $\ln(q_e - q_t) = \ln q_e - k_1 t$			Pseudo-Second Order $\frac{t}{q_t} = \frac{1}{k_2 q_e^2} + \frac{t}{q_e}$			Elovich $q_t = \frac{1}{\beta}\ln[\alpha\beta] + \frac{1}{\beta}\ln t$			
		k_1 (min^{-1})	q_e (mg/g)	R^2	k_2 (min^{-1})	q_e (mg/g)	R^2	q_e (mg/g)	α (mg/g min)	β (g/mg)	R^2
36.53	1.1903	0.0137	1.1903	0.9170	0.0146	1.1784	0.9980	1.3616	0.1605	6.3694	0.9603
55.41	2.3751	0.0102	2.3751	0.9078	0.0066	2.3493	0.9873	2.2835	0.3030	3.8610	0.9755
73.78	3.0160	0.0991	3.0160	0.8830	0.0034	2.9656	0.9895	2.9769	0.3804	3.0321	0.9824

As it is possible to observe from Table 2 and Figure 7, the absorption process can be well described by the intraparticle diffusion model. In this case, the absorption process takes place in two absorption phases—a faster initial one, that for the lower concentration lasted 90 min, while it lasted 80 min for the other two higher concentrations of Cr(VI), followed by a slowed second phase. The absorption constants of the two phases for the three tested concentrations of Cr(VI) are reported in Table 2.

Table 2. Kinetic constants and correlation factors of intraparticle diffusion model for Cr(VI) absorption by silica-based xerogel.

Cr(VI) (mg/L)	Intraparticle Diffusion $q_t = k_{id} t^{1/2} + C$					
	Phase 1			Phase 2		
	kd_1 (mg/g min$^{0.5}$)	Duration (min)	R^2	kd_2 (mg/g min$^{0.5}$)	Duration (min)	R^2
36.53	0.0891	90	0.9712	0.0069	—	0.9879
55.41	0.1462	80	0.9734	0.0158	—	0.987
73.78	0.1745	80	0.9762	0.0212	—	0.9978

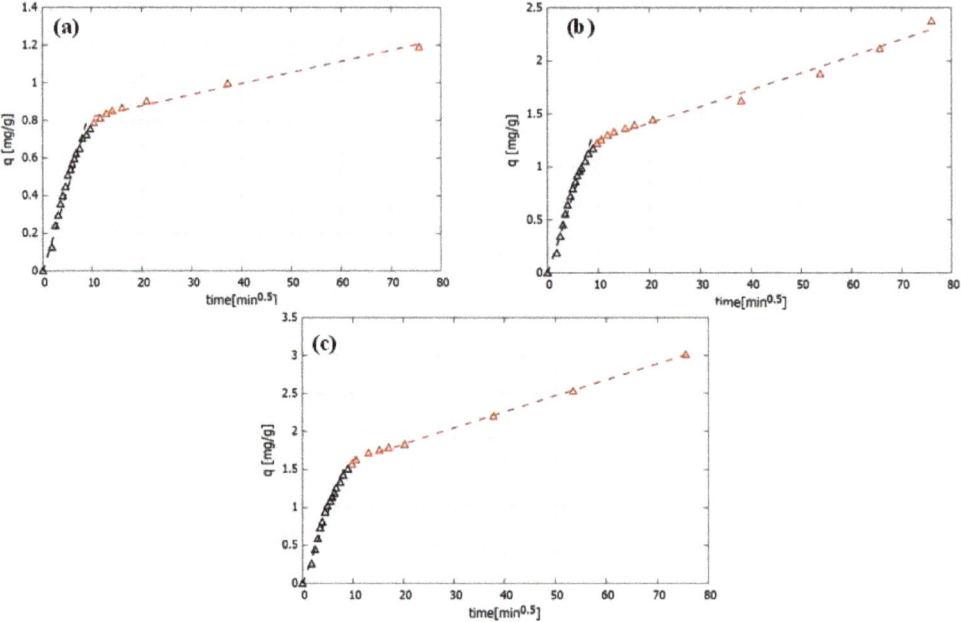

Figure 7. Intraparticle diffusion model of Cr(VI) adsorption on silica-based xerogel of (**a**) 36.53, (**b**) 55.41 and (**c**) 73.7 mg/L starting concentrations; black triangles: faster first absorption phase, red triangles: slower second absorption phase.

2.4. Absorption Equilibrium Study

The influence of the Cr(VI) concentration on the adsorption process has been evaluated by studying the equilibrium data with Freundlich and Langmuir isotherms, reported as Equations (8) and (9), respectively [33]:

$$\ln Q_e = \ln K_F + \frac{1}{n} \ln C_e \tag{8}$$

$$\frac{C_e}{Q_e} = \frac{1}{K_L} + \frac{a_L}{K_L} C_e \tag{9}$$

where C_e is the Cr(VI) concentration (mg/L), Q_e is the adsorbed amount on the monolith xerogel in mg/g, K_F is the Freundlich constant that represent the absorption capacity and $1/n$ is the absorption intensity. For the Langmuir model, K_L and a_L are the Langmuir constants and the ratio between them represents the theoretical saturation capacity of the monolith xerogel. Table 3 reports the concentration of the starting solution and the relative absorbed amount of Cr(VI), Q_e, evaluated for the absorption equilibrium study.

Table 3. Starting concentration and absorbed concentration at equilibrium of evaluated Cr(VI).

Cr(VI) (mg/L)	Q_e (mg/g)
36.53	1.1903
55.41	2.3751
73.78	3.0160
103.21	5.4302
151.45	6.4223
211.84	9.0002
269.20	10.8608

Figure 8 reports the isotherm plots for the Langmuir and Freundlich model, respectively; as it is possible to observe that the absorption process follows a Freundlich model, with high correlation factor R^2 confirming that the adsorption takes place as multiple layers on the porous surface of the silica-based xerogel monolith. Table 4 reports the equilibrium data for the Freundlich isotherm model.

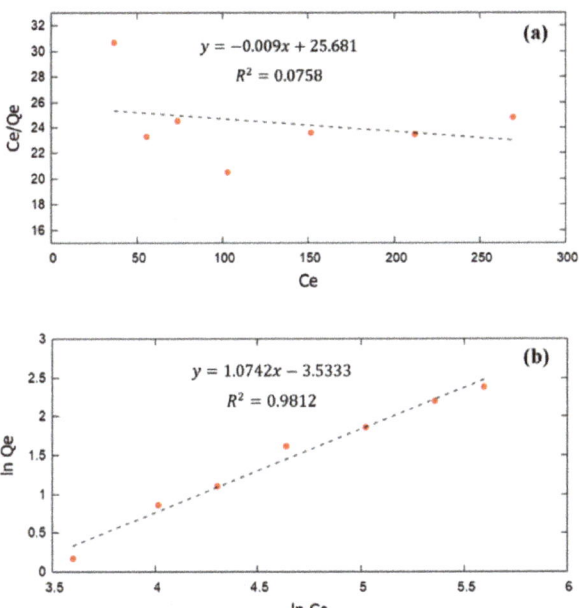

Figure 8. (a) Langmuir isotherm and (b) Freundlich isotherm plots.

Table 4. Equilibrium data and correlation factor of Freundlich isotherm model.

	Freundlich Isotherm Model
$1/n$	1.0742
lnK_F	3.5333
K_F	0.0292
R^2	0.9812

The absorption spectrum of Cr(VI) absorbed on the silica-based xerogel monolith can be easily detected by fiber optic spectrophotometer as reported in Figure 9. The Cr(VI) is adsorbed as $HCrO_4^-$ and remains stable in this form after the absorption inside the silica matrix confirmed by the unchanged UV-Vis spectra, with respect to the liquid phase. In this case, the calculated molar extinction coefficient of $HCrO_4^-$ on the xerogel is 383.22 L·mol^{-1}·cm^{-1}.

In this case, the adsorption properties of silica in acidic conditions can be explained considering the high surface area and porosity that characterizes this adsorbent material and the presence of protonated hydroxyl groups on the silicates surface (S_{surf}) that, during the adsorption process, can favor the adsorption of $HCrO_4^-$ ions [34] (Scheme 1).

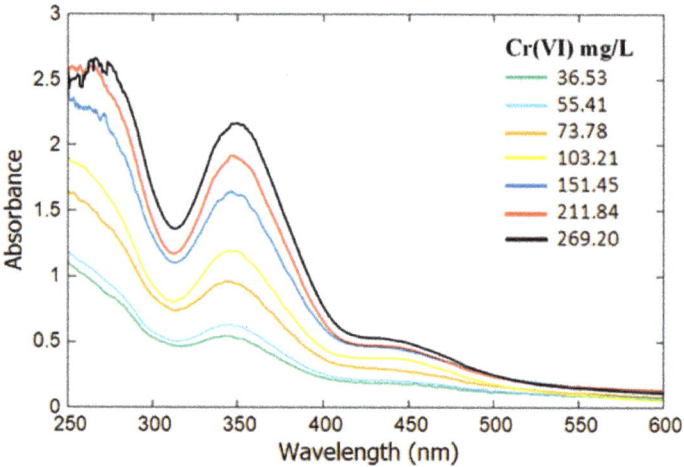

Figure 9. Absorption spectra of Cr(VI) adsorbed on silica-based xerogel at different initial concentrations (mg/L) from 36.53 to 269.20 mg/L.

$$S_{surf}\text{—OH} + H_3O^+ \rightleftharpoons S_{surf}\text{—OH}_2^+ + H_2O \xrightarrow{\text{Cr(VI) solution}} S_{surf}\text{—OH}_2^+/HCrO_4^-$$

Scheme 1. Absorption mechanism of Cr(VI) on silica-based xerogel in acidic pH condition.

Raman spectrum reported in Figure 10 also confirms the actual presence of Cr(VI) onto the matrix of the silica-based xerogel. In fact, the comparison between silica-based xerogel and adsorbed Cr(VI) xerogel, evidences the presence of a new peak at 860 cm^{-1} attributable to the Cr^{VI}-O bonds, which is also confirmed by the Raman spectrum of Cr(VI) salt [35].

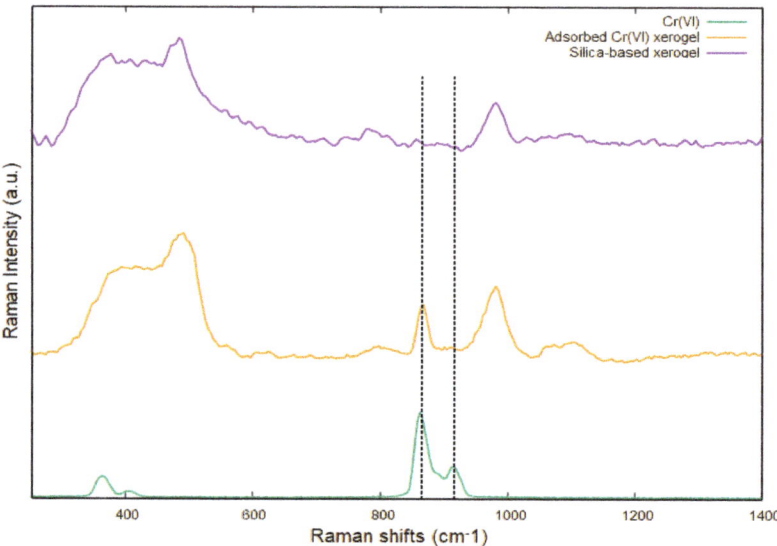

Figure 10. Raman spectra of silica-based xerogel before (purple) and after Cr(VI) absorption (yellow) and K$_2$CrO$_4$ powder (green).

2.5. Material Regeneration

The xerogel monolith was treated with 1,5-diphenylcarbazide (DPC) and, after this process, the Cr(VI), highly toxic and dangerous, was reduced directly on the material to the less toxic Cr(III). In this case, the DPC reduces the Cr(VI) to Cr(III), oxidizing itself to 1,5-diphenylcarbazone (DPCA), the latter form a purple complex with Cr(III), named as Cr(III)-DPCA [36]. To regenerate the material, the Cr(III)-DPCA xerogel was then treated with HCl (4M); in this case, as it is possible to observe in Figure 11, the xerogel returns transparent and can be reused for another cycle of absorption for Cr(VI). In Figure 11, the schematized method for the regeneration of the xerogel after Cr(VI) absorption treatment is reported.

Figure 11. Regeneration method of silica-based xerogel.

To confirm the reduction of Cr(VI) to Cr(III), XPS analysis on dried and pulverized sample was performed; the results are reported in Figure 12.

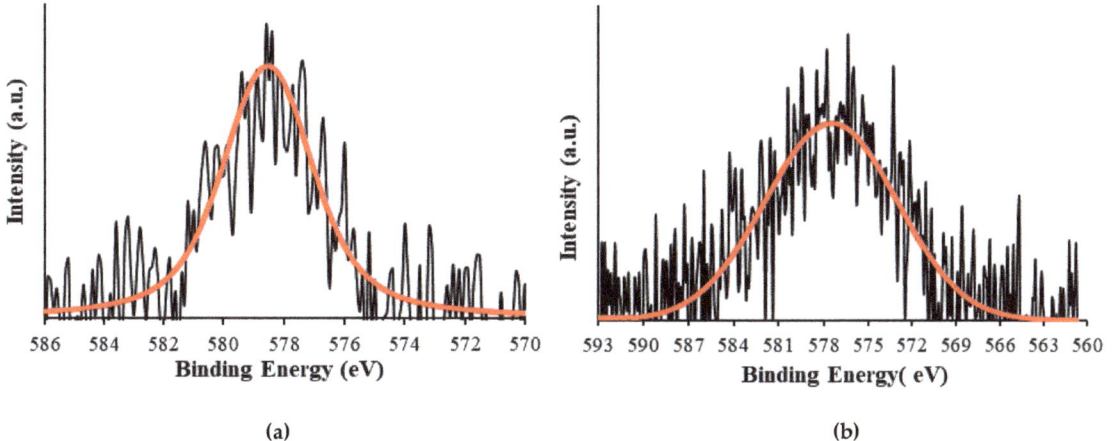

Figure 12. XPS spectra of Cr2p$_{3/2}$ for silica-based xerogel (**a**) before and (**b**) after treatment with DPC.

The XPS spectra show the peak relative to Cr2p$_{3/2}$; when Cr(VI) is absorbed on the silica-based xerogel, the signal is positioned at 578.6 eV, while after the reduction with DPC and the formation of the relative complex Cr(III)-DPCA, the XPS signal of chromium is at 577.4 eV, confirming the reduction to Cr(III) on the silica-based xerogel [37].

3. Materials and Methods

Hydrochloric acid (HCl), ethanol (EtOH), tetraethyl orthosilicate (TEOS), Dimethylformamide (DMF) and 1,5-Diphenylcarbazide are purchased from Merck and used without further purification.

3.1. Silica-Based Xerogel Preparation

Two separate solutions were prepared: the first contained HCl/H$_2$O and the second was a mixture of TEOS/EtOH/DMF. The acidic aqueous solution was added to the precursor solution and stirred at room temperature for 1 h. The molar ratio between H_2O/EtOH/TEOS/DMF was 16:4:1:2.5. Successively, 1 mL of the precursor solution was transferred into a plastic container covered with a plastic cap and left in the oven at 50 °C for seven days. Afterwards, a final treatment at 70 °C for 4 h was carried out obtaining a transparent xerogel monolith disk with a diameter of around 18 mm and a thickness of around 2.5 mm.

3.2. Xerogel Characterization

The xerogel crystalline structure was evaluated by X-ray diffraction (XRD) technique. The monolith xerogel was crushed into powder and analyzed by a Debye–Scherrer diffractometer equipped with an INEL CPS 180 (INEL, Artenay, France) curved position sensitive detector; in such a way, a drastic reduction of the acquisition time for each pattern was obtained. The X-rays source was a Mo K-alpha (lambda = 0.7093 Å). X-ray generated by a Philips sealed X-ray tube and monochromatized through a graphite crystal along the 002 plane. The sample was inserted in a glass capillary tube (diameter 100 μm) and centered on the beam.

In addition, the xerogel was characterized by Raman spectroscopy by using a micro-Raman spectrometer (iHR320, Horiba). The sample was analyzed with a green laser (532 nm), 50× objective outlet at room temperature.

BET measurements were performed in which the xerogel was first dried at 80 °C for 24 h, after the porosity and the specific area were evaluated through volumetric N_2 adsorption at 77 K using an ASAP 2020 (Micrometrics) instrument. The porosity of the materials was determined by BJH method (Halsey thickness equation).

The morphology of the silica-based xerogel was studied by using Field Emission Scanning Electron Microscopy (FE-SEM, Sigma 300, Zeiss, Germany) operated at 3 kV. The dried xerogel by using a self-adhesive carbon conductive tab was deposited on aluminum stabs. To prevent charging during the analysis, the sample was sputtered with chromium (5 nm) by Quorum Q150T-ES (Quorum Technologies, Lewes, UK). In order to perform EDX analysis, a portion of the adsorbed-Cr(VI) xerogel was sputtered with graphite.

3.3. Chromium Absorption Test and Kinetics Studies

The silica-based xerogel was tested for the absorption of Cr(VI) from water solutions. Different concentrations of K_2CrO_4 from 6.8×10^{-4} to 5×10^{-3} M at a pH of about 2 were prepared and studied in the absorption on silica-based xerogel. The absorption of Cr(VI) was evaluated at different times of absorption, monitoring the UV-Vis spectra of the Cr(VI) solutions by Cary 8454 Diode Array System spectrophotometer (Agilent Technologies, Santa Clara, CA, USA). The UV-Vis absorption spectra of silica-xerogel with the absorbed Cr(VI) were directly recorded by an HR4000-UV-NIR Ocean Optics fibre optic spectrophotometer.

The spectroscopic data were used for a detailed kinetics study and for the evaluation of the better isotherm model to explain the absorption of chromium on silica-based xerogel.

3.4. Material Regeneration

After the absorption of Cr(VI) the xerogel was treated with a solution of 1,5-diphenylcarbazide (DPC), prepared by dissolving 0.04 g of DPC in 10 mL of acetone. In this case, Cr(VI) was reduced to Cr(III), then the xerogel monolith was washed several time with HCl solution (4M). An X-ray Photoelectron analysis (XPS, VG Scientific Ltd., East Grinstead, UK) of the dried silica-xerogel, before and after the treatment with DPC, was performed. The deconvolution of the XPS spectrum was carried out by Fityk software (Microsoft, GitHub, San Francisco, CA, USA).

4. Conclusions

A silica-based transparent xerogel monolith, with a disk shape, was synthesized by an easy single step sol-gel process, at a low temperature around 50 °C, by using TEOS as silica precursor and without surface modification. The material was characterized by Raman, SEM, XRD and BET, showing the typical features of silica material with high porosity and amorphous phase.

The synthetized xerogel was easy to handle and showed prominent absorption capacity towards Cr(VI), in the form of $HCrO_4^-$. The transparency of the material permitted an evaluation of the absorption properties directly at naked eyes on the material and an easy quantification by UV-Vis analysis. The absorption process was deeply investigated, resulting in an intraparticle diffusion kinetic model, that occurs in two steps, while the absorption equilibrium was regulated by Freundlich isotherm model.

In addition, the absorbed Cr(VI) after reduction to Cr(III) by the use of DPC, can be successively regenerated by HCl washing. This silica material can be therefore advantageously applied on environmental remediation of water contaminated with Cr(VI).

Author Contributions: Conceptualization, M.Z. and R.G.; data curation, M.Z., A.R., M.M., S.F. and L.P.; formal analysis, M.Z., A.R., M.M. and L.P.; investigation, M.Z., A.R., M.M., S.F. and L.P.; methodology, M.Z.; project administration, R.G.; supervision, R.G.; writing—original draft, M.Z.; writing—review and editing, M.Z., S.F. and R.G. All authors have read and agreed to the published version of the manuscript.

Funding: This research received no external funding.

Institutional Review Board Statement: Not Applicable.

Informed Consent Statement: Not applicable.

Data Availability Statement: Additional data is available on request.

Acknowledgments: The authors gratefully acknowledge the School of Science of and Technology of University of Camerino for providing important technical and scientific resources such as the field-emission SEM (Sigma 300). the customized micro-Raman spectroscopy equipment (Olympus-Horiba iHR320) and the XPS equipment, necessary for this work.

Conflicts of Interest: The authors declare no conflict of interest.

References

1. Vodyanitskii, Y.N. Standards for the contents of heavy metals in soils of some states. *Ann. Agrar. Sci.* **2016**, *14*, 257–263. [CrossRef]
2. Dinker, M.K.; Kulkarni, P.S. Recent Advances in Silica-Based Materials for the Removal of Hexavalent Chromium: A Review. *J. Chem. Eng. Data* **2015**, *60*, 2521–2540. [CrossRef]
3. Guan, X.Y.; Fan, H.J.; Yan, S.X.; Chang, J.M. Chromium(VI) concurrent detoxification and immobilization by gallate: Kinetics, equilibrium, thermodynamics, and mechanism studies. *J. Environ. Chem. Eng.* **2017**, *5*, 5762–5769. [CrossRef]
4. Chebeir, M.; Chen, G.D.; Liu, H.Z. Emerging investigators series: Frontier review: Occurrence and speciation of chromium in drinking water distribution systems. *Environ. Sci.-Water Res. Technol.* **2016**, *2*, 906–914. [CrossRef]
5. Pourbaix, M. *Atlas of Electrochemical Equilibria in Aqueous Solutions*; National Association of Corrosion Engineers: Houston, TX, USA, 1974; pp. 257–260.
6. Njoya, O.; Zhao, S.; Qu, Y.; Shen, J.; Wang, B.; Shi, H.; Chen, Z. Performance and potential mechanism of Cr(VI) reduction and subsequent Cr(III) precipitation using sodium borohydride driven by oxalate. *J. Environ. Manag.* **2020**, *275*, 111165. [CrossRef]
7. Marzouk, I.; Dammak, L.; Chaabane, L.; Hamrouni, B. Optimization of Chromium (Vi) Removal by Donnan Dialysis. *Am. J. Anal. Chem.* **2013**, *4*, 8. [CrossRef]
8. Liu, W.; Zheng, J.; Ou, X.; Liu, X.; Song, Y.; Tian, C.; Rong, W.; Shi, Z.; Dang, Z.; Lin, Z. Effective Extraction of Cr(VI) from Hazardous Gypsum Sludge via Controlling the Phase Transformation and Chromium Species. *Environ. Sci. Technol.* **2018**, *52*, 13336–13342. [CrossRef]
9. Xu, T.; Zhou, Y.; Lei, X.; Hu, B.; Chen, H.; Yu, G. Study on highly efficient Cr(VI) removal from wastewater by sinusoidal alternating current coagulation. *J. Environ. Manag.* **2019**, *249*, 109322. [CrossRef]
10. Ren, Y.; Han, Y.; Lei, X.; Lu, C.; Liu, J.; Zhang, G.; Zhang, B.; Zhang, Q. A magnetic ion exchange resin with high efficiency of removing Cr (VI). *Colloids Surf. A Physicochem. Eng. Asp.* **2020**, *604*, 125279. [CrossRef]
11. Mitra, S.; Sarkar, A.; Sen, S. Removal of chromium from industrial effluents using nanotechnology: A review. *Nanotechnol. Environ. Eng.* **2017**, *2*, 11. [CrossRef]
12. Busetty, S. *Handbook of Environmental Materials Management*; Hussain, C.M., Ed.; Springer International Publishing: Berlin/Heidelberg, Germany, 2019; pp. 1367–1397.
13. Mondal, N.K.; Chakraborty, S. Adsorption of Cr(VI) from aqueous solution on graphene oxide (GO) prepared from graphite: Equilibrium, kinetic and thermodynamic studies. *Appl. Water Sci.* **2020**, *10*, 1–10. [CrossRef]

14. Miglio, V.; Zaccone, C.; Vittoni, C.; Braschi, I.; Buscaroli, E.; Golemme, G.; Marchese, L.; Bisio, C. Silica Monolith for the Removal of Pollutants from Gas and Aqueous Phases. *Molecules* **2021**, *26*, 1316. [CrossRef]
15. Shimizu, T.; Kanamori, K.; Nakanishi, K. Silicone-Based Organic–Inorganic Hybrid Aerogels and Xerogels. *Chem.—A Eur. J.* **2017**, *23*, 5176–5187. [CrossRef]
16. Hoffmann, F.; Cornelius, M.; Morell, J.; Froba, M. Silica-based mesoporous organic-inorganic hybrid materials. *Angew. Chem. Int. Ed. Engl.* **2006**, *45*, 3216–3251. [CrossRef]
17. Bryans, T.R.; Brawner, V.L.; Quitevis, E.L. Microstructure and Porosity of Silica Xerogel Monoliths Prepared by the Fast Sol-Gel Method. *J. Sol-Gel Sci. Technol.* **2000**, *17*, 211–217. [CrossRef]
18. Lenza RF, S.; Nunes EH, M.; Vasconcelos DC, L.; Vasconcelos, W.L. Preparation of sol–gel silica samples modified with drying control chemical additives. *J. Non-Cryst. Solids* **2015**, *423–424*, 35–40. [CrossRef]
19. Einarsrud, M.-A.; Haereid, S.; Wittwer, V. Some thermal and optical properties of a new transparent silica xerogel material with low density. *Sol. Energy Mater. Sol. Cells* **1993**, *31*, 341–347. [CrossRef]
20. Wright, J.D.; Sommerdijk, N.A.J.M. *Sol-Gel Materials Chemistry and Applications*; CRC Press Taylor & Francis Group: Boca Raton, FL, USA, 2001.
21. Satvekar, R.K.; Phadatare, M.R.; Patil, R.N.; Tiwale, B.M.; Pawar, S.H. Influence of Silane Content on the Optical Properties of Sol Gel Derived Spin Coated Silica Thin Films. *Int. J. Basic Appl. Sci.* **2012**, *1*, 9. [CrossRef]
22. Saavedra, R.; León, M.; Martin, P.; Jiménez-Rey, D.; Vila, R.; Girard, S.; Boukenter, A.; Ouerdane, Y. Raman measurements in silica glasses irradiated with energetic ions. *Fundam. Appl. Silica Adv. Dielectr. (Sio2014)* **2014**, *1624*, 118–124. [CrossRef]
23. Heili, M.; Poumellec, B.; Burov, E.; Gonnet, C.; Le Losq, C.; Neuville, D.R.; Lancry, M. The dependence of Raman defect bands in silica glasses on densification revisited. *J. Mater. Sci.* **2016**, *51*, 1667. [CrossRef]
24. Varkentina, N.; Dussauze, M.; Royon, A.; Ramme, M.; Petit, Y.; Canioni, L. High repetition rate femtosecond laser irradiation of fused silica studied by Raman spectroscopy. *Opt. Mater. Express* **2016**, *6*, 79–90. [CrossRef]
25. Fan, H.; Hartshorn, C.; Buchheit, T.; Tallant, D.; Assink, R.; Simpson, R.; Kissel, D.J.; Lacks, D.J.; Torquato, S.; Brinker, C.J. Modulus-density scaling behaviour and framework architecture of nanoporous self-assembled silicas. *Nat. Mater.* **2007**, *6*, 418–423. [CrossRef]
26. Kruk, M.; Jaroniec, M. Gas Adsorption Characterization of Ordered Organic−Inorganic Nanocomposite Materials. *Chem. Mater.* **2001**, *13*, 3169–3183. [CrossRef]
27. Giovannetti, R.; Rommozzi, E.; D'Amato, C.A.; Zannotti, M. Kinetic Model for Simultaneous Adsorption/Photodegradation Process of Alizarin Red S in Water Solution by Nano-TiO2 under Visible Light. *Catalysts* **2016**, *6*, 84. [CrossRef]
28. Largitte, L.; Pasquier, R. A review of the kinetics adsorption models and their application to the adsorption of lead by an activated carbon. *Chem. Eng. Res. Des.* **2016**, *109*, 495–504. [CrossRef]
29. Kajjumba, G.W.E.S.; Öngen, A.; Özcan, H.K.; Aydin, S. *Advanced Sorption Process Applications*; Edebali, S., Ed.; IntechOpen: London, UK, 2018.
30. Bilgili, M.S.; Varank, G.; Sekman, E.; Top, S.; Özçimen, D.; Yazıcı, R. Modeling 4-Chlorophenol Removal from Aqueous Solutions by Granular Activated Carbon. *Environ. Model. Assess.* **2012**, *17*, 289–300. [CrossRef]
31. Zhang, Z.; Liba, D.; Alvarado, L.; Chen, A. Separation and recovery of Cr(III) and Cr(VI) using electrodeionization as an efficient approach. *Sep. Purif. Technol.* **2014**, *137*, 86–93. [CrossRef]
32. Sanchez-Hachair, A.; Hofmann, A. Hexavalent chromium quantification in solution: Comparing direct UV–visible spectrometry with 1,5-diphenylcarbazide colorimetry. *Comptes Rendus Chim.* **2018**, *21*, 890–896. [CrossRef]
33. Giovannetti, R.; Rommozzi, E.; Zannotti, M.; D'Amato, C.A.; Ferraro, S.; Cespi, M.; Bonacucina, G.; Minicucci, M.; Di Cicco, A. Exfoliation of graphite into graphene in aqueous solution: An application as graphene/TiO2 nanocomposite to improve visible light photocatalytic activity. *RSC Adv.* **2016**, *6*, 93048–93055. [CrossRef]
34. Lowe, B.M.; Skylaris, C.-K.; Green, N.G. Acid-base dissociation mechanisms and energetics at the silica–water interface: An activationless process. *J. Colloid Interface Sci.* **2015**, *451*, 231–244. [CrossRef]
35. Karthik, C.; Ramkumar, V.S.; Pugazhendhi, A.; Gopalakrishnan, K.; Arulselvi, P.I. Biosorption and biotransformation of Cr(VI) by novel Cellulosimicrobium funkei strain AR6. *J. Taiwan Inst. Chem. Eng.* **2017**, *70*, 282–290. [CrossRef]
36. Zarghampour, F.; Yamini, Y.; Baharfar, M.; Javadian, G.; Faraji, M. On-chip electromembrane extraction followed by sensitive digital image-based colorimetry for determination of trace amounts of Cr(vi). *Anal. Methods* **2020**, *12*, 483–490. [CrossRef]
37. Bandara, P.C.; Peña-Bahamonde, J.; Rodrigues, D.F. Redox mechanisms of conversion of Cr(VI) to Cr(III) by graphene oxide-polymer composite. *Sci. Rep.* **2020**, *10*, 9237. [CrossRef]

Disclaimer/Publisher's Note: The statements, opinions and data contained in all publications are solely those of the individual author(s) and contributor(s) and not of MDPI and/or the editor(s). MDPI and/or the editor(s) disclaim responsibility for any injury to people or property resulting from any ideas, methods, instructions or products referred to in the content.

Article

Recovery of W(VI) from Wolframite Ore Using New Synthetic Schiff Base Derivative

Rawan E. Elbshary [1], Ayman A. Gouda [2], Ragaa El Sheikh [2], Mohammed S. Alqahtani [3,4,5], Mohamed Y. Hanfi [6,7], Bahig M. Atia [6], Ahmed K. Sakr [8,*] and Mohamed A. Gado [6,*]

[1] Department of Chemistry, Faculty of Pharmacy, Heliopolis University, El Salam City, Cairo 11785, Egypt
[2] Department of Chemistry, Faculty of Science, Zagazig University, Zagazig 44519, Egypt
[3] Radiological Sciences Department, College of Applied Medical Sciences, King Khalid University, Abha 61421, Saudi Arabia
[4] BioImaging Unit, Space Research Centre, University of Leicester, Michael Atiyah Building, Leicester LE1 7RH, UK
[5] Research Center for Advanced Materials Sciences (RCAMS), King Khalid University, Abha 61413, Saudi Arabia
[6] Nuclear Materials Authority, El Maadi, Cairo P.O. Box 530, Egypt
[7] Institute of Physics and Technology, Ural Federal University, St. Mira, 19, 620002 Yekaterinburg, Russia
[8] Department of Civil and Environmental Engineering, Wayne State University, 5050 Anthony Wayne Drive, Detroit, MI 48202, USA
* Correspondence: akhchemist@gmail.com (A.K.S.); mag.nma@yahoo.com (M.A.G.)

Abstract: A new synthetic material, namely, (3-(((4-((5-(((S)-hydroxyhydrophosphoryl)oxy)-2-nitrobenzylidene) amino) phenyl) imino) methyl)-4-nitrophenyl hydrogen (R)-phosphonate)), was subjected to a quaternary ammonium salt and named (HNAP/QA). Several characterizations, such as FTIR spectrometry, ^{1}H-NMR analysis, ^{13}C-NMR analysis, ^{31}P-NMR Analysis, TGA analysis, and GC-MS analysis, were performed to ensure its felicitous preparation. HNAP/QA is capable of the selective adsorption of W(VI) ions from its solutions and from its rock leachate. The optimum factors controlling the adsorption of W(VI) ions on the new adsorbent were studied in detail. Furthermore, kinetics and thermodynamics were studied. The adsorption reaction fits the Langmuir model. The sorption process of the W(VI) ions is spontaneous due to the negative value of $\Delta G°$ calculated for all temperatures, while the positive value of $\Delta H°$ proves that the adsorption of the W(VI) ions adsorption on HNAP/QA is endothermic. The positive value of $\Delta S°$ suggests that the adsorption occurs randomly. Ultimately, the recovery of W(IV) from wolframite ore was conducted successfully.

Keywords: adsorption; Schiff base; tungsten; synthesis; kinetics

1. Introduction

The potential toxicity of heavy metals in aquatic environments varies significantly between their different chemical forms and the fact that they are neither biodegradable nor photodegradable [1]. As a lithophile element, tungsten can be found in wolframite and scheelite ores, where it is recovered as tungstate (WO_4^{2-}) and naturally occurs in soils and sediments [2,3]. The concentration of W(VI) in the lithosphere varies from 0.2 to 2.4 mg kg^{-1} [2–4], while in ocean water it is present in the range of 8.0 to 100 mg L^{-1}. On the other hand, in the surface rocks of the Northern Atlantic and the Pacific Oceans, it is in the range of 1.0 to 1.3 mg kg^{-1} [2,4]. In terms of solubility and dominance, tungstate (WO_4^{2-}) predominates over its hydroxyl counterpart. The oxidation state of tungsten ranges from −2 to +6, making it a transition metal. There is a potential for tungstate species to polymerize in mildly acidic conditions, yielding iso-polytungstates, polytungstates, and monotungstates [5]. For example, tungsten inhibits molybdenum metabolism, which, in turn, reduces the biological activity of molybdenum-containing enzymes such phosphatase and adenosine triphosphate enzymes [6]. Even in tiny quantities, tungstate species present a significant hazard. As a result, testing and treating natural water supplies for tungsten is essential [7–9]. The disposal of their waste requires proper measurements, as this waste also contains valuable metals, and tossing these heavy metals is perilous due to the depletion of their primary resources, which leads to environmental hazards because of these minerals. Manganese is an important metal that has numerous applications in industry. It should be reprocessed and not disposed of. The high melting point of tungsten has extensive applications, such as manufacturing super alloys for high-temperature tasks, catalysts for redox chemical reactions, and fertilizers and in power plants as a regulator of oxygen and nitrogen compounds to mitigate environmental damage [10–13]. A hydrometallurgical process is used for the extraction of tungsten from both its ores and industrial effluents. Tungsten is recovered using processes such as solvent extraction [14,15], ion-exchange method [16], chemical precipitation [17], filtration [18], electrolysis and electro-deposition [19].

Tungsten is one of the elements classified as a significant raw material by the European Commission, according to its economic significance and supply chain. In 2022, the average price of tungsten was around USD 270 per metric ton unit of tungsten trioxide [20]. Tungsten has many extensive applications in several industries, such as the automotive industry, defense equipment, chemical, and aviation industries [21]. However, there are stringent restrictions for the production of tungsten in which Mo, V, Si, and P are frequently found in the products as contaminants which violate the European standards of tungsten purity. When manufacturing high-purity tungsten and tungsten products, molybdenum is one of the most difficult contaminants to remove because of its chemical similarity to tungsten [22]. It is common for molybdenum minerals to be found around wolframite and scheelite minerals. According to a previous study from China, the mass ratio of Mo/WO$_3$ was about 2.0% in W-Mo-Bi-Sn scheelite crystals in the Shizhuyuan mine, while the mass ratio was above 5.0% in Mo-W scheelite middlings from the Sandaogou mine [23].

The novelty of this work is the synthesis of a new adsorbent material with a high selectivity of W(VI) ions from its aqueous solutions and rock leachate. The prepared material undergoes several characterizations to ensure its successful preparation. The novel adsorbent is used for sorption of W(VI) ions from a synthetic solution and is finally used for the recovery of W(VI) ions from wolframite ore leachate via sorption and precipitation in a pure product.

2. Result and Discussions

2.1. Characterizations of the Sorbent

2.1.1. FTIR Analysis

The FTIR spectra of the prepared materials were collected and are presented in Figure 1. The broad peaks at 3385 and 3156 cm^{-1} are due to the vibrations of –NH and –OH groups, respectively. The band observed at 1588 cm^{-1} is related to the imine bond C=N, which is the most significant identification of the Schiff base [24]. The intense assignments located at 3049–3063 and 2987–3005 cm^{-1} are attributed to aromatic and aliphatic υ(C–H), respectively [25,26]. The N–O stretching vibrations in the nitro group occur nearly at 1538 and 1328 cm^{-1} of the asymmetrical and symmetrical vibrations, respectively. After phosphorylation, the composition of the HNAP was validated by FTIR, and new bands appearing at 2349 cm^{-1} are assigned to the P–H stretching mode. The assignment at 1170 cm^{-1} corresponds to the P=O stretching vibration. Moreover, a P–OH stretching vibration is recorded at 990 cm^{-1} [27]. The quaternization reaction (a reaction with glycidyl trimethyl ammonium chloride) brings numerous –CH$_3$ groups with which a symmetric stretching vibration appears at 2875 cm^{-1} [28]. In addition, the two sharp features located at 1463 and 1409 cm^{-1} are assigned to the C–H stretching mode of the quaternary ammonium salt. The peak obtained at 964 cm^{-1} is due the asymmetrical stretching vibration of the quaternary nitrogen of quaternary ammonium salt [29].

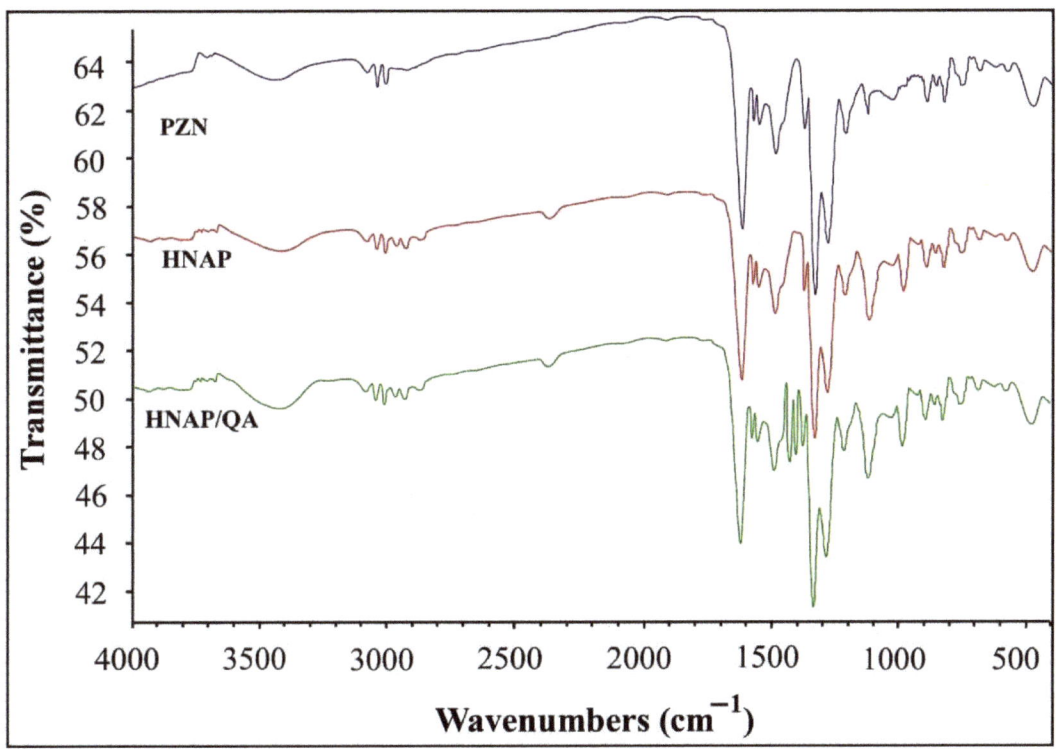

Figure 1. FTIR spectra of PZN, HNAP, and HNAP/QA.

2.1.2. BET Surface Analysis

The surface area (S_{BET}) and porosity of HNAP and HNAP/QA were detected using the Brunner–Emmett–Teller theory at 77K. As seen in Figure 2, the N_2 sorption/desorption curves of HNAP and HNAP/QA are defined as IV type isotherms with limited H_3-type hysteresis. This means that HNAP and HNAP/QA are mesoporous-type materials. The HANP/QA surface area (S_{BET}: 178.4 m^2 g^{-1}) is higher than that of the HANP surface area of 101.6 m^2 g^{-1}. In addition, there is a slight increase in the pore volume of HNAP, increasing partly from 0.205 cm^3 g^{-1} for HNAP to 0.288 cm^3 g^{-1} for HNAP/QA after a reaction with glycidyl trimethyl ammonium chloride (Table 1).

Table 1. Surface area and porosity of HNAP and HNAP/QA.

Substances	S_{BET} (m^2 g^{-1})	Pore Volume (cm^3 g^{-1})	Pore Size (nm)
HNAP	101.6	0.205	2.8
HNAP/QA	178.4	0.288	1.78

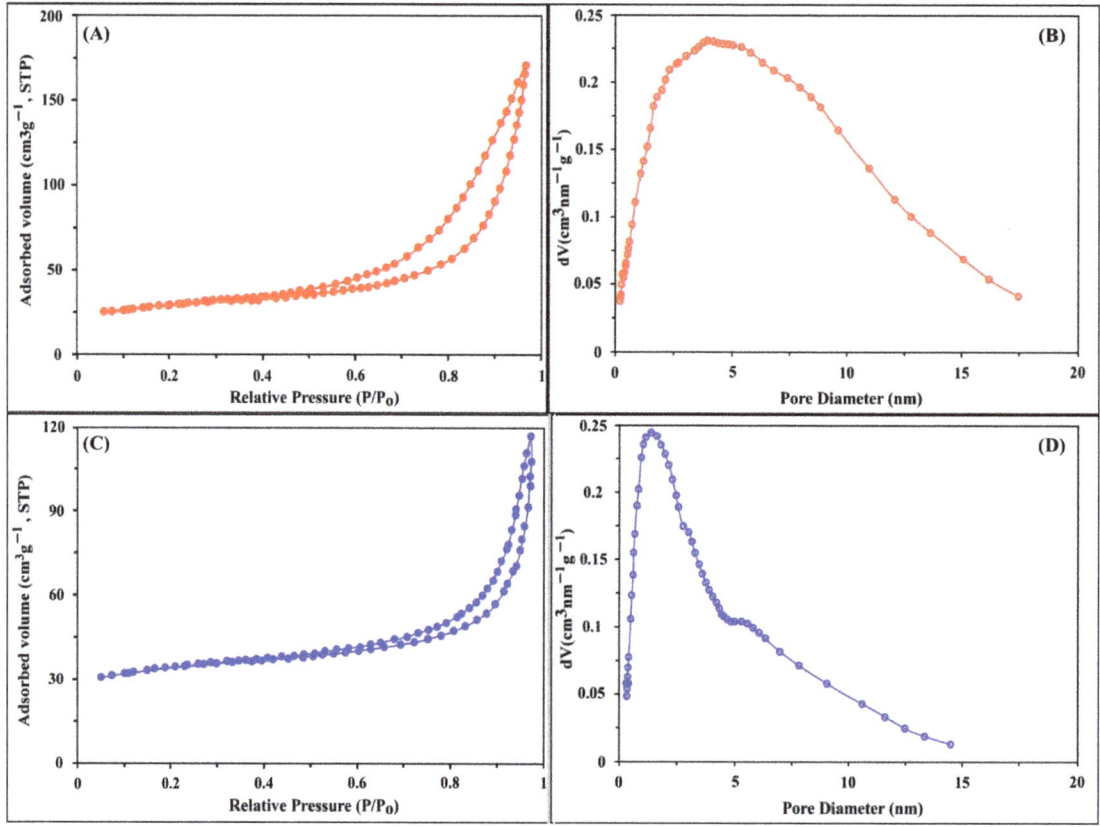

Figure 2. (**A,C**) N$_2$ sorption/desorption isotherms of HNAP and HNAP/QA, and (**B,D**) pore volume distribution of HNAP and HNAP/QA.

2.1.3. ^{1}H-NMR Analysis

^{1}H-NMR analysis is considered an efficient technique that provides significant information about the protons of the synthetic adsorbent to determine its structure. The principal assignments (δ) observed at 7.31–7.91 ppm are assigned to the protons of –CH in the benzene ring. The assignments at 9.06, 8.84, and 6.28 ppm are due to the protons of –OH, –NH, and –PH, respectively. The characterization of HNAP is illustrated in Figure 3. After modification, significant differences are observed in HNAP/QA spectrum. The chemical shift of the –OH group appears as a doublet at 6.06 ppm with a coupling constant of J = 5.74 Hz. A shift to a lower chemical shift (7.55 ppm, d, 1H, J = 14.3 Hz) is observed for –the PH group due to the modification. Furthermore, new assignments were observed for methyl groups (3.18 ppm, s, 3H, J = 8.2 Hz), methine groups (4.2 ppm, m, 1H, J = 7.46 Hz), and aliphatic methylene groups (4.44 ppm, m, 2H, J = 11.38 Hz).

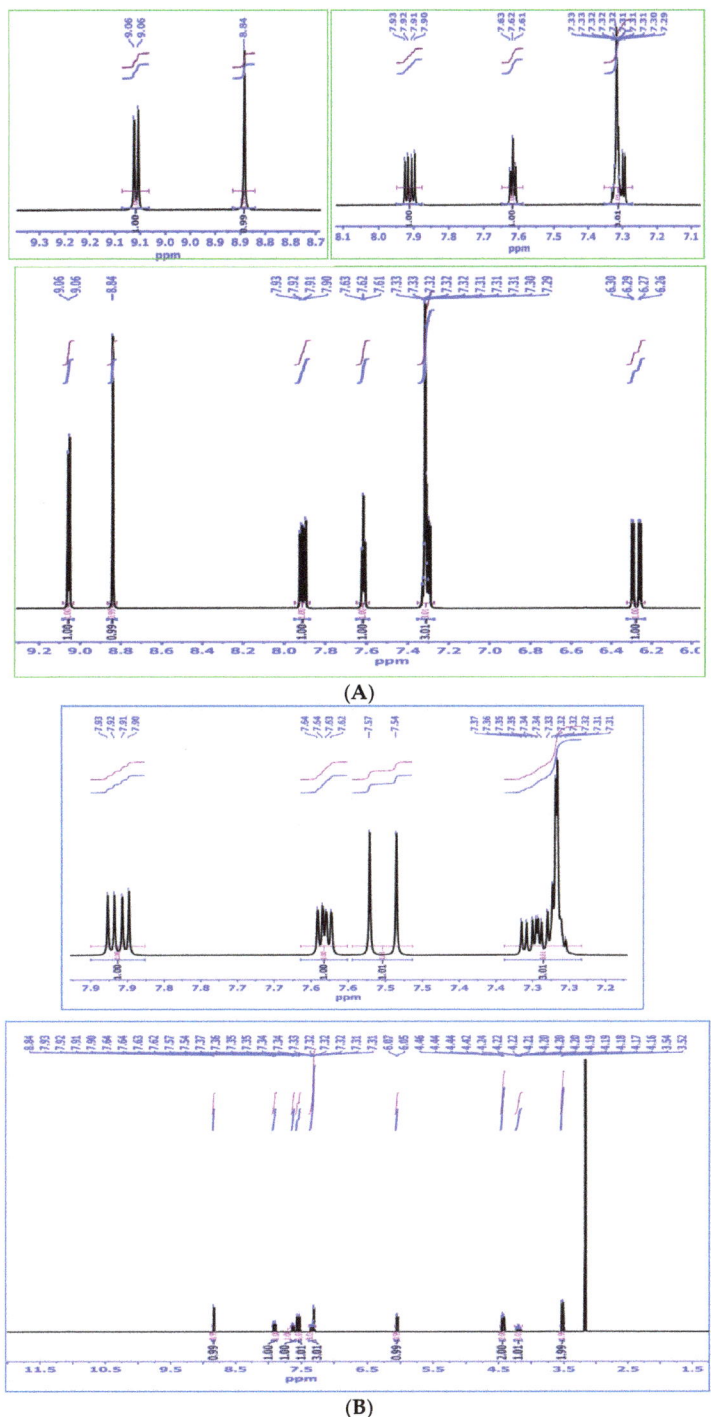

Figure 3. ^{1}H-NMR spectra of (A) HNAP and (B) HNAP/QA.

2.1.4. ^{13}C-NMR Analysis

A ^{13}C-NMR analysis of both HNAP and HNAP/QA was applied at 100.01 MHz to investigate the number of carbon atoms. As demonstrated in Figure 4A, the main assignments δ located around 126.9–127.5 ppm (s, J = 11.5, 6.9 Hz) are related to the carbon atom attached to the carbon of the imine group, the features at 146.6 ppm (s, J = 11.7, 7.9 Hz, –C=N) are related to the carbon of the imine group, the assignments at 143.8 ppm (s, J = 19.5 Hz) are attributed to the carbon attached to the nitro groups, and the assignments at 158.2 ppm (s, J = 7.5 Hz) are due to the carbon of the benzene ring attached to the phosphonic acid group, while the carbons of the benzene ring show assignments at 116.2–127 ppm (s, J = 3 Hz). Interestingly, the assignments of the de-protonated carbons were shown to be higher when de-shielded than any protonated carbons. After modification, a significant difference is observed in HNAP/QA, as illustrated in Figure 4B. A new feature appearing at 66.5 ppm (s, J = 6, 3.7 Hz, –CH$_2$) is related to the methylene group attached to the oxygen of the phosphonic acid group, the assignment at 64.2 ppm (s, J = 6.1 Hz, –CH) is related to the carbon attached to the –OH group, the assignment at 67.1 ppm (s, J = 6.1 Hz, –CH$_2$) is due to the carbon attached to the nitrogen of the quaternary nitrogen group, and finally, the carbons of the methyl groups of the quaternary nitrogen group show assignment at 53.7 ppm (s, J = 4.6 Hz, –CH$_3$).

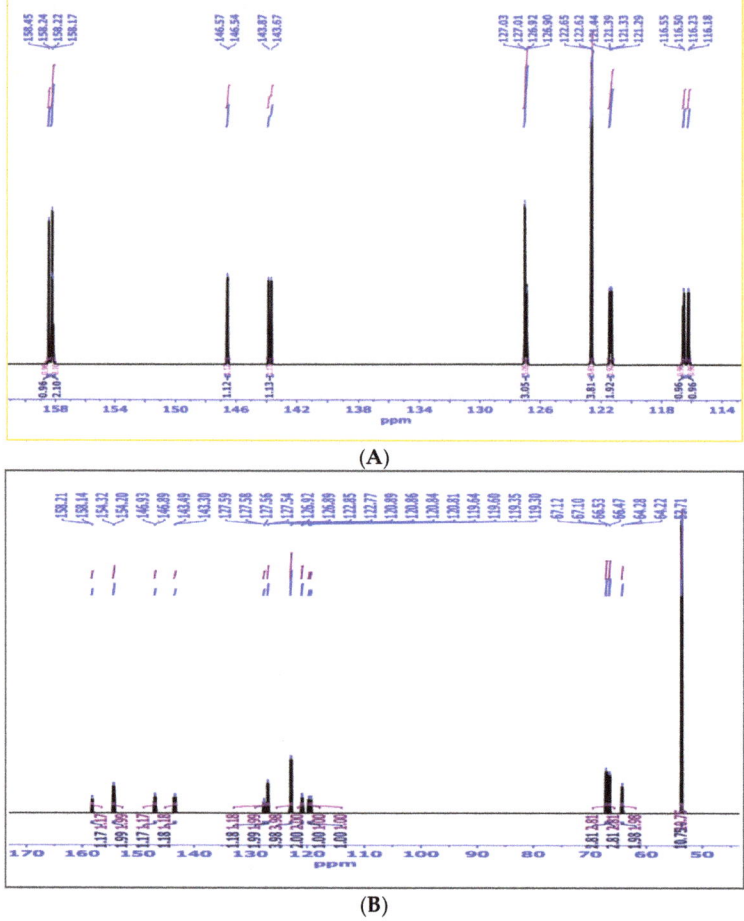

Figure 4. ^{13}C-NMR analysis of (**A**) HNAP and (**B**) HNAP/QA.

2.1.5. ³¹P-NMR Analysis

A ³¹P-NMR analysis with an energy of 200.01 MHz is presented in Figure 5. The primary assignments of the HNAP/QA phosphorus atoms are observed at 6.53 ppm as singlets.

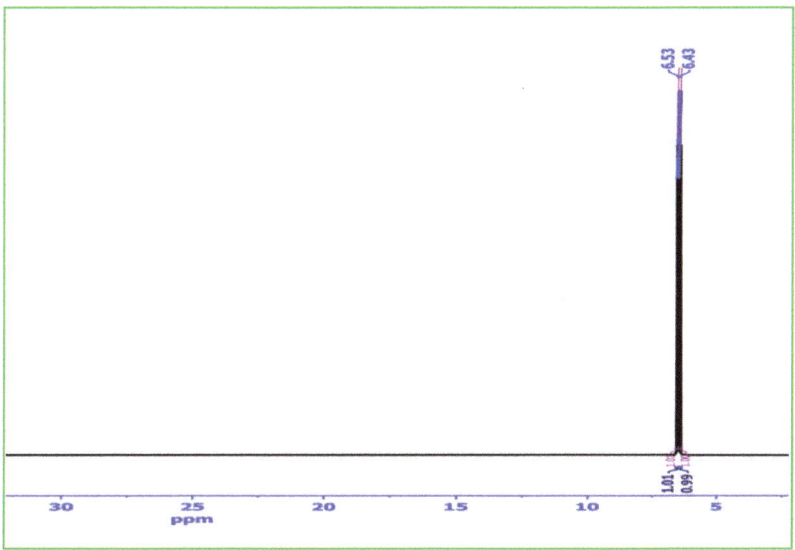

Figure 5. ³¹P-NMR analysis of HNAP/QA.

2.1.6. GC-MS Analysis

Figure 6 depicts the GC-MS chromatogram of HNAP/QA. It reveals that the most stable fragment $[m/z]^+$ is $[C_{32}H_{44}N_6P_2O_{12}Cl_2]^{\cdot}$, with a molecular weight of 837.58 g/mol and a relative abundance of 35%. The obvious peak at 78 g/mol with a relative abundance of 99% is due to the $[C_6H_6]^{\cdot}$ benzene ring $[C_6H_6]^{\cdot}$ or the $[C_3H_7Cl]^{\cdot}$ propyl chloride. The fragments of nitro benzene $[C_6H_5NO_2]^{\cdot}$, tropolium cation $[C_7H_7]^{\cdot}$, and para-phenylene diamine $[C_6H_8N_2]^{\cdot}$ have molecular weights of 123, 91, and 108 g/mol. The tiny peak at 64 g/mol with a relative abundance of 5% illustrates the formation of an ethyl chloride fragment $[C_2H_6Cl]^{\cdot}$. The molecular weight of 36 g/mol is related to the $[HCl]^{\cdot}$ fragment, which explains the liberation of hydrogen chloride gas. Moreover, the molecular weight of 95 g/mol refers to the formation of the quaternary trimethyl ammonium chloride moiety $[C_3H_{10}NCl]^{\cdot}$. The recorded fragment with a molecular weight of 140 g/mol is due to para nitro phenol $[C_6H_5NO_2]^{\cdot}$. An important note should be made for the fragmentation pattern mechanism for the compounds containing chlorine atoms. This fragmentation mechanism is most important where chlorine is a good leaving group. When one chlorine atom is present, M and M + 2 isotopic peaks become very significant, where M represents the molecular ion peak (M = 837 with an A% of 35% and M + 2 = 839 with an A% of 21). If the compound contains two chlorine atoms, a M + 4 isotopic peak should be observed, as well as an intense M + 2 peak (M + 4 = 841 with an A% of 10%). In addition, the thermal analysis of HNAP/QA is introduced in a supplementary file Figure S1.

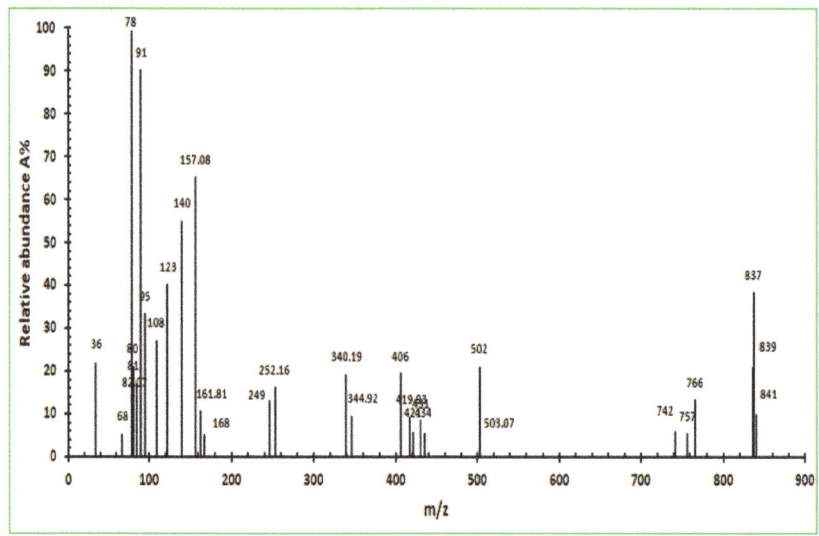

Figure 6. GC-Mass analysis of HNAP/QA.

2.2. Sorption of W(VI) Ions

Different factors were studied to optimize the sorption of W(VI) ions on the HNAP/QA ligand, as described below.

2.2.1. Influence of pH on W(VI) Ion Sorption

It is noteworthy that the pH value of the aqueous solution has a crucial impact on the adsorption of W(VI) ions. Changing the pH value has an obvious influence on not only the protonated/deprotonated active sites of the HNAP/QA but also the speciation of the tungsten ions formed. The pH value and metal content can affect the formation of W(VI) ions, as reported previously in in the literature [30]. Tungsten is a predominantly anionic species at pH values of over 2.0. Tungsten's cationic species are completely absent at these pH levels. Since most W(VI) is found in anionic species, the adsorption behavior of W(VI) ions with HNAP/QA was studied in relation to the pH value. Given that $WO_3(s)$ formation is possible at pH values below 2.0 in solutions [31], the obtained data (shown in Figure 7) demonstrate a general observed trend that the adsorption of W(VI) ions increased gradually from a pH 1.0 to 4.0 and reached a maximum adsorption efficiency at pH 4.5. The W(VI) ions sorption then decreased as the pH increased past a value of 4.5. At a pH value of 4.5, the percentage of absorbed tungsten reached a plateau of greater than 80% and then decreased gradually as a function of pH between pH 5.0 and pH 8.0. Understanding the tungsten solution species helps interpret this pH change.

For pH values between 2.0 and 8.0, W(VI) ions are present in the form of polynuclear anionic species, such as $[W_{12}O_{39}]^-$, $[W_{12}O_{41}]^{10-}$, and $[W_6O_{20}(OH)]^{5-}$, while the formation of a neutral species, WO_2, occurs at higher pH values. As the pH of a solution rose from 7.5 to 9.0, the adsorption percentage of W(VI) decreased. Furthermore, the active site of the HNAP/QA sorbent is the quaternary ammonium chloride moiety, which acts as anion exchanger. This can be explained by a shift in the species distribution. A pH of 4.5 is the most suitable pH for the adsorption of W(VI) ions in this study.

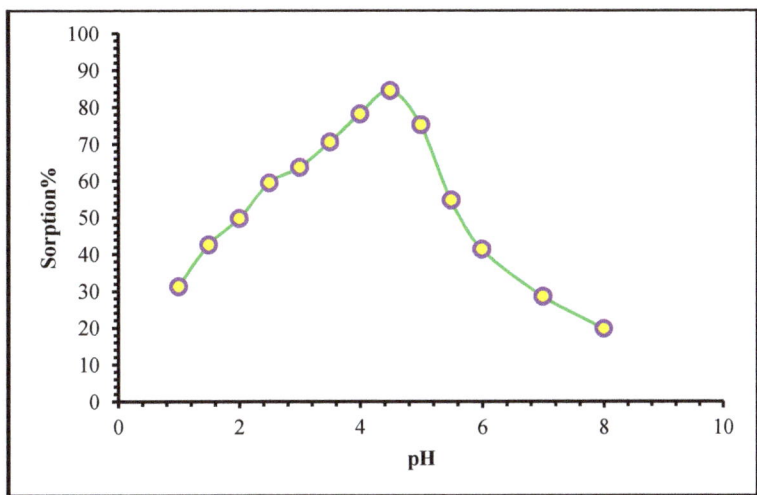

Figure 7. Effect of the solution pH on the adsorption % of W(VI) using HNAP/QA (W(VI) = 250 mg L^{-1}, 20 mL solution, 30 min, room temperature).

2.2.2. Effect of HNAP/QA Dose

Figure 8 shows the impact of HNAP/QA dosage on the sorption percentage of W(VI) ions. The experiment was applied by boosting the adsorbent quantity from 0.01 to 0.2 g while holding all other variables stable at a 250 mg L^{-1} W(VI) ion concentration and room temperature. High concentrations of HNAP/QA expose more of the tungsten's surface, leading to a higher rate of adsorption. The adsorption was improved noticeably as the dose was raised from 0.01 to 0.08 g. There was a little noticeable change in the adsorption process after adding more adsorbents. The percentage of metal adsorbed determines the sorption capacity of HNAP/QA, and the data show that the greatest adsorption efficiency of tungsten was achieved at 0.08 g of HNAP/QA.

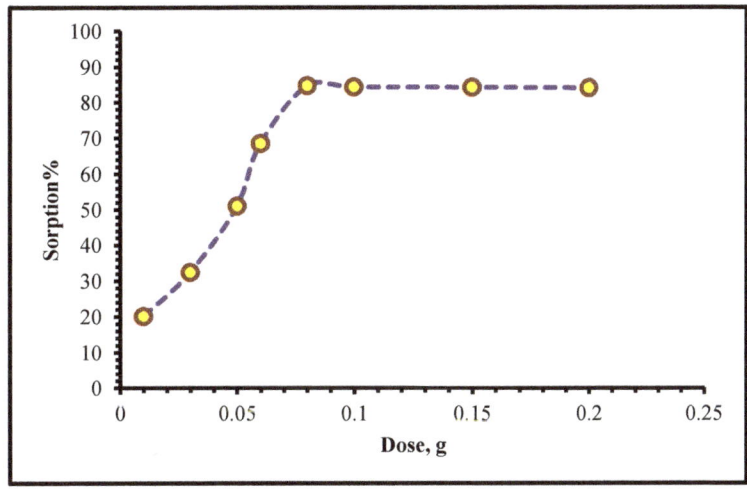

Figure 8. Effect of the HNAP/QA dose on the adsorption % of W(VI) (pH 4.5, W(VI) = 250 mg L^{-1}, 20 mL solution, 30 min, room temperature).

2.2.3. Kinetics Study

By analyzing kinetic profile, we can learn not only how long it takes to reach equilibrium but also how stable the material is, how it interacts with the target solute, and which stage in the sorption mechanism is ultimately responsible for regulating the process [32]. In fact, sorption occurs on the easily accessible active sites at the surface of the sorbent and the first exterior layers of the material, and almost 84% of total sorption occurs within the first 30 min of contact. The second stage involves a more gradual sorption caused by diffusion inside the sorbent's pores. As presented in Figure 9, after 60 min of agitation, the equilibrium time was achieved because the WO_4^{2-} ions were adsorbed continuously from the solution until equilibrium was accomplished, and then no further change was observed after the increase in time.

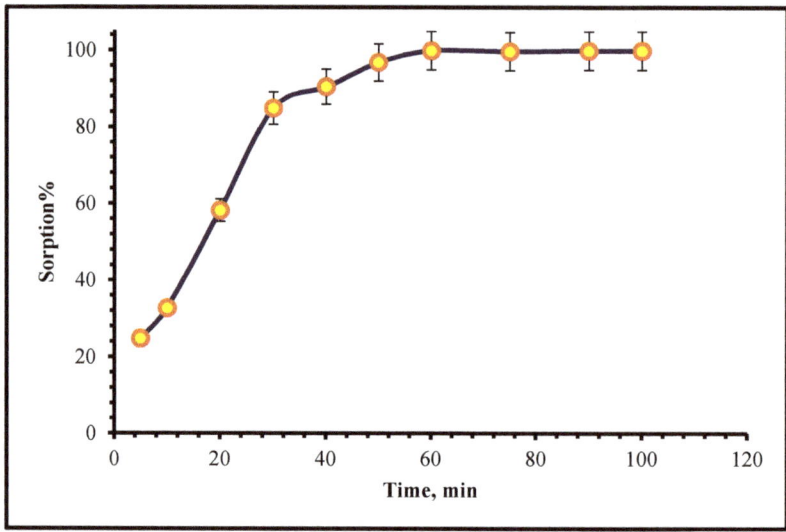

Figure 9. Effect of the agitation time on the adsorption % of W(VI) using HNAP/QA (pH 4.5, W(VI) = 250 mg L^{-1}, 20 mL solution, 0.08 g, room temperature).

The linear forms of Lagergren's pseudo-first-order (PFO) and second-order (PSO) rate expression were used to predict the order of the kinetic adsorption process of W(VI) ions on HNAP/QA, as presented in equations below [33–36]:

$$\log(q_e - q_t) = \log q_e - \frac{K_1 t}{2.303} \quad (1)$$

$$\frac{t}{q_t} = \frac{t}{q_e} + \frac{1}{K_2 q_e^2} \quad (2)$$

where K_1 represents the PFO rate constant (min^{-1}), K_2 is defined as the rate constant of the PSO (g mg^{-1} min^{-1}), and the quantities of WO_4^{2-} ions sorbed at any moment are denoted by q_t (mg g^{-1}). As provided in Figure 10, the $\log(q_e - q_t)$ vs. the agitation time (t) shows that the PFO model does not match the adsorption of WO_4^{2-} species on HNAP/QA. The calculated q_e value is 110.535 mg g^{-1} (Table 2), higher than the value of the experimental q_e (62.44 mg g^{-1}). Nonetheless, the PSO model has a significantly higher value of R^2 (0.9965), which is close to unity, as listed in Table 2. Furthermore, the calculated q_e value, 66.225 mg g^{-1}, is also very close to the experimental value of q_e, 62.44 mg g^{-1}. The data point to a better match of the PSO equation to the adsorption of WO_4^{2-} ions on HNAP/QA.

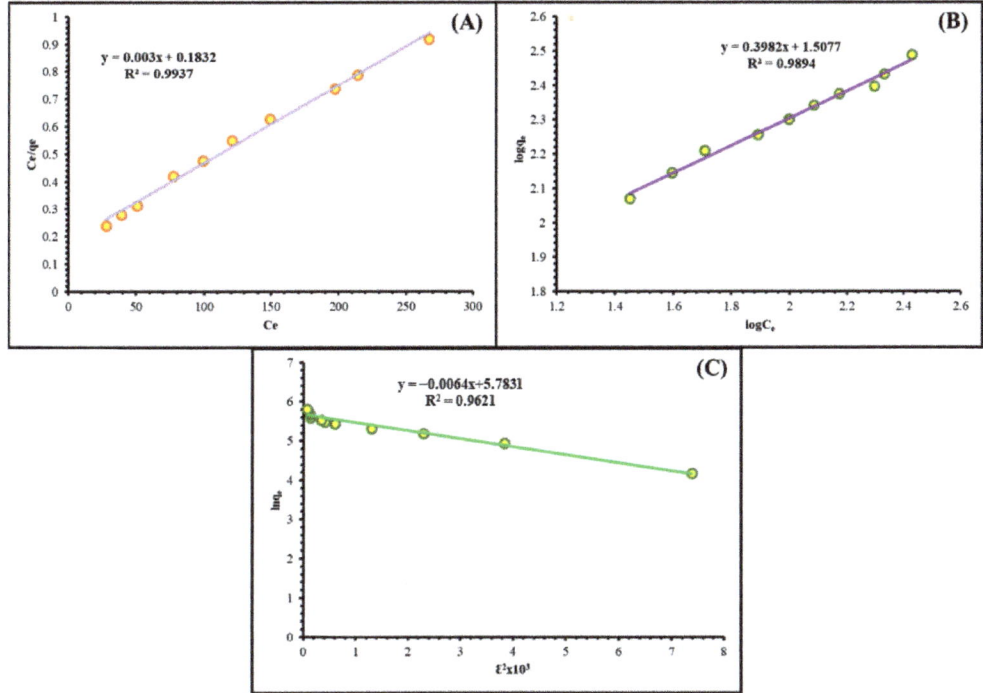

Figure 12. (**A**) Freundlich, (**B**) Freundlich, and (**C**) Dubinin–Radushkevich isotherm models of W(VI) ions adsorbed on HNAP/QA.

Table 3. Isotherms parameters of W(VI) ions adsorbed on HNAP/QA.

Models	Parameters	
Langmuir isotherm	Equation q_{max} (mg g^{-1}) K_1 R^2	y = 0.003x + 0.1832 333.33 0.0164 0.9937
Freundlich isotherm	Equation K_f (mg g^{-1}) $1/n$ (mg min g^{-1}) R^2	y = 0.3982x + 1.5077 32.189 0.3982 0.9894
D-R isotherm	Equation q_D (mg g^{-1}) B_D (Mol^2 kJ^{-2}) E (kJ mol^{-1}) R^2	y = −0.0064x + 5.7831 324.76 0.0064 8.8388 0.9621
Practical Capacity	$q_{e(exp)}$ (mg g^{-1})	326.75

Secondly, the Freundlich isotherm denotes the multilayer sorption of ions onto the surface of the sorbent material, as described in the equation below [42,43]:

$$\log q_e = \log K_f + \frac{1}{n} \log C_e \tag{4}$$

where K_f corresponds to the adsorption capacity of HNAP/QA (mg g^{-1}) and n is a constant. The smaller the $1/n$ value, the stronger the interaction between the metal ions, the adsorbent and the working adsorbate. Additionally, $1/n = 1$ signifies linear adsorption, which results in the same adsorption energies across all sites. According to the Freundlich isotherm, the binding strength decreases when the number of binding sites is increased [44]. The data in Table 3 and the plot in Figure 12B reveal that the Freundlich isotherm does not agree with the sorption of W(VI) ions on HNAP/QA, for which the K_f value (32.189 mg g^{-1}) is significantly lower when compared with the $q_{e(exp)}$.

Lastly, the Dubinin–Radushkevich isotherm model is applied to differentiate between physical and chemical sorption because of adsorption heterogeneity [45–47]. It calculates the surface energy heterogeneity. The following is a linear formula of the Dubinin–Radushkevich model, calculated from the following equations:

$$\ln q_e = \ln q_D - B_D(\varepsilon)^2 \tag{5}$$

$$\varepsilon = RT \ln(1 + \frac{1}{C}) \tag{6}$$

where B_D (mol^2 kJ^{-2}) represents the energy of sorption, and ε symbolizes the Polanyi potential. B_D offers an opportunity to determine the adsorption energy E (kJ mol^{-1}) of W(VI) ions on HNAP/QA [48].

$$E = \frac{1}{\sqrt{2B_D}} \tag{7}$$

If the calculated value of E is less than 8.0 kJ mol^{-1}, the sorption process is defined as physisorption, while if the value of E is $8.0 < E < 16.0$ kJ mol^{-1}, the sorption is believed to be chemisorption [49,50]. The data obtained from Figure 12C and listed in Table 3 show the E value is 8.839 kJ mol^{-1}. Therefore, the sorption of W(VI) ions on HNAP/QA is chemisorption process. The result is convenient with the Langmuir isotherm, which suggests a monolayer sorption.

2.2.6. Thermodynamics Study

The impact of temperature on the adsorption process was investigated using a 1250 mg L^{-1} W(VI) solution, a pH value of 4.5, and 0.08 g (dry mass) of HNAP/QA at 298, 313, 323, 333, and 343 K. As demonstrated in Figure 13, the reaction rate rose gradually when the temperature was increased. As a result, the adsorption is considered an endothermic process; therefore, increasing the ambient temperature of the system is the correct choice to speed up the adsorption.

The thermodynamic parameters controlling the sorption system ($\Delta G°$, $\Delta H°$, and $\Delta S°$) were mathematically determined from the Van 't Hoff equations as follows [51,52]:

$$\log K_d = \frac{\Delta S}{2.303R} - \frac{\Delta H}{2.303RT} \tag{8}$$

$$\Delta G° = \Delta H° - T\Delta S° \tag{9}$$

where T is a symbol of the temperature of the system (K), and R stands for the gas constant (8.314 J mol^{-1} K^{-1}). Both $\Delta H°$ and $\Delta S°$ can be calculated from the slope and the intercept of log K_d against the $1/T$ plot, as illustrated in Figure 14. The correlation coefficient is valued at $R^2 = 0.9956$. The values of the thermodynamic parameters are tabulated in Table 4; the negative value of $\Delta G°$ for all temperatures proves that the adsorption of W(VI) ions on HNAP/QA occurs spontaneously. In addition, the positive values of $\Delta H°$ and $\Delta S°$ indicate the endothermicity and randomness of the W(VI) sorption process.

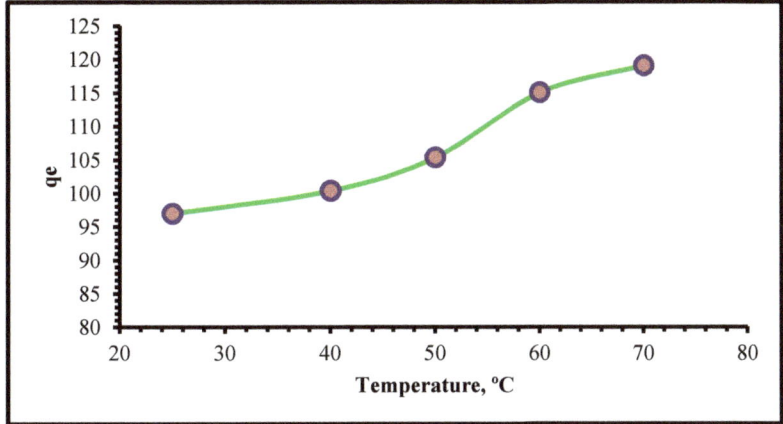

Figure 13. Effect of temperature on the adsorption % of W(VI) using HNAP/QA (pH 4.5, W(VI) = 250 mg L^{-1}, 20 mL solution, 0.08 g).

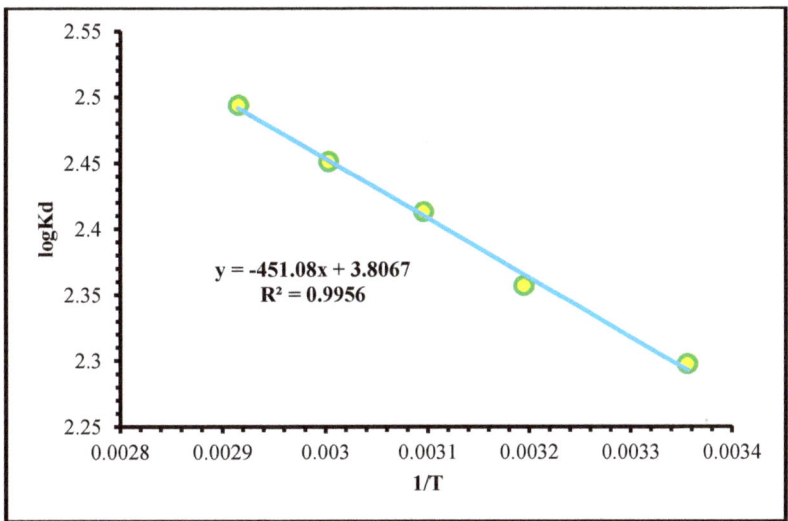

Figure 14. Plot of log K_d vs. $1/T$ of W(VI) ions adsorbed on HNAP/QA.

Table 4. Thermodynamic parameters of W(VI) ions adsorbed on HNAP/QA.

$\Delta S°$ (J mol^{-1} K^{-1})	$\Delta H°$ (kJ mol^{-1})	$\Delta G°$ (kJ mol^{-1})				
		298 K	313 K	323 K	333 K	343 K
75.887	10.36	−21.89	−22.81	−23.54	−24.27	−25.0

2.2.7. Effect of Foreign Metal Ions on Selectivity

In order to investigate the selectivity of HNAP/QA toward the uptake capacity of W(VI) ions in the presence of different existing metal ions, the sorption process of W(VI) ions was conducted employing several ion solutions of Mo(VI), V(V), Ca(II), Cu(II), and Pb(II) with different concentrations ranging between 10 and 500 mg L^{-1}. The obtained data in Figure 15 illustrate that the selectivity strength of the HNAP/QA material was slightly affected by the high concentrations of V(V), Ca(II), Cu(II), and Pb(II) ions. Impressively, the uptake of W(VI) ions was intensely influenced by the presence of Mo(VI) ions. It is more likely that the ability of the HNAP/QA material to adsorb cations through the chelation mechanism with a lone pair of electrons on the hydroxyl groups as well as the phosphine group of oxygen results in a reduction in the selectivity of the HNAP/QA sorbent to some extent. In the case of Mo(VI), both metals have the same chemical properties; therefore, both cations compete for the same binding sites, causing an obvious decrease in the adsorption efficiency. However, the HNAP/QA adsorbent is still effective in the case of the presence of several ions in the system. The result indicates that the HNAP/QA adsorbent demonstrates a high selectivity of W(VI) ions compared to other elements.

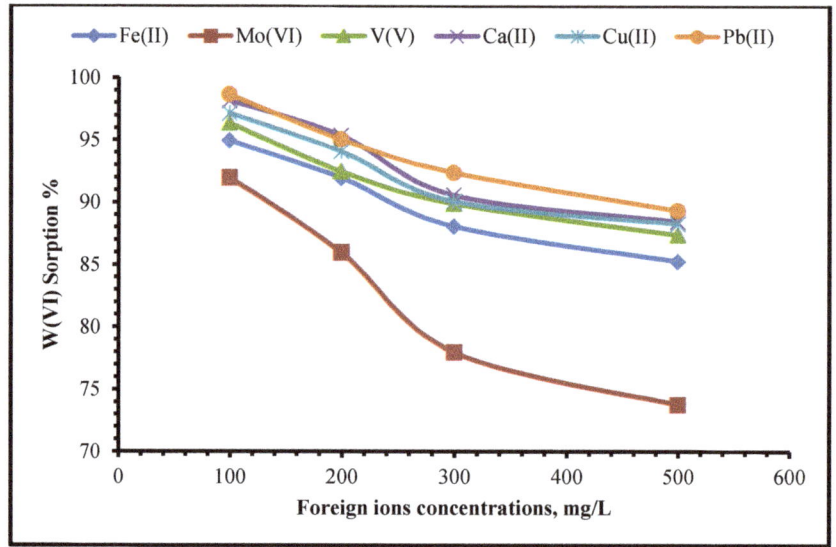

Figure 15. Effect of foreign ions on the adsorption % of W(VI) using HNAP/QA (pH 4.5, 20 mL solution, 30 min, 0.08 g, room temperature).

2.3. Desorption–Regeneration Study

The desorption of a metal-loaded sorbent was investigated not only for the possibility of reusing of the sorbent but also for the recovery of the valuable metal ion. Firstly, different types of eluents were used for the desorption of W(VI) ions, such as ascorbic acid, HCl, NH_4Cl, NaOH, and NH_4OH solutions (0.25 M), which were used as eluting agents at room temperature for 10 min. The obtained data in Figure 16A show that the best eluent for the efficient recovery of W(VI) is NH_4Cl, which is capable of recovering 79.2% of loaded W(VI) ions on HNAP/QA. Secondly, the influence of the concentration of NH_4Cl on the recovery of W(VI) ions was then studied using several concentrations varying from 0.25 to 2.0 M, as shown in Figure 16B. The obtained data show that 0.5 M NH_4Cl achieved a recovery of 92.6% of the loaded W(VI) ions on HNAP/QA. Lastly, the influence of the desorption time was assessed by ranging the elution period between 5 and 50 min. Figure 16C demonstrates that 20 min was sufficient for the recovery of almost 79.2% of the W(VI) ions from the loaded sorbent. It is completely clear that when the metal-loaded HNAP/QA was investigated before and after desorption with 0.5 M NH_4Cl for 20 min at an ambient temperature, no presence of W(VI) ions was detected after desorption; it is an evident that all W(VI) ions were recovered back to the solution, as provided in Figure 17.

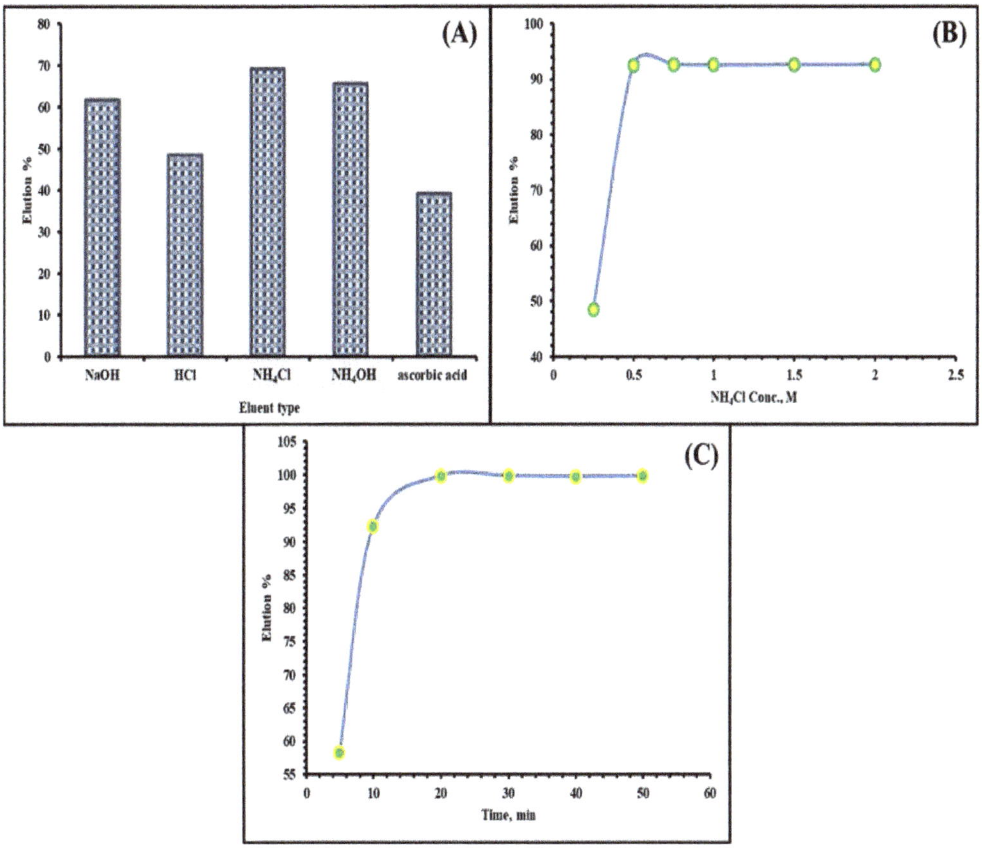

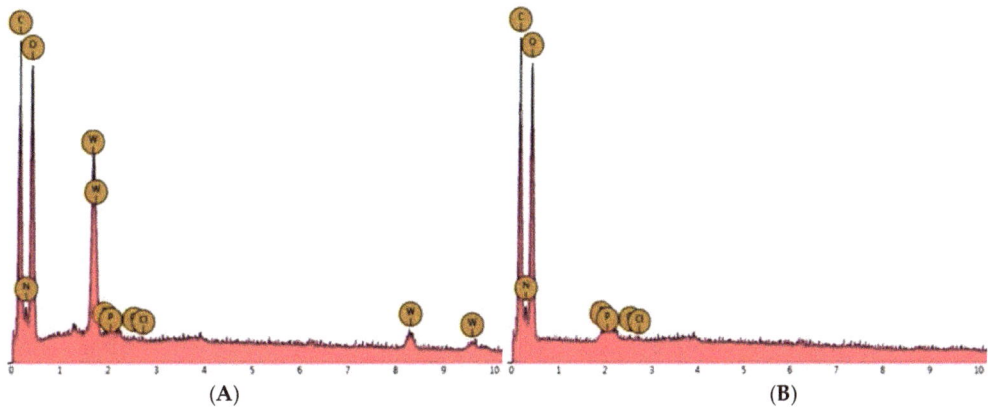

Figure 16. Factors affecting the desorption of W(VI) from HNAP/QA: (**A**) eluent type, (**B**) eluent concentration, and (**C**) elution time.

Figure 17. EDX analysis of (**A**) pregnant HNAP/QA with W(VI) and (**B**) HNAP/QA after elution.

2.4. Recovery of W(VI) Ions from Wolframite Ore

One of the most important and promising areas is Gabal (G) Qash Amir, in the extreme southeastern part of Egypt. It is considered a rich area with economic and strategic metals such as Mn, Zr, Ta, U, Nb, and W. In the last decade, many new W-rich mineralizations were discovered in the studied area of new Gabal Qash Amir [40–45]. The region known as G. Qash Amir is in the extreme southeastern corner of the Egyptian Eastern Desert; it is located approximately 28 km southwest of Abu-Ramad city and not far from the border of Sudan, bounded by longitudes 36°10′59″–36°14′24″ E and latitudes 22°14′07″–22°15′21″ N. This area is a section of the Arabian-Nubian shield zone.

2.4.1. Pre-Concentration of Wolframite Ore

A working technological ore sample was precisely collected from a mineralized invading quartz vein within G. Qash Amir wolframite granite. A bulk sample (10 kg, containing around 0.43% w/w of WO_3) was ground using a roll mill crusher and jaw crushers to diminish the mineral particles to a size of less than 1.0 mm. The gravity concentration was performed using a lab wet Wilfley shaking table (Holman-Wilfley Ltd., Redditch, UK); this operation allowed for the production of a W-rich concentrate (about 83.0 g). The next step in the pre-processing consisted of a magnetic separation utilizing a high-intensity induced magnetic roll separator (Carpco Model MLH (13) III-5, Outokumpu Technology, Inc., Jacksonville, FL, USA) to recover wolframite-rich minerals (Figure 18).

Figure 18. Photographs of selected wolframite sample and grains.

2.4.2. Characterizations of the Wolframite Ore

Wolframite is a tungsten mineral that mainly consists of manganese tungsten oxide (Fe,Mn)WO_4; it is the intermediate mineral between ferberite (Fe^{2+} rich) and hübnerite (Mn^{2+} rich). The semi-quantitative analysis of wolframite samples shows the preponderance of both W and Mn elements (88.8%, w/w) and the presence of a non-negligible fraction of silicate and iron. The wolframite sample was confirmed using SEM-EDX and XRD characterizations, as can be seen in Figure 19 and Table 5.

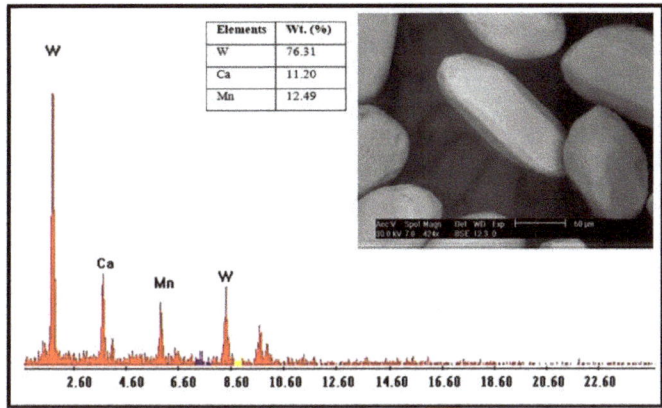

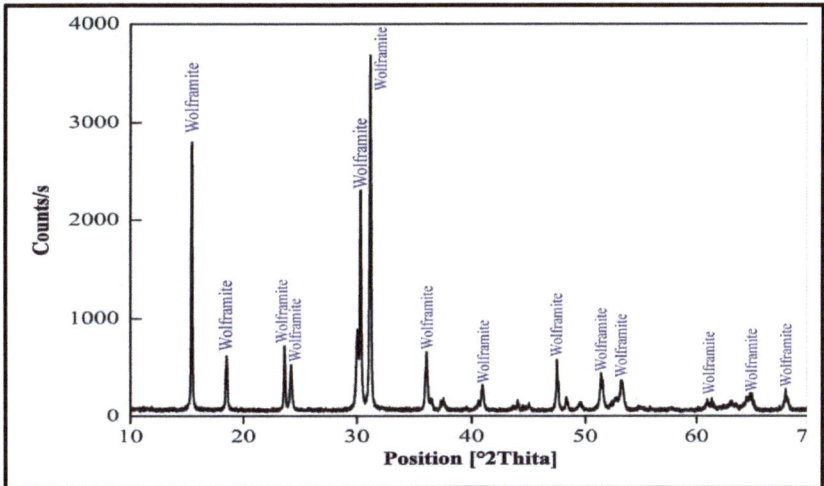

Figure 19. SEM-EDX and XRD analyses of wolframite sample.

Table 5. Chemical compositions of wolframite sample.

Element	Concentration (%)
WO_3	77.13
MnO	8.99
SiO_2	0.58
Al_2O_3	0.3
Fe_2O_3	3.9
SO_3	0.49
CaO	7.57
Others	1.04

2.4.3. Extraction of Tungsten Oxide

The first stage is the tungsten leaching from W-concentrate; this process was carried out with HCl (30% w/w HCl solution) to eliminate most of the Mn and Fe content [53]. The acid-leaching step was performed using continuous stirring at 400 rpm for 3 h at 110 °C and a 1:1.5 solid to liquid (S/L) ratio. Secondly, the wolframite sample was subjected to alkaline leaching using 300 g of NaOH (40%, w/w, NaOH solution); the leaching process was applied at 400 rpm and a S/L ratio of 1:3 for 2 h at 130 °C. The alkaline leaching process allowed for the dissolution of almost all the W-content and some traces of Mn and Fe. The obtained leachate was analyzed, and the W concentration is listed in Table 6.

Table 6. Chemical compositions of wolframite leachate.

Element	Concentration (%)
WO_3	89.46
MnO	0.004
SiO_2	0.46
Al_2O_3	0.3
Fe_2O_3	0.002
CaO	7.57

The leachate was filtered, and the pH was adjusted to 4.5. The HNAP/QA adsorbent was used to adsorb W(VI) ions from the leachate; the adsorption process was conducted based on the best controlling factors studied previously (pH 4.5, agitation time of 60 min, 2.0 g of HNAP/QA, and 250 mL of leachate at room temperature). Adsorption was accomplished, and the W(VI) content was recoverable through the separation of HNAP/QA in 0.5 M of NH_4Cl for 20 min at room temperature. Later, the W(VI) ions were precipitated from the eluate using 30% HCl. A yellow $WO_3 \cdot nH_2O$ product was formed. Finally, the precipitate was rinsed multiple times with ultrapure water and heated at 70–90 °C for 30 min to dry. This was followed by calcination at 500 °C for 2 h. The final product was characterized using SEM-EDX analysis, and the data are shown in Figure 20.

Figure 20. SEM-EDX analysis of obtained tungsten oxide product.

The EDX patterns indicate the W and O as the major elements and no notable impurities, while the SEM image shows the irregular shape of WO_3 to be a flower-like structure. A flow sheet summarizing the recovery of W(VI) ions from the wolframite ore is provided in Figure 21.

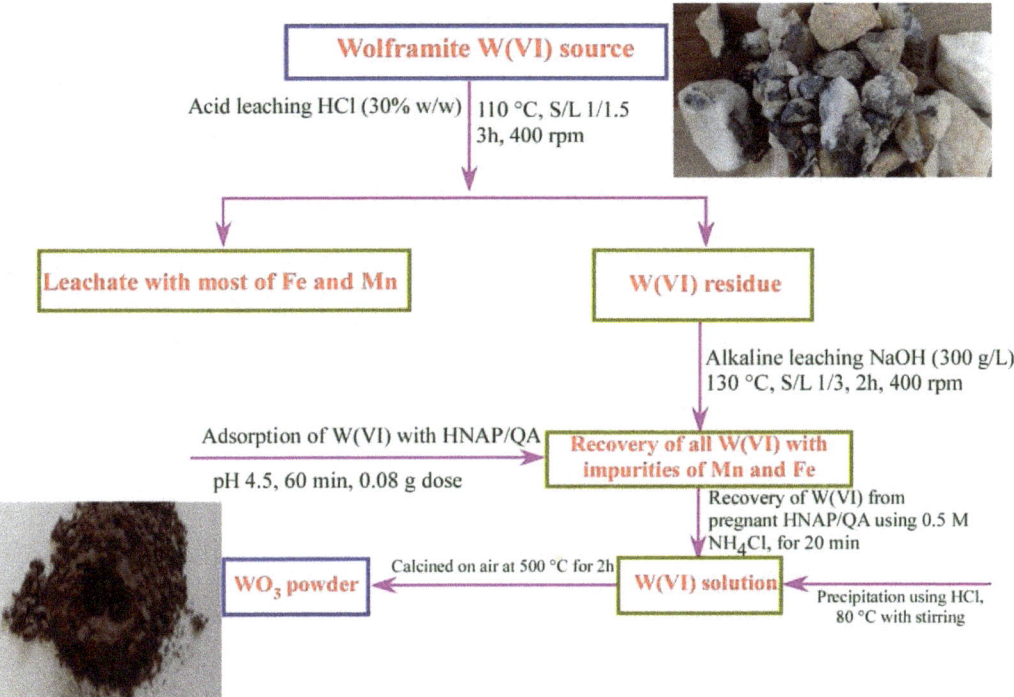

Figure 21. Flow sheet of the recovery of tungsten oxide from wolframite sample.

3. Materials and Methods

3.1. Chemicals and Reagents

P-phenylenediamine and 2-hydroxy-5-nitrobenzenaldehyde were purchased from Thermo Fisher Scientific (Morris Plains, NJ, USA). Shanghai Makclin Biochemical Co., Ltd. (Shanghai, China) provided the glycidyl trimethyl ammonium chloride (95%), $AlCl_3$, and NaOH. Phosphorus oxychloride, sodium tungstate ($Na_2WO_4 \cdot 2H_2O$), NaCl, $CHCl_3$, $CDCl_3$, dimethyl sulfoxide, and CH_3OH were purchased from Sigma-Aldrich (St. Louis, MO, USA). All chemicals were of a very high purity level and required no further processing before use. All the experimental solutions were prepared using ultrapure water of 18.2 MΩ·cm. Thin paper chromatography (PC) was utilized to explore the synthesis process. Furthermore, a basic UV lamp set to 250 nm and an eluent comprising a combination of 1:1 v/v ethanol and ethyl acetate were used to identify spots formed on the PC.

3.2. Preparation of the Adsorbent (HNAP/QA)

A methanolic solution of 2-hydroxy-5-nitrobenzenaldehyde (10 mL, 25 mmol) was added drop-wise to a 10 mL methanolic solution of p-phenylenediamine (25 mmol) in the presence of an acid (2 mL of 98% H_2SO_4) with stirring. The mixture was allowed to reflux at 70 °C for 6 h. The reaction's volume decreased by half under a vacuum. An orange-yellow precipitate was formed, which was filtered, washed out multiple times by methanol, and left to air-dry overnight. The TLC was used to monitor the progress of the condensation reaction. The produced Schiff base (3,3′-((1,4-phenylenebis(azaneylylidene))-bis(methaneylylidene))-bis(4-nitrophenol)) was provided a PZN abbreviation, and the final yield was determined as 90%. The produced material was subjected to phosphorylation, which took place through an interaction with 0.2 g of $AlCl_3$ and 25 mL phosphorus oxychloride with vigorous stirring for 30 h at a temperature of 110 °C (Scheme 1). $AlCl_3$ and $POCl_3$ were used as phosphorylating agents to introduce a phosphonic acid group. The AlCl₃ aids in the release of HCl during the reaction. A pale yellow precipitate was formed, separated through filtration, rinsed numerous times with ultrapure water and methanol, and allowed to dry overnight in the air. The produced material was named HNAP, referring to phosphorylated beads of (3-(((4-((5-(((S)-hydroxyhydrophosphoryl)oxy)-2-nitrobenzylidene) amino) phenyl) imino) methyl)-4-nitrophenyl hydrogen (R)-phosphonate)); its purity was assessed using TLC and was found to be 94.0%, (Scheme 1).

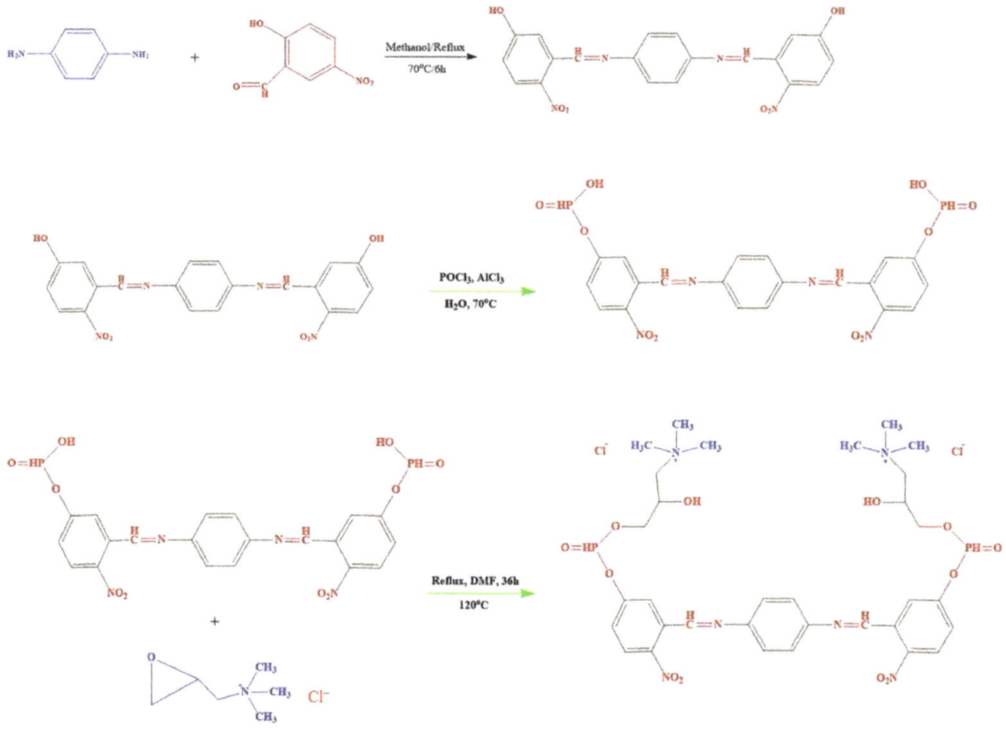

Scheme 1. Preparation of HNAP and HNAP/QA.

The phosphorylated beads (HNAP) were subjected to quaternization; this reaction occurred through the reflux of 4.5 g of HNAP in a three-necked flask containing a solution of 12.0 g of glycidyl trimethyl ammonium chloride with vigorous stirring for 36 h at 100 °C. The final product was separated by filtration and washed out several times with ultrapure water and methanol to remove the unreacted materials and solvent. The product was dried then overnight in the air. The produced material, HNAP/quaternary ammonium salt, was named (HNAP/QA). All synthesized chemicals were provided with names in accordance with the IUPAC nomenclature.

3.3. Instrumentation

We tested the adsorbent using the following spectroscopic techniques: FTIR (Nicolet™ iS50, Thermo Fischer Scientific, Morris Plains, NJ, USA), mass spectra using a Hitachi M-8000 unit, Tokyo, Japan, SEM-EDX using a Hitachi S 4160 FE SEM, Tokyo, Japan, and NMR in a 500 MHz field using a Bruker Fourier 80, Billerica, MA, USA. The concentrations of ions were measured using inductively coupled plasma mass spectrometry (ICP-MS) (NexION 1000, PerkinElmer, Waltham, MA, USA).

3.4. Adsorption and Elution Experiments

A tungstate stock solution of 1000 mg L^{-1} was prepared by dissolving a proper weight of sodium tungstate (Na$_2$WO$_4$·2H$_2$O) in a 0.01 M NaCl solution. The sorption experiment was conducted in three independent batches, each of which was made from the stock solution. HCl and NaOH, at concentration of 0.1 M each, were administered to adjust the acidity of the solution. For the standard batch sorption tests, the experiments were carried out using a 20 mL of 250 mg L^{-1} W(VI) ion solution for 30 min and 0.1 g of HNAP/QA at 25 °C, unless stated otherwise. The pH of the aqueous medium was changed from 1.0 to 6.0 to study the impact of the pH value. In addition, the experiments were performed at an amount of adsorbent between 0.01 and 0.2 g for a period of 30 min and a 250 mg L^{-1} W(VI) ion solution at a pH of 4.5 and at room temperature to investigate the effect of the adsorbent dose. The sorption time was studied likewise by varying the contact time between 5 and 90 min. Finally, the effect of temperature was investigated between 25 and 70 °C. The adsorption efficiency (S%) of the W(VI) ions and the adsorption capacity (q_e) were calculated using equations below:

$$S\% = \frac{C_o - C_e}{C_o} \times 100 \quad (10)$$

$$q_e = (C_o - C_e) \times \frac{V}{m} \quad (11)$$

where C_o and C_e (mg L^{-1}) are the W(VI) ion concentrations in the liquid phase at the beginning and end of the experiment, respectively, and q_e (mg g^{-1}) symbolizes the sorption capacity. In this equation, V (L) represents the volume of the working solution, and m (g) corresponds to the dry weight of the HNAP/QA.

3.5. Desorption Experiments

Desorption is a crucial step in recycling a sorbent material and recovering W(VI) ions. Once the adsorption experiment was completed, desorption tests were performed using the adsorbent in a batch system. NaOH, NH$_4$OH, HCl, NH$_4$Cl, and ascorbic acid were among the desorptive substances investigated in this study. The desorption time and the concentration of the desorptive substances were also studied. The concentrations of the metal ions in the supernatant were detected using ICP-MS.

4. Conclusions

HNAP/QA, which is an abbreviation of (3-(((4-((5-(((S)-hydroxyhydrophosphoryl)oxy))-2-nitrobenzylidene) amino) phenyl) imino) methyl)-4-nitrophenyl hydrogen (R)-phosphonate), is a new synthetic adsorbent material. HNAP/QA has a high selectivity for W(VI) ions from its aqueous solution. Several characterization techniques, such as FTIR, GC-MS, TGA, and NMR analyses, were used to ensure its successful preparation. The optimal factors influencing the retention of W(VI) ions on the new adsorbent were thoroughly examined and found to be a pH of 4.5 and 30 min of sorption time at room temperature. Kinetics and thermodynamics were also investigated; the adsorption reaction follows a Langmuir model. The sorption process of the W(VI) ions is spontaneous due to the negative value of $\Delta G°$ for all temperature values, while the positive value of $\Delta H°$ indicates that the adsorption on HNAP/QA is an endothermic mechanism. In addition, the positive value of $\Delta S°$ proposes an increase in the randomness of the adsorption activities in the investigated system. Finally, HNAP/QA was successfully used for the adsorption of W(VI) ions from wolframite ore, and pure W(VI) ions were separated via precipitation.

Supplementary Materials: The following supporting information can be downloaded at: https://www.mdpi.com/article/10.3390/ijms24087423/s1. References [54,55] are cited in the supplementary material.

Author Contributions: Conceptualization, A.A.G.; Methodology, R.E.E.; Software, M.A.G.; Validation, M.Y.H.; Formal analysis, A.K.S.; Investigation, A.K.S.; Resources, B.M.A.; Data curation, R.E.S.; Writing–original draft, M.A.G.; Writing–review & editing, A.K.S.; Visualization, R.E.S.; Supervision, M.A.G.; Project administration, B.M.A.; Funding acquisition, M.S.A. All authors have read and agreed to the published version of the manuscript.

Funding: The authors extend their appreciation to the Deanship of Scientific Research at King Khalid University (KKU) for funding this research through the Research Group Program Under the Grant Number: (R.G.P.2/451/44).

Institutional Review Board Statement: Not applicable.

Informed Consent Statement: Not applicable.

Data Availability Statement: Not applicable.

Acknowledgments: The authors extend their appreciation to the Deanship of Scientific Research at King Khalid University (KKU) for funding this research.

Conflicts of Interest: The authors declare no conflict of interest.

References

1. Liu, H.; Liu, H.; Nie, C.; Zhang, J.; Steenari, B.; Ekberg, C. Comprehensive treatments of tungsten slags in China: A critical review. *J. Environ. Manag.* **2020**, *270*, 110927. [CrossRef] [PubMed]
2. Shen, L.; Li, X.; Lindberg, D.; Taskinen, P. Tungsten extractive metallurgy: A review of processes and their challenges for sustainability. *Miner. Eng.* **2019**, *142*, 105934. [CrossRef]
3. Datta, S.; Vero, S.E.; Hettiarachchi, G.M.; Johannesson, K. Tungsten contamination of soils and sediments: Current state of science. *Curr. Pollut. Rep.* **2017**, *3*, 55–64. [CrossRef]
4. Koutsospyros, A.; Strigul, N.; Braida, W.; Christodoulatos, C. Tungsten: Environmental Pollution and Health Effects. In *Encyclopedia of Environmental Health*; Elsevier: Amsterdam, The Netherlands, 2011; pp. 418–426. [CrossRef]
5. Ünal, Ü.; Somer, G. A new and very simple procedure for the differential pulse polarographic determination of ultra trace quantities of tungsten using catalytic hydrogen wave and application to tobacco sample. *J. Electroanal. Chem.* **2012**, *687*, 64–70. [CrossRef]
6. Seiler, R.L.; Stollenwerk, K.G.; Garbarino, J.R. Factors controlling tungsten concentrations in ground water Carson Desert, Nevada. *Appl. Geochem.* **2005**, *20*, 423–441. [CrossRef]
7. Bednar, A.; Mirecki, J.; Inouye, L.; Winfield, L.; Larson, S.; Ringelberg, D. The determination of tungsten, molybdenum, and phosphorus oxyanions by high performance liquid chromatography inductively coupled plasma mass spectrometery. *Talanta* **2007**, *72*, 1828–1832. [CrossRef]
8. Li, S.; Deng, N.; Zheng, F.; Huang, Y. Spectrophotometric determination of tungsten (VI) enriched by nanometer-size titanium dioxide in water and sediment. *Talanta* **2003**, *60*, 1097–1104. [CrossRef]
9. Bazel, Y.; Lešková, M.; Rečlo, M.; Šandrejová, J.; Simon, A.; Fizer, M.; Sidey, V. Structural and spectrophotometric characterization of 2-[4-(dimethylamino) styryl]-1-ethylquinolinium iodide as a reagent for sequential injection determination of tungsten. *Spectrochim. Acta Part A Mol. Biomol. Spectrosc.* **2018**, *196*, 398–405. [CrossRef]
10. Chen, W.; Yang, W.Z.; Liu, X.Q.; Chen, X.M. Structural, dielectric and magnetic properties of Ba3SrLn2Fe2Nb8O30(Ln = La, Nd, Sm) filled tungsten bronze ceramics. *J. Alloys Compd.* **2016**, *675*, 311–316. [CrossRef]
11. Chong, X.; Jiang, Y.; Zhou, R.; Feng, J. Stability, chemical bonding behavior, elastic properties and lattice thermal conductivity of molybdenum and tungsten borides under hydrostatic pressure. *Ceram. Int.* **2016**, *42*, 2117–2132. [CrossRef]
12. Ravi, U.K.; Kumar, J.; Kumar, V.; Sankaranarayana, M.; Nageswara Rao, G.V.S.; Nandy, T.K. Effect of cyclic heat treatment and swaging on mechanical properties of the tungsten heavy alloys. *Mater. Sci. Eng. A* **2016**, *656*, 256–265. [CrossRef]
13. Stanciu, V.I.; Vitry, V.; Delaunois, F. Tungsten carbide powder obtained by direct carburization of tungsten trioxide using mechanical alloying method. *J. Alloys Compd.* **2016**, *659*, 302–308. [CrossRef]
14. Medici, S.; Costa, M.; Brocato, J.; Peana, M.; Oksuz, B.A.; Cartulari, L.; Vaughan, J.; Laulicht, F.; Wu, F.; Kluz, T.; et al. Tungsten-induced carcinogenesis in human bronchial epithelial cells. *Toxicol. Appl. Pharmacol.* **2015**, *288*, 33–39. [CrossRef]
15. Chen, Y.; Huo, G.; Guo, X.; Wang, Q. Sustainable extraction of tungsten from the acid digestion product of tungsten concentrate by leaching-solvent extraction together with raffinate recycling. *J. Clean. Prod.* **2022**, *375*, 133924. [CrossRef]
16. Huo, G.; Peng, C.; Song, Q.; Lu, X. Tungsten removal from molybdate solutions using ion exchange. *Hydrometallurgy* **2014**, *147–148*, 217–222. [CrossRef]
17. Fu, F.; Wang, Q. Removal of heavy metal ions from wastewaters: A review. *J. Environ. Manag.* **2011**, *92*, 407–418. [CrossRef]
18. Kim, J.W.; Lee, W.G.; Hwang, I.S.; Lee, J.Y.; Han, C. Recovery of tungsten from spent selective catalytic reduction catalysts by pressure leaching. *J. Ind. Eng. Chem.* **2015**, *28*, 73–77. [CrossRef]
19. Qin, W.; Xi, X.; Zhang, L.; Nie, Z.; Liu, C.; Li, R. The effect of MOx (M = Zr, Ce, La, Y) additives on the electrochemical preparation of tungsten in eutectic Na_2WO_4–WO_3 melt. *J. Electroanal. Chem.* **2023**, *935*, 117343. [CrossRef]

20. Han, Z.; Golev, A.; Edraki, M. A Review of Tungsten Resources and Potential Extraction from Mine Waste. *Minerals* **2021**, *11*, 701. [CrossRef]
21. Wu, X.; Zhang, G.; Zeng, L.; Zhou, Q.; Li, Z.; Zhang, D.; Cao, Z.; Guan, W.; Li, Q.; Xiao, L. Study on removal of molybdenum from ammonium tungstate solutions using solvent extraction with quaternary ammonium salt extractant. *Hydrometallurgy* **2019**, *186*, 218–225. [CrossRef]
22. Zhao, Z.; Cao, C.; Chen, X. Separation of macro amounts of tungsten and molybdenum by precipitation with ferrous salt. *Trans. Nonferrous Met. Soc.* **2011**, *21*, 2758–2763. [CrossRef]
23. Cao, F.; Wang, W.; Wei, D.; Liu, W. Separation of tungsten and molybdenum with solvent extraction using functionalized ionic liquid tricaprylmethylammonium bis(2,4,4-trimethylpentyl) phosphinate. *Int. J. Miner. Metall. Mater.* **2021**, *28*, 1769–1776. [CrossRef]
24. Xu, C.; Zhan, W.; Tang, X.; Mo, F.; Fu, L.; Lin, B. Self-healing chitosan/vanillin hydrogels based on Schiff-base bond/hydrogen bond hybrid linkages. *Polym. Test.* **2018**, *66*, 155–163. [CrossRef]
25. Pandya, H.J.; Ganatra, K.J. Synthesis, characterization and biological evaluation of bis-bidentate Schiff base metal complexes. *Inorg. Chem. Indian J.* **2008**, *3*, 182–187.
26. Li, L.; Sakr, A.K.; Schlöder, T.; Klein, S.; Beckers, H.; Kitsaras, M.-P.; Snelling, H.V.; Young, N.A.; Andrae, D.; Riedel, S. Searching for Monomeric Nickel Tetrafluoride: Unravelling Infrared Matrix Isolation Spectra of Higher Nickel Fluorides. *Angew. Chem. Int. Ed.* **2021**, *60*, 6391–6394. [CrossRef] [PubMed]
27. Cai, Y.; Wu, C.; Liu, Z.; Zhang, L.; Chen, L.; Wang, J.; Wang, X.; Yang, S.; Wang, S. Fabrication of a phosphorylated graphene oxide–chitosan composite for highly effective and selective capture of U(VI). *Environ. Sci. Nano* **2017**, *4*, 1876–1886. [CrossRef]
28. Belfer, S.; Fainchtain, R.; Purinson, Y.; Kedem, O. Surface characterization by FTIR-ATR spectroscopy of polyethersulfone membranes-unmodified, modified and protein fouled. *J. Membr. Sci.* **2000**, *172*, 113–124. [CrossRef]
29. Lin-Vien, D.; Colthup, N.B.; Fateley, W.G.; Grasselli, J.G. CHAPTER 10—Compounds Containing –NH$_2$, –NHR, and –NR$_2$ Groups. In *The Handbook of Infrared and Raman Characteristic Frequencies of Organic Molecules*; Lin-Vien, D., Colthup, N.B., Fateley, W.G., Grasselli, J.G., Eds.; Academic Press: San Diego, CA, USA, 1991; pp. 155–178.
30. Hamza, M.F.; Salih, K.A.M.; Zhou, K.; Wei, Y.; Abu Khoziem, H.A.; Alotaibi, S.H.; Guibal, E. Effect of bi-functionalization of algal/polyethyleneimine composite beads on the enhancement of tungstate sorption: Application to metal recovery from ore leachate. *Sep. Purif. Technol.* **2022**, *290*, 120893. [CrossRef]
31. Sethuraman, P.R.; Leparulo, M.A.; Pope, M.T.; Zonnevijlle, F.; Brevard, C.; Lemerle, J. Heteropolypentatungstobisphosphonates—Cyclopentane-like pseudorotation of an oxometalate structure. *J. Am. Chem. Soc.* **1981**, *103*, 7665–7666. [CrossRef]
32. Bajaber, M.A.; Ragab, A.H.; Sakr, A.K.; Atia, B.A.; Fathy, W.M.; Gado, M.A. Application of a new derivatives of traizole Schiff base on chromium recovery from its wastewater. *Sep. Sci. Technol.* **2023**, *58*, 737–758. [CrossRef]
33. Lagergren, S. About the theory of so-called adsorption of soluble substances, Zur theorie der sogenannten adsorption geloster stoffe. *K. Sven. Vetensk. Handl.* **1898**, *24*, 1–39.
34. Sakr, A.K.; Abdel Aal, M.M.; Abd El-Rahem, K.A.; Allam, E.M.; Abdel Dayem, S.M.; Elshehy, E.A.; Hanfi, M.Y.; Alqahtani, M.S.; Cheira, M.F. Characteristic Aspects of Uranium (VI) Adsorption Utilizing Nano-Silica/Chitosan from Wastewater Solution. *Nanomaterials* **2022**, *12*, 3866. [CrossRef] [PubMed]
35. Negm, S.H.; Abd El-Hamid, A.A.M.; Gado, M.A.; El-Gendy, H.S. Selective uranium adsorption using modifed acrylamide resins. *J. Radioanal. Nucl. Chem.* **2018**, *319*, 327–337. [CrossRef]
36. Washahy, A.R.; Sakr, A.K.; Gouda, A.A.; Atia, B.M.; Somaily, H.H.; Hanfi, M.Y.; Sayyed, M.I.; El Sheikh, R.; El-Sheikh, E.M.; Radwan, H.A.; et al. Selective Recovery of Cadmium, Cobalt, and Nickel from Spent Ni-Cd Batteries using Adogen®464 and Mesoporous Silica Derivatives. *Int. J. Mol. Sci.* **2022**, *23*, 8677. [CrossRef]
37. Wang, J.; Guo, X. Adsorption kinetic models: Physical meanings, applications, and solving methods. *J. Hazard. Mater.* **2020**, *390*, 122156. [CrossRef]
38. Ibrahium, H.A.; Atia, B.M.; Awwad, N.S.; Nayl, A.A.; Radwan, H.A.; Gado, M.A. Efficient preparation of phosphazene chitosan derivatives and its applications for the adsorption of molybdenum from spent hydrodesulfurization catalyst. *J. Dispers. Sci. Technol.* **2022**, *43*, 1–16. [CrossRef]
39. Ruiping, L.; Chunye, L.; Xitao, L. Adsorption of tungstate on kaolinite: Adsorption models and kinetics. *RSC Adv.* **2006**, *6*, 19872–19877. [CrossRef]
40. Alharbi, A.; Gouda, A.A.; Atia, B.M.; Gado, M.A.; Alluhaybi, A.A.; Alkabli, J. The Role of Modified Chelating Graphene Oxide for Vanadium Separation from Its Bearing Samples. *Russ. J. Inorg. Chem.* **2022**, *67*, 560–575. [CrossRef]
41. Atia, B.M.; Sakr, A.K.; Gado, M.A.; El-Gendy, H.S.; Abdelazeem, N.M.; El-Sheikh, E.M.; Hanfi, M.Y.; Sayyed, M.I.; Al-Otaibi, J.S.; Cheira, M.F. Synthesis of a New Chelating Iminophosphorane Derivative (Phosphazene) for U (VI) Recovery. *Polymers* **2022**, *14*, 1687. [CrossRef]
42. Atia, B.M.; Khawassek, Y.M.; Hussein, G.M.; Gado, M.A.; El-Sheify, M.A.; Cheira, M.F. One-pot synthesis of pyridine dicarboxamide derivative and its application for uranium separation from acidic medium. *J. Environ. Chem. Eng.* **2021**, *9*, 105726. [CrossRef]
43. Atia, B.M.; Gado, M.A.; Cheira, M.F.; El-Gendy, H.S.; Yousef, M.A.; Hashem, M.D. Direct synthesis of a chelating carboxamide derivative and its application for thorium extraction from Abu Rusheid ore sample, South Eastern Desert, Egypt. *Int. J. Environ. Anal. Chem.* **2021**, *101*, 1–24. [CrossRef]

44. Gado, M.; Rashad, M.; Kassab, W.; Badran, M. Highly Developed Surface Area Thiosemicarbazide Biochar Derived from Aloe Vera for Efficient Adsorption of Uranium. *Radiochemistry* **2021**, *63*, 353–363. [CrossRef]
45. Akar, T.; Kaynak, Z.; Ulusoy, S.; Yuvaci, D.; Ozsari, G.; Akar, S.T. Enhanced Biosorption of Nickel (II) Ions by Silica-GelImmobilized Waste Biomass: Biosorption Characteristics in Batch and Dynamic Flow Mode. *J. Hazard Mater.* **2009**, *163*, 1134–1141. [CrossRef] [PubMed]
46. Chen, A.H.; Chen, S.M. Biosorption of Azo Dyes from Aqueous Solution by Glutaraldehyde-Crosslinked Chitosans. *J. Hazard Mater.* **2009**, *172*, 1111–1121. [CrossRef] [PubMed]
47. Dubinin, M.M. The Potential Theory of Adsorption of Gases and Vapors for Adsorbents with Energetically Nonuniform Surfaces. *Chem. Rev.* **1960**, *60*, 235–241. [CrossRef]
48. Mittal, A.; Kaur, D.; Mittal, J. Batch and Bulk Removal of a Triarylmethane Dye, Fast Green FCF, from Wastewater by Adsorption over Waste Materials. *J. Hazard Mater.* **2009**, *163*, 568–577. [CrossRef]
49. Chen, A.H.; Yang, C.Y.; Chen, C.Y.; Chen, C.Y.; Chen, C.W. The Chemically Crosslinked Metal-Complexed Chitosans for Comparative Adsorptions of Cu (II), Zn (II), Ni (II) and Pb (II) Ions in Aqueous Medium. *J. Hazard Mater.* **2009**, *163*, 1068–1075. [CrossRef]
50. Kundu, S.; Gupta, A.K. Arsenic Adsorption onto Iron OxideCoated Cement (IOCC): Regression Analysis of Equilibrium Data with Several Isotherm Models and Their Optimization. *Chem. Eng. J.* **2006**, *122*, 93–106. [CrossRef]
51. Hassanin, M.A.; Negm, S.H.; Youssef, M.A.; Sakr, A.K.; Mira, H.I.; Mohammaden, T.F.; Al-Otaibi, J.S.; Hanfi, M.Y.; Sayyed, M.I.; Cheira, M.F. Sustainable Remedy Waste to Generate SiO_2 Functionalized on Graphene Oxide for Removal of U(VI) Ions. *Sustainability* **2022**, *14*, 2699. [CrossRef]
52. Gado, M.; Atia, B. Adsorption of Thorium Ions Using Trioctylphosphine Oxide Impregnated Dowex 1 × 4 Powder. *Radiochemistry* **2019**, *61*, 168–176. [CrossRef]
53. Weshahy, A.R.; Gouda, A.A.; Atia, B.M.; Sakr, A.K.; Al-Otaibi, J.S.; Almuqrin, A.; Hanfi, M.Y.; Sayyed, M.I.; El Sheikh, R.; Radwan, H.A.; et al. Efficient Recovery of Rare Earth Elements and Zinc from Spent Ni–Metal Hydride Batteries: Statistical Studies. *Nanomaterials* **2022**, *12*, 2305. [CrossRef] [PubMed]
54. Borodina, E.; Karpov, S.I.; Selemenev, V.F.; Schwieger, W.; Maracke, S.; Fröba, M.; Rößner, F. Surface and texture properties of mesoporous silica materials modified by silicon-organic compounds containing quaternary amino groups for their application in base-catalyzed reactions. *Microporous Mesoporous Mater.* **2015**, *203*, 224–231. [CrossRef]
55. Tarasova, N.P.; Smetannikov, Y.V.; Zanin, A.A. Radiation-chemical transformation of elemental phosphorus in the presence of ionic liquids. *Dokl. Chem.* **2013**, *449*, 111–113. [CrossRef]

Disclaimer/Publisher's Note: The statements, opinions and data contained in all publications are solely those of the individual author(s) and contributor(s) and not of MDPI and/or the editor(s). MDPI and/or the editor(s) disclaim responsibility for any injury to people or property resulting from any ideas, methods, instructions or products referred to in the content.

Article

Facile Synthesis of Sulfur-Containing Functionalized Disiloxanes with Nonconventional Fluorescence by Thiol–Epoxy Click Reaction

Jing Tang [1,2], Shengyu Feng [2] and Dengxu Wang [1,2,*]

[1] Institute of Novel Semiconductors, State Key Laboratory of Crystal Materials, Shandong University, Jinan 250100, China
[2] National Engineering Research Center for Colloidal Materials & Key Laboratory of Special Functional Aggregated Materials, Ministry of Education, Shandong Key Laboratory of Advanced Organosilicon Materials and Technologies, School of Chemistry and Chemical Engineering, Shandong University, Jinan 250100, China
* Correspondence: dxwang@sdu.edu.cn

Abstract: Herein, a series of novel sulfur-containing functionalized disiloxanes based on a low-cost and commercially available material, i.e., 1,3-bis(3-glycidoxypropyl)-1,1,3,3-tetramethyldisiloxane, and various thiol compounds were prepared by thiol–epoxy click reaction. It was found that both lithium hydroxide (LiOH) and tetrabutylammonium fluoride (TBAF) have high catalytic activity after optimizing the reaction condition, and the reaction can be carried out with high yields, excellent regioselectivity, mild reaction condition, and good tolerance of functional groups. These compounds exhibit excellent nonconventional fluorescence due to the formation of coordination bonds between Si atoms and heteroatoms (e.g., S or N) and can emit blue fluorescence upon ultraviolet (UV) irradiation. These results demonstrate that the thiol–epoxy click reaction could promisingly act as an efficient organosilicon synthetic methodology to construct various organosilicon materials with novel structures and functionality, and thus their application scope will be significantly expanded.

Keywords: thiol–epoxy reaction; functionalized disiloxanes; organosilicon synthetic methodology; nonconventional fluorescence

1. Introduction

In recent years, intense efforts have been devoted to developing new organosilicon materials because of their unique properties and indispensable roles in emerging areas, such as aerospace, new energy, electronics, and biomedicine [1–5]. However, the development lags far behind the increasing requirements. One crucial reason is that organosilicon synthetic methodologies are relatively few compared to numerous synthetic strategies to build carbon-based materials. Up to now, three traditional methodologies, including direct synthesis, hydrolysis and polycondensation, and hydrosilylation reactions, are still the most widely used techniques in the laboratory and in industry [6–9], although these traditional methodologies were found more than 80 years ago. Thus, developing new efficient synthetic methodologies to design and construct organosilicon materials is still highly desirable.

Thanks to the well-developed and emerging organic synthetic methodologies, diversified organosilicon materials with novel structures and functionality have been designed, prepared, and new applications found, and thus organosilicon synthetic methodologies are increasingly growing [10–16]. Click reactions have attracted specific attention because of their intriguing features, such as mild reaction conditions, fast speed, high yield, and brilliant tolerance of functional groups [17–19]. The azide–alkyne reaction [20], thiol–ene reaction [21,22], and aza-Michael addition reaction [23,24] with the "click" feature have all been widely used to prepare organosilicon compounds [21], functionalized polysiloxanes [20,25–27], and silicone elastomers [28] by us and other groups, and thus have become mature or semi-mature organosilicon synthetic methodologies. From the standpoint of existing raw materials, these strategies require azide- (or alkyl), thiol-, vinyl-, and amine-containing silanes or (poly)siloxanes, most of which have relatively high costs except vinyl-containing materials. However, epoxy-based materials (e.g., epoxy-containing disiloxane and polysiloxanes), which have a much lower cost than thiol- or amine-based ones, have not been fully exploited. It is of great interest to develop an organosilicon synthetic methodology using epoxy groups.

Herein, we introduce the thiol–epoxy reaction as the new organosilicon synthetic methodology and choose the preparation of functionalized disiloxanes as the representative examples. The selection of the thiol–epoxy reaction is motivated by its click feature and wide usage in the construction of functional molecules and polymers [29,30]. Although this reaction has been used to prepare some organosilicon materials [31], such as silsesquioxane-based materials [32–36], silicone/epoxy composites [37,38], and epoxy/silanes hybrid coating [39], exploration in this area is still limited and systematical investigation is rare. Moreover, the nonconventional properties of organosilicon materials constructed by this reaction have still not been studied despite the presence of unique S→Si coordination bonds within the structures. Thus, a systematic investigation on the synthesis of new sulfur-containing functionalized disiloxanes by the thiol–epoxy reaction is reported in the present study. The catalytic system and reaction condition for the synthesis of organosilicon compounds were optimized by studying the effects of catalysts, substituents, reaction temperature, and reaction times on the reaction efficiency. Using the optimal reaction, eleven new functionalized disiloxanes were prepared and their nonconventional fluorescent properties were explored.

2. Results and Discussion

2.1. Optimization of Reaction Condition

The thiol–epoxy reaction is considered a click reaction [40] when using the base as the catalyst, and previous reports proved that tetrabutylammonium fluoride (TBAF) and lithium hydroxide (LiOH) were the most efficient catalysts [41]. When this reaction was used to prepare organosilicon materials, the present sulfur-containing functionalized disiloxanes, the base catalyst is required to evaluate the suitability because of the sensitivity of Si–O or Si–C bonds under basic conditions. Thus a model reaction between 1,3-bis(3-glycidoxypropyl)tetramethyldisiloxane (E_1) and 1-dodecanethiol (T_1) was conducted to find out the optimal condition using LiOH or TBAF as catalysts under solvent-free condition or in tetrahydrofuran (THF) (Scheme 1) [41]. Table 1 demonstrates the thiol–epoxy reaction data of E_1 and T_1.

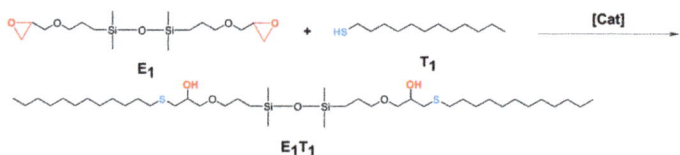

Scheme 1. The model reaction between 1,3-bis(3-glycidoxypropyl)tetramethyldisiloxane (E_1) and 1-dodecanethiol (T_1).

LiOH was initially used as the catalyst (5 mol%) without solvent and with a reaction time of 6 h at room temperature. The conversion rate of the epoxy group is 63.1% (Table 1, entry 1). It is a good sign that prolonging the reaction time to 12 h led to the quantitative conversion of the epoxy group (Table 1, entry 2). However, when the amount of catalyst was reduced to 3 mol% and 1 mol%, the conversion rates were significantly decreased to 80.3% and 39.8%, respectively (Table 1, entries 3–4). Then TBAF was used as a catalyst in THF. It was found that the conversion rate of the epoxy group was 81.3% with 8 mol% of TBAF and a reaction time of 12 h at room temperature (Table 1, entry 5). When the temperature was increased to 50 °C, the epoxy groups could be quantitatively consumed (Table 1, entry 6). Then, the amount of catalyst was reduced to 5 mol%, 3 mol%, 1 mol%, and 0.8 mol% under the same condition (Table 1, entries 7–10) and it was found that 3 mol% of catalyst could still provide high catalytic activity with a high conversion rate of the epoxy group (98.6%) (Table 1, entry 8). Based on this finding, the reaction time was further optimized with 3 mol% of catalyst loading (Table 1, entries 11–12). The results demonstrated that 6 h was sufficient to achieve almost quantitative conversion (Table 1, entry 11).

Table 1. Optimization of the reaction condition of the thiol–epoxy reaction based on E_1 and T_1.

Entry [a]	Catalyst	Cat. Loading (mol%)	Time (h)	Temp (°C)	Conversion (%) [c]
1	LiOH	5	6	25	63.1
2	LiOH	5	12	25	99.5
3	LiOH	3	12	25	80.3
4	LiOH	1	12	25	39.8
5	TBAF [b]	8	12	25	81.3
6	TBAF [b]	8	12	50	99.5
7	TBAF [b]	5	12	50	99.4
8	TBAF [b]	3	12	50	98.6
9	TBAF [b]	1	12	50	92.3
10	TBAF [b]	0.8	12	50	80.0
11	TBAF [b]	3	6	50	96.7
12	TBAF [b]	3	3	50	91.8

[a] Reaction conditions: E_1 (1 mmol), T_1 (1.25 mmol), and catalyst. The amount of catalyst was calculated based on epoxy group. [b] The reactions were carried out in THF. [c] Determined by ^1H NMR.

In addition, no obvious by-products, which may be generated by the cleavage of Si–O or Si–C bonds, were found in the final products using TBAF and LiOH as the catalysts, as evidenced by the high yields (98% and 96% for entries 2 and 11) after purification and the ^{29}Si NMR spectra of the reaction mixture, in which no peaks in the ranges of −30 ppm to −20 ppm and 20 ppm to 40 ppm, which are assigned to the silicon atoms from Si–OH or Si-F bonds, respectively, were observed during the reaction (Figure S1). These results indicate that both TBAF and LiOH can act as efficient catalysts for the synthesis of new functionalized disiloxanes. It is known that LiOH is soluble in water but insoluble in most of organic solvents; thus, it is considered a heterogeneous catalyst [30]. Although it can efficiently catalyze this reaction of E_1T_1 without the solvent, its catalytic activity is relatively low when thiol compounds (e.g., methimazole) appear as the solid, and thus the condition (e.g., additional solvents) needs to be further optimized [30]. In contrast, this possibility will not occur when selecting TBAF in THF because THF can dissolve most thiol compounds. In addition, the alkalinity of TBAF is weaker than LiOH, and the residual LiOH in the system may cause the instability of the products (e.g., crosslinking networks) [42]. Thus, considering the universality of TBAF in THF, herein we choose it for further synthesis of various sulfur-containing functionalized disiloxanes.

2.2. Synthesis and Characterization of Sulfur-Containing Functionalized Disiloxanes

The capability of the optimized reaction condition, that is TBAF (3 mol%) in THF at 50 °C for 6 h, was tested by preparing sulfur-containing functionalized disiloxanes from E_1 and various thiol compounds (Scheme 2). The results revealed that eleven new disiloxanes were successfully achieved with excellent conversion rates of epoxy groups (≥99%) and isolated yields (>89%) and high regioselectivity (Table 2, entries 1–11) regardless of aliphatic thiols (e.g., 1-butanethiol, a yield of 95.5%, Table 2, entry 2) or aromatic thiols (e.g., 3-methylbenzenethiol, a yield of 96.5%, Table 2, entry 9). As expected, in using aromatic thiols, the products can be generated with high yields, even shortening the reaction times due to their stronger nucleophile ability of the formed thiolate anion than aliphatic thiols [30,43]. For example, when 3-methylbenzenethiol (T_9) reacted with E_1, 3 h was sufficient to afford a nearly quantitative conversion of the epoxy group (Table 2, entry 16). In addition, the yield of E_1T_5 (Table 2, entry 5) was slightly lower than that of E_1T_2 (Table 2, entry 2), and this finding could be explained by higher steric hindrance of 3-methyl-2-butanethiol (T_5) than 1-butanethiol (T_2). This speculation can also explain the fact that E_1T_2 and E_1T_4 can also be afforded with high yields, even shortening the reaction time to 3 h or 5 h (Table 2, entries 12 and 14), in comparison to the requirement of 6 h for E_1T_1 (Table 2, entry 1), concerning the higher steric hindrance of T_1 than T_2 and T_4. These results demonstrated that the electronic effect and steric hindrance of substituent groups influenced the activity of thiol groups, resulting in the different reactivity of thiol compounds and further different yields.

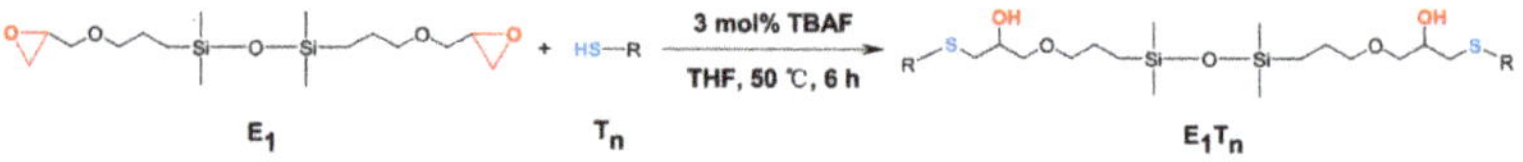

Scheme 2. Synthetic routes of sulfur-containing functionalized disiloxanes from 1,3-bis(3-glycidoxypropyl)tetramethyldisiloxane (E_1) and various thiol compounds (T_n).

Table 2. The performance of optimized conditions in the thiol–epoxy reaction of E_1 and various thiol compounds (T_n).

Entry [a]	Substrates	Products	Time (h)	Conversion (%) [b]	Yields (%) [c]
1	T_1	 E_1T_1	 6	96.0	96.0
2	T_2	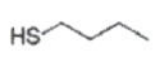 E_1T_2	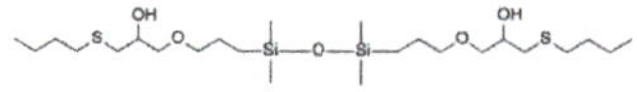 6	99.9	95.5
3	T_3	 E_1T_3	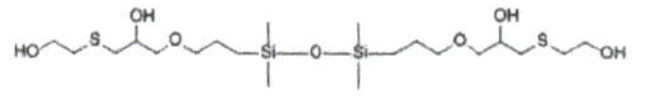 6	99.9	95.4
4	T_4	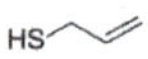 E_1T_4	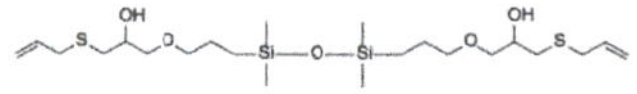 6	99.9	94.3
5	T_5	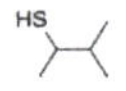 E_1T_5	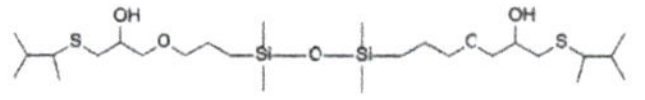 6	99.0	90.3
6	T_6	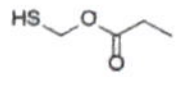 E_1T_6	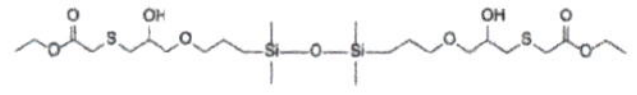 6	98.0	95.0
7	T_7	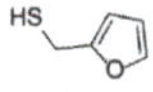 E_1T_7	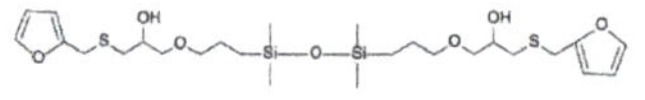 6	99.9	94.7
8	T_8	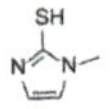 E_1T_8	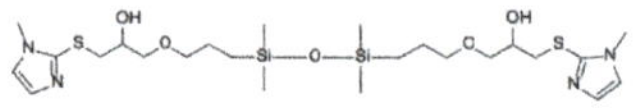 6	99.9	92.3
9	T_9	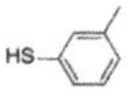 E_1T_9	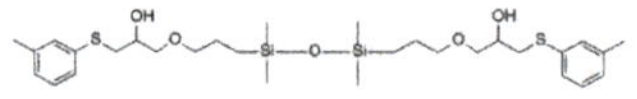 6	99.9	96.5
10	T_{10}	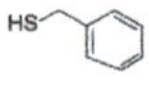 E_1T_{10}	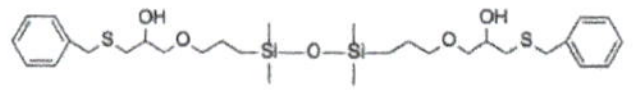 6	99.9	93.8
11	T_{11}	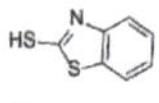 E_1T_{11}	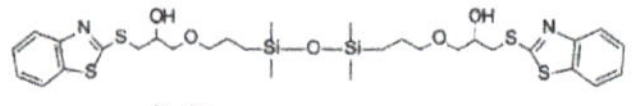 6	98.0	89.3
12	T_2	E_1T_2	3	99.9	95.5
13	T_3	E_1T_3	5	99.9	95.3
14	T_4	E_1T_4	5	99.9	93.8
15	T_8	E_1T_8	5	99.9	92.1
16	T_9	E_1T_9	3	99.9	96.4

[a] Reaction conditions: E_1 (1 mmol), substrates (1.25 mmol), TBAF (3 mol%), THF (0.6 mL), 50 °C, 6 h. [b] The conversion rate of epoxy groups determined by ^1H NMR. [c] Isolated yields and confirmed by FT-IR, NMR, HR-MS, and elemental analysis.

The successful achievement of these sulfur-containing functionalized disiloxanes can be determined by FT-IR, $^1H/^{13}C/^{29}Si$ NMR, HR-MS, and elemental analysis. Considering their similar structures, E_1T_2 and E_1T_9 were selected as the representative examples (the original spectra of all the products can be found in the Figures S2–S19 of the Supporting Information). Figure 1a illustrates the IR spectrum of E_1, E_1T_2, and E_1T_9. After the reaction, the peaks assigned to the epoxy group at 910 cm^{-1} disappeared in the products E_1T_2 and E_1T_9, while a new strong and broad peak attributed to the generated β-hydroxyl group at 3464 cm^{-1} was observed. Figure 1b illustrates the ^{1}H NMR of E_1, E_1T_2, and E_1T_9, and the assignments of the characteristic protons. The peaks assigned to epoxy protons at 2.55 ppm ($H^{c'}$), 2.74 ppm (H^c), and 3.09 ppm (H^b) in E_1 completely disappeared in E_1T_2 and E_1T_9, while two new signals at 2.76 ppm and 3.82 ppm, which are attributed to the protons from β-hydroxyl groups (H^4 in E_1T_2 and H^D in E_1T_9) and −CH groups adjacent to it (H^2 in E_1T_2 and H^B in E_1T_9), were observed (Figure 1b). In the ^{29}Si NMR, a single peak was found at 7.73 ppm and 7.75 ppm for E_1T_2 and E_1T_9, in comparison to a single peak of 7.46 ppm for E_1, indicating that the chemical environment of the Si atom has been changed (Figure 1c). It was proven that the Si atom could interact with heteroatoms (e.g., N or S) by accepting a lone pair of p electrons, resulting in S→Si or N→Si coordination bonds [44]. Thus, the present variation is apparently due to the formation of S→Si coordination.

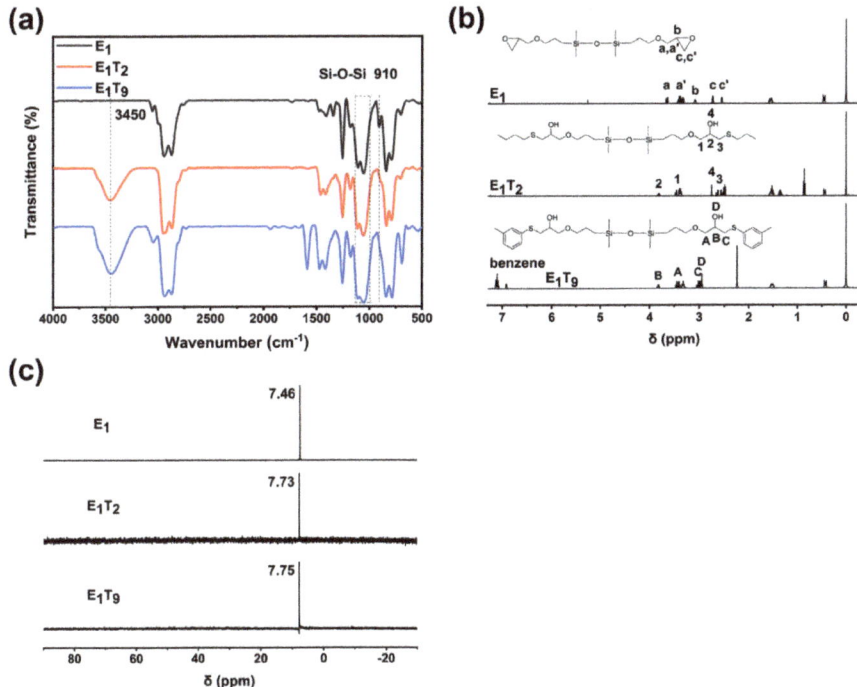

Figure 1. (a) FT-IR spectra of epoxy-containing groups disiloxane E_1 and functionalized disiloxanes E_1T_2, E_1T_9. (b) ^{1}H NMR spectrum of E_1, E_1T_2, and E_1T_9 in CDCl$_3$. (c) ^{29}Si NMR spectrum of E_1, E_1T_2, and E_1T_9 in CDCl$_3$.

2.3. Fluorescent Properties

Previous reports have demonstrated that the existence of heteroatom in organosilicon molecules or polymers could exhibit interesting nonconventional fluorescence [44]. The present sulfur-containing functionalized disiloxanes expectedly show nonconventional fluorescence. All the samples in ethanol (EtOH) solutions emit similar and strong blue fluorescence with the maximum emission wavelength (λ_{max}) at ca. 410 nm when excited by 365 nm (Figure 2). Moreover, this phenomenon can be visualized under 365 nm ultraviolet (UV) light (Figure 3). Similar to previous reports [44], this nonconventional fluorescence was generated by through-space charge transfer from lone pairs of electrons in S atoms to the 3d empty orbital of Si atoms and aggregation of chromophores, including ester groups in E_1T_6 and aromatic units in E_1T_7 to E_1T_{11}. It is worth noting that E_1T_{11} has the highest fluorescent intensities. Previous reports demonstrated that benzothiazole and its derivatives have excellent fluorescence due to the large delocalized π bond and rigid planar construction. Hence, the aromatic π–π conjugation contributes to the blue emission region with the λ_{max} at ca. 400 nm, in addition to the contribution from S→Si and N→Si coordination [45]. It is worth noting that the coordination is not only contributed by the thioether group but also by the S and N from thiazole units. These combined contributions lead to the highest fluorescence intensity. In addition, a weak emission peak at 497 nm was found. This finding is probably due to higher electronic conjugation induced by the π–π interaction between aromatic thiazole groups [46], while S→Si and N→Si coordination facilitates this interaction. The combined contribution can also explain higher fluorescence intensities found in E_1T_6 and E_1T_9 than in other samples, because additional aggregation of ester groups and N→Si coordination contributed by imidazole groups [24,47,48], exist within E_1T_6 and E_1T_9, respectively.

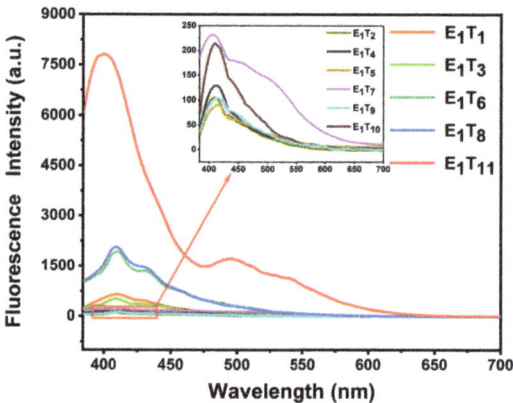

Figure 2. Fluorescence emission spectra of functionalized disiloxanes E_1T_1 to E_1T_{11} in ethanol solutions excited at 365 nm (0.01 mol L^{-1}).

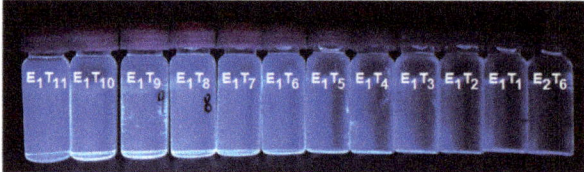

Figure 3. The photographs of E_1T_1 to E_1T_{11} and E_2T_6 in ethanol solutions under 365 nm UV light.

The nonconventional fluorescent properties of these polysiloxanes can also be determined by the fluorescence variation in different concentrations in solutions. Figure 4 shows the fluorescence spectra of E_1T_6 and E_1T_{11} in ethanol at different concentrations as the examples (see Figures S20–S22 for other samples). Their fluorescence intensities were progressively enhanced with increasing concentrations. This phenomenon is a typical example of aggregation-induced emission enhancement (AIEE), which is apparently induced by the aggregation of nonconventional chromophores, including X→Si (X = S or N) coordination bonds in these compounds and ester groups in E_1T_6 [47,49]. In addition, a new peak at ca. 500 nm could be observed at a high concentration of E_1T_{11}, which could be assigned to the extended conjugation formed by the aggregation of the thiazole groups.

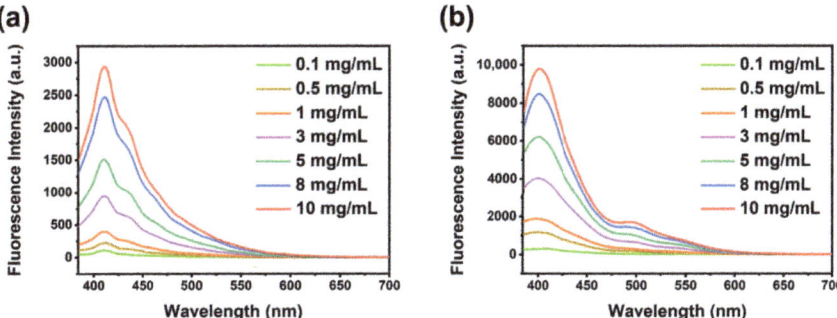

Figure 4. (a) Fluorescence emission spectra of E_1T_6 with different concentrations excited at 365 nm. (b) Fluorescence emission spectra of E_1T_{11} with different concentrations excited at 365 nm.

Generally, the fluorescence property depends on the environment. Therefore, the photophysical characteristics of these sulfur-containing functionalized disiloxanes were investigated in various organic solvents. Figure 5 shows the fluorescence spectra of E_1T_6 and E_1T_{11} in different solvents excited at 365 nm (Figures S23–S25 for other samples). For E_1T_6, the emission regions were similar in different solvents, but their fluorescence intensity varied. The fluorescence emission intensity became weaker in the order of dimethylformamide (DMF), THF, dichloromethane (DCM), ethanol, and acetone. The same variation trend of emission intensity was also found in E_1T_{11} (Figure 5b). This finding may be due to the participation of the electron-rich solvents (lone pairs) in the aggregation of nonconventional chromophores and consequently physically altered the extended electronic conjugation. In addition, the polarity of the solvents can also influence the fluorescence. As a result, these factors affect the emission behavior [50].

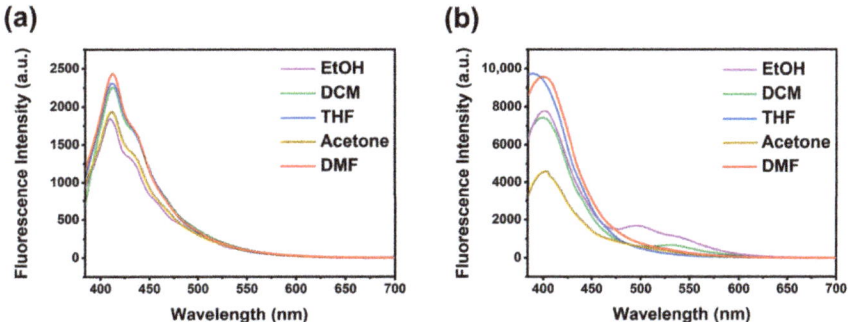

Figure 5. (a) Fluorescence emission spectra of E_1T_6 in different solvents excited at 365 nm (0.01 mol/L). (b) Fluorescence emission spectra of E_1T_{11} in different solvents excited at 365 nm (0.01 mol/L).

To further prove the contribution of S→Si or Si–O to the nonconventional fluorescence, a model compound E_2T_6, which has similar functional groups with E_1T_6 but without Si–O–Si linker, was synthesized by the thiol–epoxy reaction of diethylene glycol diglycidyl ether (E_2) and ethyl thioglycolate (T_6) (Scheme 3).

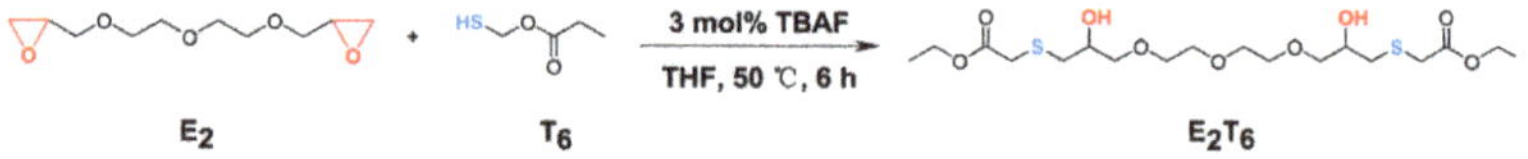

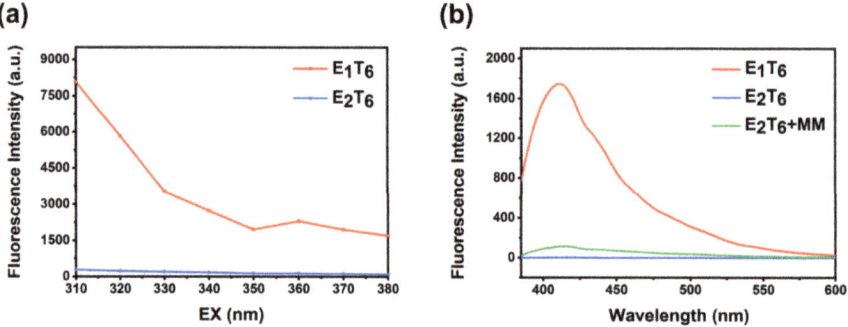

Scheme 3. Synthesis route of the model compound E_2T_6.

As illustrated in Table 3, although E_2T_6 can also emit blue fluorescence and have similar emission wavelengths to E_1T_6, its fluorescence intensity is much lower than that of E_1T_6 under the same excitation wavelength (Figure 6a), indicating the important role of Si–O–Si in the enhancement of fluorescence intensity. This Si-O-Si induced enhancement can be found even without direct introduction in the structures. As shown in Figure 6b, the fluorescence intensity of E_2T_6 was obviously enhanced after adding equimolar hexamethyl-disiloxane (MM), although it was still lower than that of E_1T_6. These results indicate that the interaction of S→Si coordination bonds can increase the aggregation of ester groups and thus enhance the fluorescence intensity.

Table 3. The excitation and emission of E_2T_6 and E_1T_6.

Entry	E_x/nm	E_2T_6		E_1T_6	
		E_m/nm	Intensity	E_m/nm	Intensity
1	310	345	299.1	349	8069
2	320	357	245	363	5824
3	330	370	212.3	366	3534
4	340	381	184.8	386	2730
5	350	392	147.9	410	1959
6	360	404	139.7	409	2310
7	370	417	131.8	409	1951
8	380	430	102	412	1700

Figure 6. (a) The relative fluorescence intensity of E_2T_6 and E_1T_6 in THF solution (0.01 mol L^{-1}). (b) Fluorescence emission spectra of E_1T_6, E_2T_6, and E_2T_6/MM mixture in THF solution excited at 365 nm (MM was insoluble in ethanol at room temperature, 0.01 mol L^{-1}).

3. Materials and Methods

3.1. Materials

All reagents and solvents were obtained from commercial suppliers and used without further purification unless otherwise noted. 1,3-bis(3-glycidoxypropyl)tetramethyldisiloxane (≥98%) was supplied by Hangzhou Sloan Materials Technology Co., Ltd. (Hangzhou, China). Diethylene glycol diglycidyl ether (68%) was purchased from Jinan Jing Shui Biotechnology Co., Ltd. (Jinan, China). Tetrabutylammonium fluoride

(TBAF), lithium hydroxide (LiOH), 1-dodecanethiol (≥98%), 2-mercaptoethanol (≥99%), 1-butanethiol (≥97%), 2-propenthiol (50%), 3-methyl-2-butanethiol (≥98%), ethyl thioglycolate (≥98%), furfuryl mercaptan (≥98%), methimazole (≥98%), 3-methylbenzenethiol (≥97%), benzyl mercaptan (≥98%), and 2-mercaptobenzothiazole (≥98%) were purchased from MACKLIN Chemical Reagent Co., Ltd. (Shanghai, China). Tetrahydrofuran (THF), dichloromethane (DCM), ethyl acetate (EA), petroleum ether (PE), acetone, ethanol, and dimethylformamide (DMF) all with Grade AR were purchased from Tianjin Fuyu Chemical Reagent Co., Ltd. (Tianjin, China).

3.2. Methods

Fourier transform infrared (FT-IR) spectra were conducted on a Bruker TENSOR-27 FT-IR spectrophotometer (Germany) using the KBr pellet technique and recorded in the range of 400 cm^{-1} to 4000 cm^{-1}. ^{1}H NMR, ^{13}C NMR, and ^{29}Si NMR were recorded on a Bruker AVANCE 400 NMR spectrometer (Rheinstetten, Germany) operating at 25 °C using CDCl$_3$ as solvent and without tetramethylsilane as an interior label, and using Software MestReNova-14.0.0-23239 to process the data. High-resolution mass spectra (HR-MS) were recorded on UPLC–QTOF–MS (Bruker, Bremen, Germany) with ESI$^+$, source spray voltage 2.6 kV and capillary temperature 180.00 °C, and data processing was performed using Bruker Compass Data Analysis 4.4.2. The fluorescent emission spectra of the samples were determined at room temperature with a F-7000 fluorescence spectrophotometer (Hitachi, Tokyo, Japan) using a monochromated Xe lamp as an excitation source. The contents of C, H, N, and S were characterized by the Elementar Vario EL III elemental analyzer (Munich, Germany).

3.3. General Procedure for the Synthesis of Sulfur-Containing Functionalized Disiloxanes E_1T_1 to E_1T_{11}

1,3-bis(3-glycidoxypropyl)-1,1,3,3-tetramethyldisiloxane (E_1, 362.6 mg, 1 mmol), thiol compounds (T_n, 2.5 mmol), and tetrahydrofuran (THF, 0.6 mL) were placed into a 25 mL round-bottom flask with a magnetic stirrer. Then, tetrabutylammonium fluoride (TBAF, 7.9 mg, 0.03 mmol) was added to the mixture at 0 °C. After removing the cooling, the mixture was heated at 50 °C for 6 h under stirring. Then, the mixture was cooled to room temperature and dichloromethane DCM (100 mL) was added. The mixture was washed with water and the organic layer was separated and dried over anhydrous magnesium sulfate. After the filtration, the solvent was removed under reduced pressure to yield the crude product. The pure products E_1T_1 to E_1T_{11} were then purified by column chromatography on silica gel using PE/EA as an eluent.

1,3-bis(3-(dodecylthiopropoxy)-2-ol-propyl)-1,1,3,3-tetramethyldisiloxane (E_1T_1): The synthesis was performed according to general procedure using 1-dodecanethiol (T_1, 506.0 mg) as a thiol compound. E_1T_1 was afforded as a colorless viscous oil (736.0 mg, yield: 96.0%). FT-IR (KBr pellet cm^{-1}): ν = 3460, 2928, 2858, 1464, 1255, 1119, 1065, 843, 795. ^{1}H NMR (400 MHz, CDCl$_3$): [δ, ppm] = 0.00 (s, 12H, −SiCH$_3$), 0.39–0.48 (t, 4H, −SiCH$_2$), 0.83 (m, 6H, −CH$_2$CH$_3$), 1.20–1.32 (m, 36H, −(CH$_2$)$_9$CH$_3$), 1.53 (m, 8H, −SiCH$_2$CH$_2$, −SCH$_2$CH$_2$), 2.44–2.71 (m, 10H, −CH(OH)CH$_2$SCH$_2$), 3.31–3.51 (m, 8H, −CH$_2$OCH$_2$), 3.81 (m, 2H, −CH(OH)). ^{13}C NMR (101 MHz, CDCl$_3$): [δ, ppm] = 0.00, 13.83, 13.91, 22.38, 23.09, 28.49–29.47, 31.61, 32.30, 35.50, 68.85, 73.08, 73.95. ^{29}Si NMR (79 MHz, CDCl$_3$): [δ, ppm] = 7.70. HR-MS (FAB) calcd for C$_{40}$H$_{86}$O$_5$S$_2$Si$_2$ (MH$^+$): 767.5533; Found: 767.5535. Anal. Calcd for C$_{40}$H$_{86}$O$_5$S$_2$Si$_2$ (%): C, 62.61; H, 11.30; S, 8.36. Found: C, 61.69; H, 10.18; S, 9.52.

1,3-bis(3-(butylthiopropoxy)-2-ol-propyl)-1,1,3,3-tetramethyldisiloxane (E_1T_2): The synthesis was performed according to general procedure using 1-butanethiol (T_2, 225.5 mg) as a thiol compound. E_1T_2 was obtained as a colorless viscous oil (518.0 mg). Yield: 95.5%. FT-IR (KBr pellet cm^{-1}): ν = 3456, 2949, 2868, 1464, 1418, 1259, 1188, 1121, 1063, 845, 795. ^{1}H NMR (400 MHz, CDCl$_3$): [δ, ppm] = 0.00 (s, 12H, −SiCH$_3$), 0.40–0.49(t, 4H, −SiCH$_2$), 0.86 (t, 6H, −CH$_2$CH$_3$), 1.28–1.42 (m, 4H, −CH$_2$CH$_3$), 1.46–1.60 (m, 8H, −SiCH$_2$CH$_2$,

−SCH$_2$CH$_2$), 2.45–2.68 (m, 8H, −CH$_2$SCH$_2$), 2.76 (s, 2H, −CH(OH)), 3.33–3.50 (m, 8H, −CH$_2$OCH$_2$), 3.77–3.87 (m, 2H, −CH(OH)). ^{13}C NMR (101 MHz, CDCl$_3$): [δ, ppm] = 0.11, 13.69, 16.44, 22.00, 24.63, 32.07, 32.20, 36.21, 70.16, 73.49, 74.09. ^{29}Si NMR (79 MHz, CDCl$_3$): [δ, ppm] = 7.73. HR-MS (FAB) calcd for C$_{24}$H$_{54}$O$_5$S$_2$Si$_2$ (MH$^+$): 543.3029; Found: 543.3031. Anal. Calcd for C$_{24}$H$_{54}$O$_5$S$_2$Si$_2$ (%): C, 53.09; H, 10.02; S, 11.81. Found: C, 52.58; H, 10.03; S, 12.08.

1,3-bis(3-((2-hydroxyethyl)-thiopropoxy)-2-ol-propyl)-1,1,3,3-tetramethyldisiloxane (E$_1$T$_3$): The synthesis was performed according to general procedure using 2-mercaptoethanol (T$_3$, 195.0 mg) as a thiol compound. E$_1$T$_3$ was obtained as a colorless viscous oil (494.4 mg). Yield: 95.4%. FT-IR (KBr pellet cm^{-1}): ν = 3410, 2941, 2869, 1651, 1410, 1256, 1182, 1115, 1057, 845, 791. ^{1}H NMR (400 MHz, CDCl$_3$): [δ, ppm] = 0.00 (s, 12H, −SiCH$_3$), 0.39–0.48 (m, 4H, −SiCH$_2$), 1.54 (dq, 4H, −SiCH$_2$CH$_2$), 2.52–2.74 (m, 8H, −CH$_2$SCH$_2$), 3.29–3.47 (m, 8H, −CH$_2$OCH$_2$), 3.56 (s, 4H, −CH$_2$OH, −CH(OH)), 3.70 (t, 4H, −CH$_2$OH), 3.86 (m, 2H, −CH(OH)). ^{13}C NMR (101 MHz, CDCl$_3$): [δ, ppm] = 0.00, 13.81, 22.98, 35.39, 35.41, 60.87, 69.49, 73.03, 73.90. ^{29}Si NMR (79 MHz, CDCl$_3$): [δ, ppm] = 7.70. HR-MS (FAB) calcd for C$_{20}$H$_{46}$O$_7$S$_2$Si$_2$ (MH$^+$): 519.2302; Found: 519.2305. Anal. Calcd for C$_{20}$H$_{46}$O$_7$S$_2$Si$_2$ (%): C, 46.30; H, 8.94; S, 12.36. Found: C, 45.64; H, 8.93; S, 12.40.

1,3-bis(3-(allylthiopropoxy)-2-ol-propyl)-1,1,3,3-tetramethyldisiloxane (E$_1$T$_4$): The synthesis was performed according to general procedure using 2-propenthiol (T$_4$, 370.0 mg) as a thiol compound. E$_1$T$_4$ was obtained as a colorless viscous oil (480.1 mg). Yield: 94.2%. FT-IR (KBr pellet cm^{-1}): ν = 3456, 2924, 2868, 1639, 1414, 1256, 1188, 1121, 1065, 835, 791. ^{1}H NMR (400 MHz, CDCl$_3$): [δ, ppm] = 0.00 (s, 12H, −SiCH$_3$), 0.40–0.49 (m, 4H, −SiCH$_2$), 1.48–1.60 (m, 4H, −SiCH$_2$CH$_2$), 2.44–2.65 (m, 4H, −CH(OH)CH$_2$SCH$_2$), 2.85 (s, 2H, −CH(OH)), 3.10 (dd, 4H, −CH(OH)CH$_2$SCH$_2$), 3.30–3.50 (m, 8H, −CH$_2$OCH$_2$), 3.82 (q, 2H, −CH(OH)), 4.99–5.16 (m, 4H, −CH=CH$_2$), 5.66–5.81 (m, 2H, −CH=CH$_2$). ^{13}C NMR (101 MHz, CDCl$_3$): [δ, ppm] = 0.00, 13.90, 23.07, 33.79, 34.78, 68.87, 73.07, 73.90, 117.11, 133.80. ^{29}Si NMR (79 MHz, CDCl$_3$): [δ, ppm] = 7.73. HR-MS (FAB) calcd for C$_{22}$H$_{46}$O$_5$S$_2$Si$_2$ (MH$^+$): 511.2403; Found: 511.2402. Anal. Calcd for C$_{22}$H$_{46}$O$_5$S$_2$Si$_2$ (%): C, 51.72; H, 9.08; S, 12.55. Found: C, 51.27; H, 8.47; S, 13.05.

1,3-bis(3-((3-methylbutan-2-methyl)-thiopropoxy)-2-ol-propyl)-1,1,3,3-tetramethyldisiloxane (E$_1$T$_5$): The synthesis was performed according to general procedure using 3-methyl-2-butanethiol (T$_5$, 260.3 mg) as a thiol compound. E$_1$T$_5$ was obtained as a colorless viscous oil (514.4 mg). Yield: 90.2%. FT-IR (KBr pellet cm^{-1}): ν = 3456, 2964, 2869, 1462, 1379, 1249, 1109, 1061, 843, 795. ^{1}H NMR (400 MHz, CDCl$_3$): [δ, ppm] = 0.00 (s, 12H, −SiCH$_3$), 0.39–0.51 (m, 4H, −SiCH$_2$), 0.91 (dd, 12H, −CH(CH$_3$)$_2$), 1.17 (d, 6H, −SCH(CH$_3$)), 1.48–1.61 (m, 4H, −SiCH$_2$CH$_2$), 1.72–1.86 (m, 2H, −CH(CH$_3$)$_2$), 2.52–2.71 (m, 6H, −CH$_2$SCH), 2.85 (m, 2H, −CH(OH)), 3.32–3.50 (m, 8H, −CH$_2$OCH$_2$), 3.80 (m, 2H, −CH(OH)). ^{13}C NMR (101 MHz, CDCl$_3$): [δ, ppm] = 0.01, 13.93, 17.35, 18.14, 19.55, 23.09, 32.58, 34.58, 47.22, 69.07, 73.08, 73.95. ^{29}Si NMR (79 MHz, CDCl$_3$): [δ, ppm] = 7.74. HR-MS (FAB) calcd for C$_{26}$H$_{58}$O$_5$S$_2$Si$_2$ (MH$^+$): 571.3342; Found: 571.3345. Anal. Calcd for C$_{26}$H$_{58}$O$_5$S$_2$Si$_2$ (%): C, 54.69; H, 10.24; S, 11.23. Found: C, 52.85; H, 10.09; S, 9.99.

1,3-bis(3-(ethyl-3-acetate-thiopropoxy)-2-ol-propyl)-1,1,3,3-tetramethyldisiloxane (E$_1$T$_6$): The synthesis was performed according to general procedure, ethyl thioglycolate (T$_6$, 300.0 mg) as a thiol compound. E$_1$T$_6$ was obtained as a colorless viscous oil (572.1 mg). Yield: 95.0%. FT-IR (KBr pellet cm^{-1}): ν = 3469, 2939, 2870, 1738, 1466, 1408, 1265, 1123, 1042, 843, 787. ^{1}H NMR (400 MHz, CDCl$_3$): [δ, ppm] = 0.00 (s, 12H, −SiCH$_3$), 0.44 (m, 4H, −SiCH$_2$), 1.23 (m, 6H, −CH$_2$CH$_3$), 1.53 (m, 4H, −SiCH$_2$CH$_2$), 2.62–2.83 (m, 4H, −SCH$_2$CH(OH)), 2.88 (s, 2H, −CH(OH)), 3.25 (t, 4H, −SCH$_2$C(O)), 3.31–3.51 (m, 8H, −CH$_2$OCH$_2$), 3.88 (m, 2H, −CH(OH)), 4.14 (m, 4H, −OCH$_2$CH$_3$). ^{13}C NMR (101 MHz, CDCl$_3$): [δ, ppm] = 0.00, 13.83, 13.88, 23.05, 33.94, 36.14, 61.18, 69.10, 72.94, 73.93, 170.47. ^{29}Si NMR (79 MHz, CDCl$_3$): [δ, ppm] = 7.65. HR-MS (FAB) calcd for C$_{24}$H$_{50}$O$_9$S$_2$Si$_2$ (MH$^+$): 603.2513; Found: 603.2520. Anal. Calcd for C$_{24}$H$_{50}$O$_9$S$_2$Si$_2$ (%): C, 47.81; H, 8.36; S, 10.63. Found: C, 46.86; H, 8.40; S, 10.68.

1,3-bis(3-((furanyl-methyl)-thiopropoxy)-2-ol-propyl)-1,1,3,3-tetramethyldisiloxane (E_1T_7): The synthesis was performed according to general procedure using furfuryl mercaptan (T_7, 285.0 mg) as a thiol compound. E_1T_7 was obtained as a pale-yellow viscous oil (559.0 mg). Yield: 94.7%. FT-IR (KBr pellet cm^{-1}): ν = 3452, 3119, 2930, 2870, 1506, 1414, 1254, 1119, 1065, 868, 796. ^{1}H NMR (400 MHz, CDCl$_3$): [δ, ppm] = 0.01 (s, 12H, −SiCH$_3$), 0.39–0.49 (m, 4H, −SiCH$_2$), 1.47–1.60 (m, 4H, −SiCH$_2$CH$_2$), 2.50–2.68 (m, 4H, −SCH$_2$CH(OH)), 2.81 (t, 2H, −CH(OH)), 3.31–3.48 (m, 8H, −CH$_2$OCH$_2$), 3.68–3.74 (m, 4H, −CHCH$_2$SCH$_2$), 3.75–3.84 (m, 2H, −CH(OH)), 6.11–6.27 (m, 4H, −OCH=CHCH=C, Furan ring), 7.27–7.33 (m, 2H, −OCH=CHCH=C, Furan ring). ^{13}C NMR (101 MHz, CDCl$_3$): [δ, ppm] = 0.00, 13.88, 23.05, 28.37, 34.71, 69.03, 72.97, 73.87, 107.41, 110.10, 141.86, 151.05. ^{29}Si NMR (79 MHz, CDCl$_3$): [δ, ppm] = 7.74. HR-MS (FAB) calcd for C$_{26}$H$_{46}$O$_7$S$_2$Si$_2$ (MH$^+$): 591.2297; Found: 591.2292. Anal. Calcd for C$_{26}$H$_{46}$O$_7$S$_2$Si$_2$ (%): C, 52.85; H, 7.85; S, 10.85. Found: C, 51.16; H, 7.23; S, 10.13.

1,3-bis(3-((1-methylimidazol)-thiopropoxy)-2-ol-propyl)-1,1,3,3-tetramethyldisiloxane (E_1T_8): The synthesis was performed according to general procedure, methimazole (T_8, 285.0 mg) as a thiol compound. E_1T_8 was obtained as a pale-yellow viscous oil (545.0 mg). Yield: 92.3%. FT-IR (KBr pellet cm^{-1}): ν = 3390, 3198, 3117, 2937, 2865, 1674, 1514, 1466, 1258, 1117, 1055. ^{1}H NMR (400 MHz, CDCl$_3$): [δ, ppm] = 0.00 (s, 12H, −SiCH$_3$), 0.39–0.49 (m, 4H, −SiCH$_2$), 1.46–1.59 (m, 4H, −SiCH$_2$CH$_2$), 3.05–3.27 (m, 4H, −SCH$_2$CH(OH)), 3.34–3.52 (m, 8H, −CH$_2$OCH$_2$), 3.54 (s, 6H, −Imidazole−CH$_3$), 4.05 (m, 2H, −CH(OH)), 5.43–6.04 (s, 2H, -CH(OH)), 6.85 (m, 4H, −Imidazole). ^{13}C NMR (101 MHz, CDCl$_3$): [δ, ppm] = 0.00, 13.90, 23.08, 33.16, 37.25, 70.25, 72.89, 73.86, 121.83, 127.73, 142.69. ^{29}Si NMR (79 MHz, CDCl$_3$): [δ, ppm] = 7.73. HR-MS (FAB) calcd for C$_{24}$H$_{46}$N$_4$O$_5$S$_2$Si$_2$ (MH$^+$): 591.2526; Found: 591.2534. Anal. Calcd for C$_{24}$H$_{46}$N$_4$O$_5$S$_2$Si$_2$ (%): C, 48.78; H, 7.85; N, 9.48; S, 10.85. Found: C, 47.94; H, 8.07; N, 10.43; S, 10.39.

1,3-bis(3-(m-tolyl-thiopropoxy)-2-ol-propyl)-1,1,3,3-tetramethyldisiloxane (E_1T_9): The synthesis was performed according to general procedure using 3-methylbenzenethiol (T_9, 310.0 mg) as a thiol compound. E_1T_9 was obtained as a colorless viscous oil (589.0 mg). Yield: 96.5%. FT-IR (KBr pellet cm^{-1}): ν = 3454, 3051, 2939, 2864, 1587, 1475, 1412, 1258, 1180, 1111, 1053. ^{1}H NMR (400 MHz, CDCl$_3$): [δ, ppm] = 0.00 (s, 12H, −SiCH$_3$), 0.39–0.50 (m, 4H, −SiCH$_2$), 1.46–1.58 (m, 4H, −SiCH$_2$CH$_2$), 2.24 (s, 6H, −Phenyl−CH$_3$), 2.91–3.08 (m, 6H, −SCH$_2$CH(OH)), 3.32–3.48 (m, 8H, −CH$_2$OCH$_2$), 3.82 (m, 2H, −CH(OH)), 6.91–7.15 (m, 8H, −Phenyl). ^{13}C NMR (101 MHz, CDCl$_3$): [δ, ppm] = 0.00, 13.82, 20.93, 22.99, 36.85, 68.68, 72.62, 73.76, 125.96, 126.64, 128.39, 129.56, 135.27, 138.18. ^{29}Si NMR (79 MHz, CDCl$_3$): [δ, ppm] = 7.74. HR-MS (FAB) calcd for C$_{30}$H$_{50}$O$_5$S$_2$Si$_2$ (MH$^+$): 611.2716; Found: 611.2725. Anal. Calcd for C$_{30}$H$_{50}$O$_5$S$_2$Si$_2$ (%): C, 58.97; H, 8.25; S, 10.49. Found: C, 58.38; H, 8.31; S, 10.49.

1,3-bis(3-(benzyl-thiopropoxy)-2-ol-propyl)-1,1,3,3-tetramethyldisiloxane (E_1T_{10}): The synthesis was performed according to general procedure using, benzyl mercaptan (T_{10}, 310.0 mg) as a thiol compound. E_1T_{10} was obtained as a colorless viscous oil (572.4 mg). Yield: 93.8%. FT-IR (KBr pellet cm^{-1}): ν = 3447, 3022, 2928, 2868, 1603, 1495, 1416, 1258, 1123, 1049, 706. ^{1}H NMR (400 MHz, CDCl$_3$): [δ, ppm] = 0.00 (s, 12H, −SiCH$_3$), 0.38–0.48 (m, 4H, −SiCH$_2$), 1.46–1.58 (m, 4H, −SiCH$_2$CH$_2$), 2.50 (m, 4H, −SCH$_2$CH(OH)), 2.68–2.74 (m, 2H, −CH(OH)), 3.30–3.44 (m, 8H, −CH$_2$OCH$_2$), 3.68 (s, 4H, -CHCH$_2$SCH$_2$), 3.79 (m, 2H, −CH(OH)), 7.14–7.28 (m, 10H, −Phenyl). ^{13}C NMR (101 MHz, CDCl$_3$): [δ, ppm] = 0.00, 13.84, 23.01, 34.36, 36.19, 68.84, 72.99, 73.80, 126.68, 128.13, 128.55, 137.79. ^{29}Si NMR (79 MHz, CDCl$_3$): [δ, ppm] = 7.78. HR-MS (FAB) calcd for C$_{30}$H$_{50}$O$_5$S$_2$Si$_2$ (MH$^+$): 611.2716; Found: 611.2729. Anal. Calcd for C$_{30}$H$_{50}$O$_5$S$_2$Si$_2$ (%): C, 58.97; H, 8.25; S, 10.49. Found: C, 57.95; H, 8.32; S, 10.65.

1,3-bis(3-(benzothiazolyl-thiopropoxy)-2-ol-propyl)-1,1,3,3-tetramethyldisiloxane (E_1T_{11}): The synthesis was performed according to general procedure using 2-mercaptobenzothiazole (T_{11}, 417.5 mg) as a thiol compound. E_1T_{11} was obtained as a yellow viscous oil (622.3 mg). Yield: 89.4%. FT-IR (KBr pellet cm^{-1}): ν = 3418, 3068, 2939, 2872, 1568, 1462, 1425, 1256, 1186, 1065, 845. ^{1}H NMR (400 MHz, CDCl$_3$): [δ, ppm] = 0.00 (s, 12H, −SiCH$_3$), 0.40–0.50 (m, 4H,

−SiCH$_2$), 1.54 (dq, 4H, −SiCH$_2$CH$_2$), 3.34–3.55 (dt, 12H, −CH$_2$OCH$_2$CH(OH)CH$_2$), 4.14 (m, 2H, −CH(OH)), 4.64 (d, 2H, −CH(OH)), 7.15–7.69 (m, 8H, −Phenyl). ^{13}C NMR (101 MHz, CDCl$_3$): [δ, ppm] = 0.00, 13.86, 23.05, 37.04, 69.54, 72.68, 73.88, 120.62, 120.87, 124.09, 125.78, 134.96, 152.10, 167.45. ^{29}Si NMR (79 MHz, CDCl$_3$): [δ, ppm] = 7.80. HR-MS (FAB) calcd for C$_{30}$H$_{44}$N$_2$O$_5$S$_4$Si$_2$ (MH$^+$): 697.1750; Found: 697.1760. Anal. Calcd for C$_{30}$H$_{44}$N$_2$O$_5$S$_4$Si$_2$ (%): C, 51.69; H, 6.36; N, 4.02; S, 18.40. Found: C, 49.95; H, 6.39; N, 4.99; S, 17.65.

3.4. Synthesis of bis(2-(ethyl-3-acetate-thiopropoxy)-2-ol-ethyl)-ether (E$_2$T$_6$)

The synthesis of E$_2$T$_6$ was performed according to general procedure based on diethylene glycol diglycidyl ether (E$_2$, 218.1 mg, 1 mmol) and ethyl thioglycolate (T$_6$, 300.0 mg, 2.5 mmol). E$_2$T$_6$ was obtained as a colorless viscous oil (435.3 mg) with the same synthesis procedure of E$_1$T$_1$. Yield: 95.0%. FT-IR (KBr pellet cm^{-1}): ν = 3446.67, 2912.41, 2881, 1732.01, 1467.78, 1284.55, 1132.17, 1031.88. ^{1}H NMR (400 MHz, CDCl$_3$): [δ, ppm] = 1.23 (t, 6H, −CH$_2$CH$_3$), 2.71 (m, 4H, −SCH$_2$CH(OH)), 3.25 (d, 4H, −SCH$_2$COO), 3.43–3.66 (m, 14H, −OCH$_2$CH$_2$OCH$_2$CH(OH)), 3.85–3.93 (m, 2H, −CH(OH)), 4.14 (m, 4H, −CH$_2$CH$_3$). ^{13}C NMR (101 MHz, CDCl$_3$): [δ, ppm] = 14.15, 34.30, 35.75, 61.46, 69.50, 70.41, 70.53, 74.28, 170.71.

4. Conclusions

In summary, we developed an efficient synthetic strategy, i.e., a thiol–epoxy reaction, for the synthesis of novel sulfur-containing functionalized disiloxanes from 1,3-bis(3-glycidoxypropyl)-1,1,3,3-tetramethyldisiloxane and different thiol compounds. The study found that both lithium hydroxide (LiOH) and tetrabutylammonium fluoride (TBAF) have high catalytic activity, and TBAF in THF is universal for the synthesis of these compounds because THF can dissolve most thiol compounds. The reactions can be conducted under mild reaction conditions with high reaction yields and regioselectivity, no by-products, and good functional group tolerance. Furthermore, these new disiloxanes exhibit typical aggregation-induced emission enhancement (AIEE) characteristics and can emit strong blue fluorescence under the action of the unconventional fluorescent chromophores S→Si or N→Si coordination bonds. In addition, the S→Si coordination bonds can also increase the aggregation of the nonconventional fluorescent clusters and thus enhance the fluorescence intensity. These results demonstrate that the thiol–epoxy reaction could be developed as an efficient organosilicon synthetic methodology for constructing various organosilicon compounds or polymers with new structures and functionalities from the low-cost epoxy-containing organosilicon raw materials. It is worth noting that although TBAF is an efficient catalyst for the synthesis of functionalized disiloxanes, it may be not adaptable for the synthesis of other organosilicon compounds, especially silsesquioxanes, because fluoride ions could react rapidly with silicon atoms and it may be a cause of complicated and unpredictable synthetic results. Further synthesis of other organosilicon materials and the reaction methods may be also optimized. Our group is currently utilizing this reaction to construct various novel organosilicon polymers and materials and find their applications in emerging fields, such as sensing and catalysis.

Supplementary Materials: The following supporting information can be downloaded at: https://www.mdpi.com/article/10.3390/ijms24097785/s1.

Author Contributions: Conceptualization, D.W.; methodology, J.T. and D.W.; validation, J.T. and D.W.; investigation, J.T. and D.W.; writing—original draft preparation, J.T. and D.W.; writing—review and editing, J.T., D.W. and S.F.; supervision, D.W.; project administration, D.W.; funding acquisition, D.W. and S.F. All authors have read and agreed to the published version of the manuscript.

Funding: This research was supported by the National Natural Science Foundation of China (22271175 and 52173102), Fluorine Silicone Materials Collaborative Fund of Shandong Provincial Natural Science Foundation (ZR2020LFG011 and ZR2021LFG001), Basic Research Foundation of Institute of Silicon-Based High-End New Materials (YJG202103), and Young Scholars Program of Shandong University.

Institutional Review Board Statement: Not applicable.

Informed Consent Statement: Not applicable.

Data Availability Statement: Not applicable.

Conflicts of Interest: The authors declare no conflict of interest.

References

1. Yilgör, E.; Yilgör, I. Silicone containing copolymers: Synthesis, properties and applications. *Prog. Polym. Sci.* **2014**, *39*, 1165–1195. [CrossRef]
2. Rücker, C.; Kümmerer, K. Environmental Chemistry of Organosiloxanes. *Chem. Rev.* **2015**, *115*, 466–524. [CrossRef] [PubMed]
3. Liu, J.; Yao, Y.; Li, X.; Zhang, Z. Fabrication of advanced polydimethylsiloxane-based functional materials: Bulk modifications and surface functionalizations. *Chem. Eng. J.* **2021**, *408*, 127262. [CrossRef]
4. Cazacu, M.; Dascalu, M.; Stiubianu, G.-T.; Bele, A.; Tugui, C.; Racles, C. From passive to emerging smart silicones. *Rev. Chem. Eng.* **2022**. [CrossRef]
5. Qi, D.; Zhang, K.; Tian, G.; Jiang, B.; Huang, Y. Stretchable Electronics Based on PDMS Substrates. *Adv. Mater.* **2021**, *33*, 2003155. [CrossRef]
6. Wang, D.; Klein, J.; Mejía, E. Catalytic Systems for the Cross-Linking of Organosilicon Polymers. *Chem. Asian J.* **2017**, *12*, 1180–1197. [CrossRef]
7. Brook, M.A. *Silicon in Organic, Organometallic, and Polymer Chemistry*; Wiley: New York, NY, USA, 2000; Volume 123.
8. Pawlenko, S. *Organosilicon Chemistry*; De Gruyter: Berlin, Germany, 1986.
9. Chojnowski, J.; Marciniec, B. *Progress in Organosilicon Chemistry*; Gordon and Breach: Philadelphia, PA, USA, 1995.
10. Wang, D.; Cao, J.; Hang, D.; Li, W.; Feng, S. Novel Organosilicon Synthetic Methodologies. *Prog. Chem.* **2019**, *31*, 110–120.
11. Jones, R.; Ando, W.; Chojnowski, J. *The Science and Technology of Their Synthesis and Applications. Silicon-Containing Polymers*; Springer: Berlin/Heidelberg, Germany, 2000.
12. LaRonde, F.J.; Brook, M.A.; Hu, G. Amino acid and peptide chemistry on silicones. *Silicon Chem.* **2002**, *1*, 215–222. [CrossRef]
13. Fan, X.; Zhang, M.; Gao, Y.; Zhou, Q.; Zhang, Y.; Yu, J.; Xu, W.; Yan, J.; Liu, H.; Lei, Z.; et al. Stepwise on-demand functionalization of multihydrosilanes enabled by a hydrogen-atom-transfer photocatalyst based on eosin Y. *Nat. Chem.* **2023**. [CrossRef]
14. Fuchise, K.; Sato, K.; Igarashi, M. Precise Synthesis of Linear Polysiloxanes End-Functionalized with Alkynylsilyl Groups by Organocatalytic Ring-Opening Polymerization of Cyclotrisiloxanes. *Macromolecules* **2021**, *54*, 5765–5773. [CrossRef]
15. Siripanich, P.; Bureerug, T.; Chanmungkalakul, S.; Sukwattanasinitt, M.; Ervithayasuporn, V. Mono and Dumbbell Silsesquioxane Cages as Dual-Response Fluorescent Chemosensors for Fluoride and Polycyclic Aromatic Hydrocarbons. *Organometallics* **2022**, *41*, 201–210. [CrossRef]
16. Yi, B.; Wang, S.; Hou, C.; Huang, X.; Cui, J.; Yao, X. Dynamic siloxane materials: From molecular engineering to emerging applications. *Chem. Eng. J.* **2021**, *405*, 127023. [CrossRef]
17. Golas, P.L.; Matyjaszewski, K. Marrying click chemistry with polymerization: Expanding the scope of polymeric materials. *Chem. Soc. Rev.* **2010**, *39*, 1338–1354. [CrossRef]
18. Qin, A.; Lam, J.W.Y.; Tang, B.Z. Click polymerization. *Chem. Soc. Rev.* **2010**, *39*, 2522–2544. [CrossRef] [PubMed]
19. Franc, G.; Kakkar, A.K. "Click" methodologies: Efficient, simple and greener routes to design dendrimers. *Chem. Soc. Rev.* **2010**, *39*, 1536–1544. [CrossRef]
20. Rambarran, T.; Bertrand, A.; Gonzaga, F.; Boisson, F.; Bernard, J.; Fleury, E.; Ganachaud, F.; Brook, M.A. Sweet supramolecular elastomers from α,ω-(β-cyclodextrin terminated) PDMS. *Chem. Commun.* **2016**, *52*, 6681–6684. [CrossRef]
21. Tucker-Schwartz, A.K.; Farrell, R.A.; Garrell, R.L. Thiol-ene Click Reaction as a General Route to Functional Trialkoxysilanes for Surface Coating Applications. *J. Am. Chem. Soc.* **2011**, *133*, 11026–11029. [CrossRef]
22. Zuo, Y.; Cao, J.; Feng, S. Sunlight-Induced Cross-Linked Luminescent Films Based on Polysiloxanes and d-Limonene via Thiol-ene "Click" Chemistry. *Adv. Funct. Mater.* **2015**, *25*, 2754–2762. [CrossRef]
23. Genest, A.; Binauld, S.; Pouget, E.; Ganachaud, F.; Fleury, E.; Portinha, D. Going beyond the barriers of aza-Michael reactions: Controlling the selectivity of acrylates towards primary amino-PDMS. *Polym. Chem.* **2017**, *8*, 624–630. [CrossRef]
24. Lu, H.; Feng, L.; Li, S.; Zhang, J.; Lu, H.; Feng, S. Unexpected Strong Blue Photoluminescence Produced from the Aggregation of Unconventional Chromophores in Novel Siloxane–Poly(amidoamine) Dendrimers. *Macromolecules* **2015**, *48*, 476–482. [CrossRef]
25. Zuo, Y.; Zhang, Y.; Gou, Z.; Lin, W. Facile construction of imidazole functionalized polysiloxanes by thiol-ene "Click" reaction for the consecutive detection of Fe^{3+} and amino acids. *Sensors Actuat. Chem.* **2019**, *291*, 235–242. [CrossRef]
26. Bezlepkina, K.A.; Ardabevskaia, S.N.; Klokova, K.S.; Ryzhkov, A.I.; Migulin, D.A.; Drozdov, F.V.; Cherkaev, G.V.; Muzafarov, A.M.; Milenin, S.A. Environment Friendly Process toward Functional Polyorganosiloxanes with Different Chemical Structures through CuAAC Reaction. *ACS Appl. Polym. Mater.* **2022**, *4*, 6770–6783. [CrossRef]
27. Milenin, S.A.; Drozdov, F.V.; Bezlepkina, K.A.; Majorov, V.Y.; Muzafarov, A.M. Acid-Catalyzed Rearrangement of Azidopropyl-Siloxane Monomers for the Synthesis of Azidopropyl-Polydimethylsiloxane and Their Carboxylic Acid Derivatives. *Macromolecules* **2021**, *54*, 2921–2935. [CrossRef]

28. Zuo, Y.; Gou, Z.; Zhang, C.; Feng, S. Polysiloxane-Based Autonomic Self-Healing Elastomers Obtained through Dynamic Boronic Ester Bonds Prepared by Thiol–Ene "Click" Chemistry. *Macromol. Rapid Commun.* **2016**, *37*, 1052–1059. [CrossRef] [PubMed]
29. Xu, Q.; Huang, T.; Li, S.; Li, K.; Li, C.; Liu, Y.; Wang, Y.; Yu, C.; Zhou, Y. Emulsion-Assisted Polymerization-Induced Hierarchical Self-Assembly of Giant Sea Urchin-like Aggregates on a Large Scale. *Angew. Chem.* **2018**, *130*, 8175–8179. [CrossRef]
30. Stuparu, M.C.; Khan, A. Thiol-epoxy "click" chemistry: Application in preparation and postpolymerization modification of polymers. *J. Polym. Sci. Polym. Chem.* **2016**, *54*, 3057–3070. [CrossRef]
31. Hoang, M.V.; Chung, H.-J.; Elias, A.L. Irreversible bonding of polyimide and polydimethylsiloxane (PDMS) based on a thiol-epoxy click reaction. *J. Microme. Microeng.* **2016**, *26*, 105019. [CrossRef]
32. Shevchenko, V.; Bliznyuk, V.; Gumenna, M.; Klimenko, N.; Stryutsky, A.; Wang, J.; Binek, C.; Chernyakova, M.; Belikov, K. Coordination Polymers Based on Amphiphilic Oligomeric Silsesquioxanes and Transition Metal Ions (Co^{2+}, Ni^{2+}): Structure and Stimuli-Responsive Properties. *Macromol. Mater. Eng.* **2021**, *306*, 2100085. [CrossRef]
33. Shibasaki, S.; Sasaki, Y.; Nakabayashi, K.; Mori, H. Synthesis and metal complexation of dual-functionalized silsesquioxane nanoparticles by sequential thiol–epoxy click and esterification reactions. *React. Funct. Polym.* **2016**, *107*, 11–19. [CrossRef]
34. Lin, H.; Chen, L.; Ou, J.; Liu, Z.; Wang, H.; Dong, J.; Zou, H. Preparation of well-controlled three-dimensional skeletal hybrid monoliths via thiol–epoxy click polymerization for highly efficient separation of small molecules in capillary liquid chromatography. *J. Chromatog. A* **2015**, *1416*, 74–82. [CrossRef]
35. Lungu, A.; Ghitman, J.; Cernencu, A.I.; Serafim, A.; Florea, N.M.; Vasile, E.; Iovu, H. POSS-containing hybrid nanomaterials based on thiol-epoxy click reaction. *Polymer* **2018**, *145*, 324–333. [CrossRef]
36. Bekin Acar, S.; Ozcelik, M.; Uyar, T.; Tasdelen, M.A. Polyhedral oligomeric silsesquioxane-based hybrid networks obtained via thiol-epoxy click chemistry. *Iran. Polym. J.* **2017**, *26*, 405–411. [CrossRef]
37. Lin, G.; Yin, J.; Lin, Z.; Zhu, Y.; Li, W.; Li, H.; Liu, Z.; Xiang, H.; Liu, X. Facile thiol-epoxy click chemistry for transparent and aging-resistant silicone/epoxy composite as LED encapsulant. *Prog. Org. Coat.* **2021**, *156*, 106269. [CrossRef]
38. Murphy, Z.; Kent, M.; Freeman, C.; Landge, S.; Koricho, E. Halloysite nanotubes functionalized with epoxy and thiol organosilane groups to improve fracture toughness in nanocomposites. *SN Appl. Sci.* **2020**, *2*, 213. [CrossRef]
39. Bera, S.; Rout, T.K.; Udayabhanu, G.; Narayan, R. Water-based & eco-friendly epoxy-silane hybrid coating for enhanced corrosion protection & adhesion on galvanized steel. *Prog. Org. Coat.* **2016**, *101*, 24–44.
40. De, S.; Khan, A. Efficient synthesis of multifunctional polymersviathiol–epoxy "click" chemistry. *Chem. Commun.* **2012**, *48*, 3130–3132. [CrossRef]
41. Gadwal, I.; Stuparu, M.C.; Khan, A. Homopolymer bifunctionalization through sequential thiol–epoxy and esterification reactions: An optimization, quantification, and structural elucidation study. *Polym. Chem.* **2015**, *6*, 1393–1404. [CrossRef]
42. From, M.; Larsson, P.T.; Andreasson, B.; Medronho, B.; Svanedal, I.; Edlund, H.; Norgren, M. Tuning the properties of regenerated cellulose: Effects of polarity and water solubility of the coagulation medium. *Carbohyd. Polym.* **2020**, *236*, 116068. [CrossRef]
43. Azizi, N.; Khajeh-Amiri, A.; Ghafuri, H.; Bolourtchian, M. LiOH-Catalyzed simple ring opening of epoxides under solvent-free conditions. *Phosphorus Sulfur* **2010**, *185*, 1550–1557. [CrossRef]
44. Zuo, Y.; Gou, Z.; Quan, W.; Lin, W. Silicon-assisted unconventional fluorescence from organosilicon materials. *Coord. Chem. Rev.* **2021**, *438*, 213887. [CrossRef]
45. Huang, Z.; Zhou, C.; Chen, W.; Li, J.; Li, M.; Liu, X.; Mao, L.; Yuan, J.; Tao, L.; Wei, Y. A polymerizable Aggregation Induced Emission (AIE)-active dye with remarkable pH fluorescence switching based on benzothiazole and its application in biological imaging. *Dye. Pigment.* **2021**, *196*, 109793. [CrossRef]
46. Razi, S.S.; Gupta, R.C.; Ali, R.; Dwivedi, S.K.; Srivastava, P.; Misra, A. A new D–π–A type intramolecular charge transfer Dyad System to detect F^-: Anion induced CO_2 sensing. *Sensors Actuat. Chem.* **2016**, *236*, 520–528. [CrossRef]
47. Cao, J.; Zuo, Y.; Lu, H.; Yang, Y.; Feng, S. An unconventional chromophore in water-soluble polysiloxanes synthesized via thiol-ene reaction for metal ion detection. *J. Photo. Photo. A Chem.* **2018**, *350*, 152–163. [CrossRef]
48. Lu, H.; Zhang, J.; Feng, S. Controllable photophysical properties and self-assembly of siloxane-poly(amidoamine) dendrimers. *Phys. Chem. Chem. Phys.* **2015**, *17*, 26783–26789. [CrossRef]
49. Fu, P.-Y.; Li, B.-N.; Zhang, Q.-S.; Mo, J.-T.; Wang, S.-C.; Pan, M.; Su, C.-Y. Thermally Activated Fluorescence vs Long Persistent Luminescence in ESIPT-Attributed Coordination Polymer. *J. Am. Chem. Soc.* **2022**, *144*, 2726–2734. [CrossRef]
50. Lu, H.; Hu, Z.; Feng, S. Nonconventional Luminescence Enhanced by Silicone-Induced Aggregation. *Chem. Asian J.* **2017**, *12*, 1213–1217. [CrossRef]

Disclaimer/Publisher's Note: The statements, opinions and data contained in all publications are solely those of the individual author(s) and contributor(s) and not of MDPI and/or the editor(s). MDPI and/or the editor(s) disclaim responsibility for any injury to people or property resulting from any ideas, methods, instructions or products referred to in the content.

Article

Flexible Curcumin-Loaded Zn-MOF Hydrogel for Long-Term Drug Release and Antibacterial Activities

Jiaxin Li [1,†], Yachao Yan [1,†], Yingzhi Chen [1,2,*], Qinglin Fang [1], Muhammad Irfan Hussain [1] and Lu-Ning Wang [1,2,*]

1. School of Materials Science and Engineering, University of Science and Technology Beijing, Beijing 100083, China
2. School of Shunde Graduate, University of Science and Technology Beijing, Foshan 528399, China
* Correspondence: chenyingzhi@ustb.edu.cn (Y.C.); luning.wang@ustb.edu.cn (L.-N.W.)
† These authors contributed equally to this work.

Abstract: Management of chronic inflammation and wounds has always been a key issue in the pharmaceutical and healthcare sectors. Curcumin (CCM) is an active ingredient extracted from turmeric rhizomes with antioxidant, anti-inflammatory, and antibacterial activities, thus showing significant effectiveness toward wound healing. However, its shortcomings, such as poor water solubility, poor chemical stability, and fast metabolic rate, limit its bioavailability and long-term use. In this context, hydrogels appear to be a versatile matrix for carrying and stabilizing drugs due to their biomimetic structure, soft porous microarchitecture, and favorable biomechanical properties. The drug loading/releasing efficiencies can also be controlled via using highly crystalline and porous metal-organic frameworks (MOFs). Herein, a flexible hydrogel composed of a sodium alginate (SA) matrix and CCM-loaded MOFs was constructed for long-term drug release and antibacterial activity. The morphology and physicochemical properties of composite hydrogels were analyzed by scanning electron microscopy (SEM), Fourier transform infrared spectroscopy (FT-IR), X-ray diffraction (XRD), ultraviolet-visible spectroscopy (UV-Vis), Raman spectroscopy, and mechanical property tests. The results showed that the composite hydrogel was highly twistable and bendable to comply with human skin mechanically. The as-prepared hydrogel could capture efficient CCM for slow drug release and effectively kill bacteria. Therefore, such composite hydrogel is expected to provide a new management system for chronic wound dressings.

Keywords: curcumin; MOF; sodium alginate hydrogel; drug release; antibacterial activity

1. Introduction

Chronic wounds and the accompanying persistent bacterial inflammation have always been a vital challenge in clinical treatment. For the treatment of wounds, curcumin (CCM) is highly recognized for benefitting human health [1]. It is a natural lipophilic polyphenol and natural antioxidant compound that has been widely used as a food additive, food coloring [2], and flavoring agent [3] in the past decades, and in recent years researchers have revealed the pleiotropic nature of the biological effects of this molecule. Much research has clarified its antioxidant, anti-inflammatory, and antibacterial activities, which are beneficial for wound healing [4]. However, poor water solubility (under acidic and neutral conditions), chemical instability (especially under neutral and alkaline conditions), rapid-metabolic-rate-limited bioavailability, and delivery efficiency severely affect its application in wound healing [5]. In this regard, nanotechnology-based delivery systems (encapsulating the drug into specific nanocarriers) are considered a promising technology for circumventing these obstacles.

To date, various drug delivery platforms have been developed, including polymeric nanoparticles [6], carbon-based nanostructures, lipid-based nanoparticles [7], and inorganic nanostructures [8]. Among them, carbon-based nanostructures display high surface area,

functionalization versatility, and drug-loading capacity for drug delivery [9]. Biomolecules such as lipids or proteins are good choices as biocompatible platforms [10]. Inorganic or organic nanoparticles are appealing for their altered pharmacokinetics and biodistribution profiles, sometimes presenting unique photonic/thermal/electronic/magnetic effects in guiding the accumulation or release of drugs [11,12]. To combine the merits of loading efficiency, tunability, and biocompatibility, metal-organic frameworks (MOFs) stand out as promising antibacterial platforms. MOFs are a class of highly crystalline and nanoporous materials that can be built from various metal-ion metal-clusters and organic linkers for tunable chemical and topological structures [13]. Recently, increasing use of MOFs has been made in drug storage and release for their combined properties of high adsorption capacity, tunable host-guest interaction and release efficiency, biocompatibility, and nontoxicity [14].

To be properly used as a wound dressing, it should also be able to fill wound spaces and provide mechanical stability as well as good penetrability to water vapor and metabolites. Hydrogels with three-dimensional (3D) hydrophilic polymeric structures can meet these requirements [15] by mimicking the native extracellular matrix (flexible and stable geometry, high plasticity, and not sticking to the wound) and creating a moist microenvironment for wounds [16]. Simultaneously, hydrogels can host various bioactive ingredients [17], and their high porosity and controllable crosslinking allow these ingredients to be delivered suitably for accelerating wound healing [18]. Hydrogels' water absorption and penetrability arise from their flexible polymer network with many hydrophilic groups [19] (e.g., R-COOH, R-CONH$_2$, R-NH$_2$, R-OH, R-SO$_3$H). These functional groups in hydrogels facilitate the adsorption of metal ions by binding with various oxygen-, nitrogen-, or sulfur-containing functional groups [20]. This also provides anchoring sites for the nucleation and growth of MOFs, highlighting the rational design of an efficient wound-dressing system.

Numerous studies have been conducted to combine the above components into complexes for wound healing. As for the typical binary system, (1) functional MOFs carrying drugs have excellent antibacterial properties via the controllable release of antibacterial drugs and metal ions for efficient wound healing [21,22]. Recent research has suggested that modifying the surface of MOFs could enhance their adherence to bacterial cells, inhibiting the growth of bacteria [23]. However, it is in powder, making it challenging to adhere to the skin stably. (2) MOF-anchored hydrogel dressings are an alternative that could achieve antibacterial action by destroying the integrity of bacterial cell membranes and decomposing H$_2$O$_2$ to improve wound hypoxia due to the peroxidase performance of MOF [24–26]. However, without drug loading, antimicrobial efficiency is limited. (3) The drug-included hydrogel delivery system, which encapsulates drugs physically, has excellent deformability and oxygen transmission rates as wound dressings [27,28]. However, it is difficult to regulate the drug release rate. Recently, several studies have demonstrated the excellent antimicrobial activity of drug–MOF–biomolecular carrier systems (ternary systems) [29–31]. Despite the improved performance for wound healing, its preparation process was relatively complicated. This is because synthesizing the precursors requires long preparation cycles, harsh preparation conditions, and cumbersome preparation processes. Therefore, simplifying the preparation methods for ternary systems is currently an issue that needs to be addressed.

In this study, CCM ligands and Zn^{2+}-centered MOFs with a similar structure to the zeolitic imidazolate framework (ZIF-8) were facilely prepared on sodium alginate (SA) hydrogel to improve the stability and administration efficiency of CCM, affording controlled release of CCM and improved bioavailability. Zn^{2+}, introduced as a crosslinking agent and presented on the surface of the prepared hydrogels, could also be used as a metal-ligand center linked to both 2-methylimidazole (2-mIM) ligand and CCM, resulting in a one-step synthesis of CCM-loaded Zn-MOF hydrogels (CCM@ZIF-8@SA), which simplified preparation process to a great extent. The composite hydrogel displayed controlled antibacterial and anti-inflammatory effects by changing the amount of hosted CCM. The composite hydrogel also exhibited reasonable flexibility that could be twisted and bent on human skin, which is promising for use as a wound dressing. Such a system fully uses the performance advantages and structural characteristics of each conventional single component, circumventing the shortcomings of CCM, traditional dressings, and MOFs in biomedical applications, thus providing a rational design guideline for an antibacterial CCM-based platform for wound healing.

2. Results and Discussion

2.1. Characterization of CCM@ZIF-8@SA-Composite Hydrogels

Zn^{2+} was used as the cross-linking agent for the preparation of flexible sodium SA hydrogels, and the surficial Zn^{2+} on the as-prepared hydrogel could also work as a metal-coordinated center to simultaneously connect with the 2-mIM ligand and CCM, leading to the formation of CCM@ZIF-8@SA, as shown in Figure 1a. With increasing CCM amounts (1.0, 3.0, and 5.0 mg), three kinds of composite hydrogels were obtained, 1CCM@ZIF-8@SA, 3CCM@ZIF-8@SA, and 5CCM@ZIF-8@SA, respectively. Optical and scanning electron microscopy (SEM) images of the as-prepared SA, ZIF-8@SA, and 1/3/5CCM@ZIF-8@SA were depicted in Figure 1. The optical images showed that the SA hydrogel after Zn^{2+}-crosslinking was transparent (Figure 1b), and after loading the ZIF-8 particles, the sample changed from transparent to white (Figure 1c). When CCM was added together with the coordination ligand, the sample color turned orange and gradually deepened as the CCM amount increased (Figure 1d–f). As shown in the SEM image (Figure 1g–k), they all had a porous sponge structure, which contributed to water retention/penetration and sustained drug release inside the hydrogel. As shown in the magnified insets in Figure 1g,h, the surface of SA hydrogel was smooth, while coordination of Zn^{2+} with 2-mIM gave rise to regular dodecahedral ZIF-8 particles homogeneously dispersed on SA (ZIF-8@SA). From Figure 1i, it can be seen that a small dosage of CCM did not change the surface state slightly, but when the added CCM amount was increased to 3 mg, most of the regular dodecahedral structure was replaced by a sheet structure (Figure 1j), meaning the CCM could work as the regulator in the coordination reaction and even as another ligand to coordinate with Zn^{2+}, thus rebuilding the MOF structure [32,33]. With CCM increasing to 5 mg (Figure 1k), the surface dodecahedral structures were changed to lamellar structures, verifying the similar role of CCM as ligands. The lamellar structures are expected to serve as effective channels for drug release.

The X-ray diffraction (XRD) patterns (Figure 2a) showed that ZIF-8 exhibited characteristic diffraction peaks at 7.39, 10.45, 12.84, and 16.51°, which corresponded to the (110), (200), (211), and (222) planes, respectively [34]. These typical peaks appeared in ZIF-8@SA and continued to endure in the 5CCM@ZIF-8@SA (containing the most CCM) samples, indicating that ZIF-8 crystals were successfully formed onto the hydrogel [35], and the participation of CCM did not affect the packing of ZIF-8. The Fourier transform infrared spectroscopy (FT-IR) spectra are shown in Figure 2b. The samples displayed distinct absorption peaks at 3435, 2927, 1620, and 1567 cm^{-1}. The peaks at 2927 and 1567 cm^{-1} were attributed to the stretching vibrations of the C-H and C-N bonds of the aliphatic group in 2-methylimidazole. In the spectrum of 1/3/5CCM@ZIF-8@SA, the peak at 1620 cm^{-1} corresponded to the superposition vibrations of the C=C and C=O bonds in CCM. Pure CCM exhibited a characteristic absorption peak (phenolic O-H stretching vibration) at 3490 cm^{-1}. Still, this peak was blue-shifted in the 3490–3435 cm^{-1} band in 1/3/5CCM@ZIF-8@SA, indicating that the hydrogel surface was gradually covered by CCM [36]. UV-Vis absorption spectra were provided to analyze the composition. As shown in Figure 2c, bare ZIF-8 has a characteristic absorption peak at 226 nm, which also appeared in 1/3/5CCM@ZIF-8@SA, indicating that coating of ZIF-8 on the SA surface. Similarly, the peak at approximately 427 nm, corresponding to CCM, was also found in 1/3/5CCM@ZIF-8@SA. This indicated that CCM had been successfully loaded into the ZIF-8 framework. In contrast to pure CCM, the composite showed a redshift of approximately 13 nm. This further revealed the strong interaction between CCM and Zn^{2+} in either ZIF-8 or the hydrogel, which decreased the band gap between the $\pi-\pi*$ [37] electronic transition of CCM. It is supposed that CCM contains highly conjugated 1, 3-diketone moieties (1, 3-diketones and two enols) in the tautomer, which could be connected with Zn^{2+} to form porous skeleton compounds with stable structures. In the Raman spectra (Figure 2d), no obvious bond was found in SA; ZIF-8@SA showed the characteristic bonds of ZIF-8 at 286 cm^{-1} (Zn-N vibrations in the ZnN_4 tetrahedron) [38], 1123 cm^{-1} (C-N stretching), and 1461 cm^{-1} (C-H) [39], indicating the successful loading of ZIF-8 onto SA; the extracted bending vibration of the imidazole ring at 694 cm^{-1} was assigned to Zn-N vibrations [40]. These distinct Raman bands of CCM were also observed in 1/3/5CCM@ZIF-8@SA, further confirming the incorporation of CCM [41].

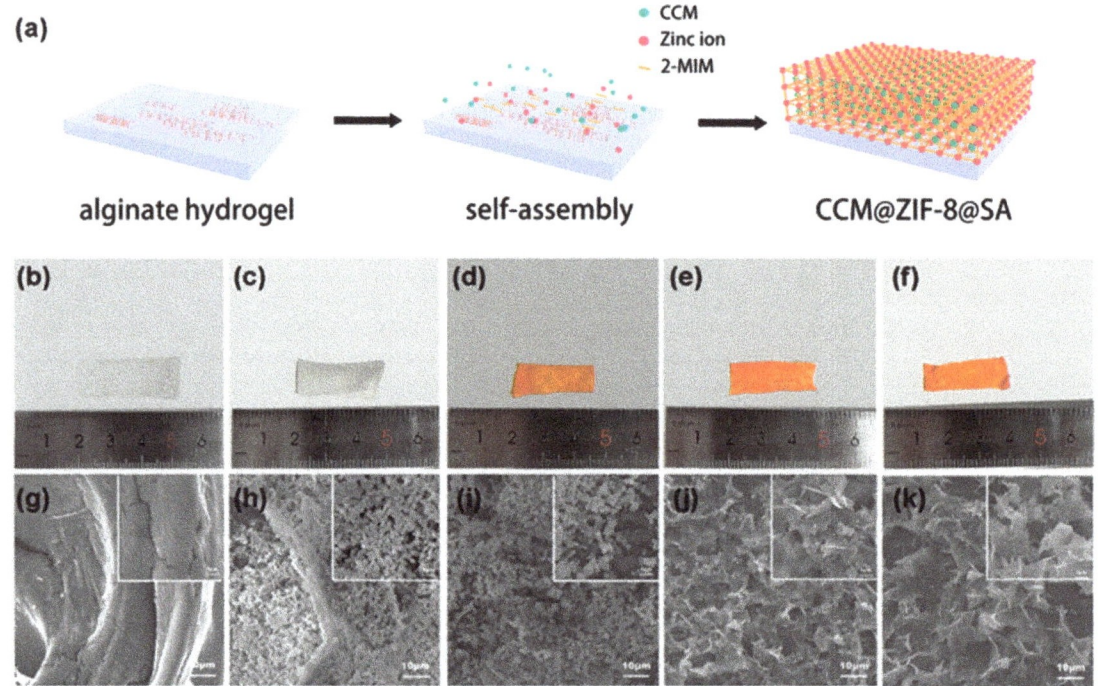

Figure 1. (a) Schematic diagram of the hydrogel synthesis. Optical images of SA (b), ZIF-8@SA (c), 1CCM@ZIF-8@SA (d), 3CCM@ZIF-8@SA (e), 5CCM@ZIF-8@SA (f). SEM images of SA (g), ZIF-8 @SA (h), 1CCM@ZIF-8@SA (i), 3CCM@ZIF-8@SA (j), 5CCM@ZIF-8@SA (k).

Based on the thermogravimetric (TGA) analysis, the thermal stability of the sample was determined. It has been revealed that at elevated temperatures, CCM begins to degrade weightlessly at 200 °C, and ZIF-8 starts to collapse weightlessly and structurally at 580 °C [41]. As shown in Figure 3a, the weightlessness curve of SA exhibited a stable decreasing trend, but ZIF-8@SA experienced a rapid weight loss at 544 °C caused by the collapse and degradation of the ZIF-8 frame loaded on the SA. Meanwhile, the 1/3/5 CCM@ZIF-8@SA samples underwent rapid weight loss at 243 and 544 °C. The first prompt weight loss that occurred at 243 °C was due to the degradation of the encapsulated CCM. The increase in weight loss temperature compared to the pure CCM weight loss temperature was caused by the interaction of the CCM with the metal part in the ZIF-8 framework. And the collapse degradation of the ZIF-8 framework triggered the second rapid weightlessness at 544 °C. The pore structure parameters of the sample as determined by the BET test, in Figure 3b, ZIF-8 and 1/3/5CCM@ZIF-8 exhibited I-type adsorption curve, with significant increases in N_2 uptake at low relative pressure (<0.01 MPa) due to the presence of micropores. As seen in Table 1, the specific surface area of 1/3/5CCM@ZIF-8 was decreased by

1.63%, 1.98%, and 9.51%, the pore volume was reduced by 3.2%, 15.8%, and 31.6%, and the average pore size was also reduced by 11.9%, 12.3%, and 14.8%, respectively, while the pure ZIF-8 had the specific surface area of 968.3 m^2/g, and the pore volume and pore size were 0.461 cm^3/g and 3.302 nm. The decrease in the porosity of 1/3/5CCM@ZIF-8 was ascribed to the encapsulating of CCM into the ZIF-8 framework. The pore size distributions of ZIF-8 and 1/3/5CCM@ZIF-8 (Figure 3c) were mainly concentrated in the microporous range, and the pore size distribution of ZIF-8 and 1/3/5CCM@ZIF-8 is mainly in the microporous range (Figure 3c). However, the average pore size was mesoporous, which may have been caused by the gap error between particles in the test.

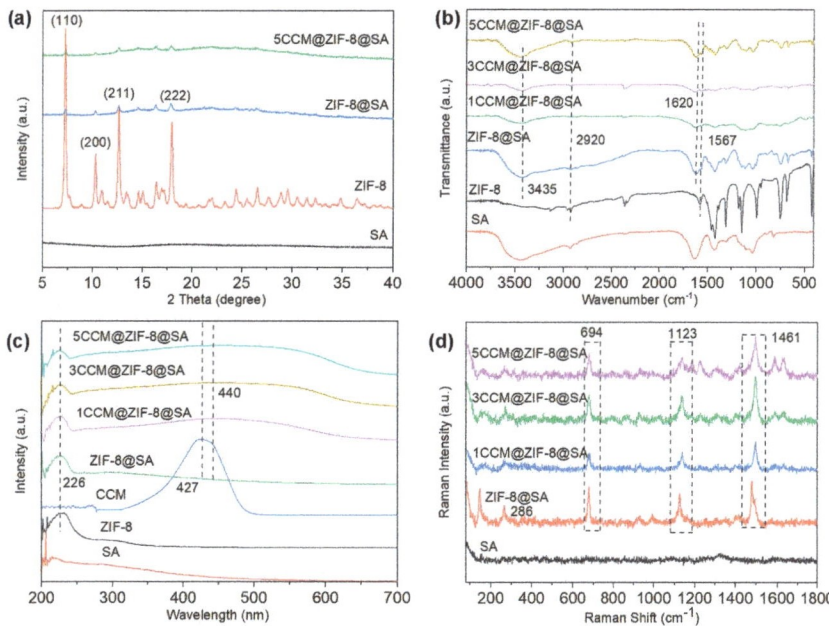

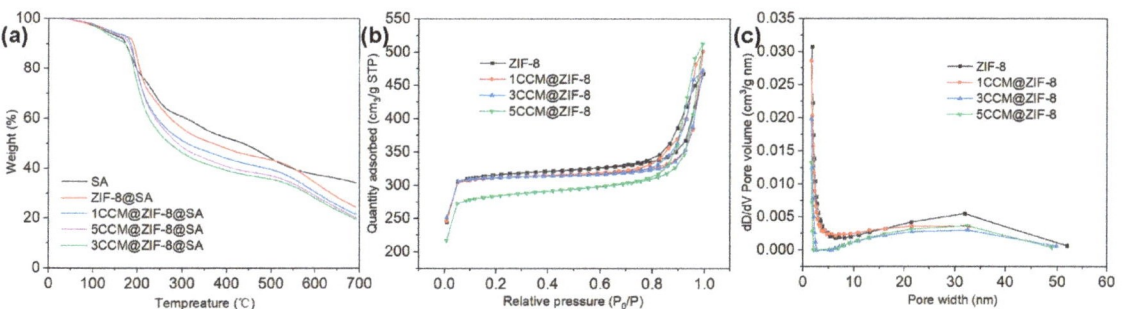

Figure 2. (**a**) XRD spectra of SA, ZIF-8, ZIF-8@SA and 5CCM@ZIF-8@SA. (**b**) FTIR spectra of CCM, ZIF-8, ZIF-8@SA, and 1/3/5CCM@ZIF-8@SA. (**c**) UV-Vis absorption spectra of SA, ZIF-8, ZIF-8@SA, and 1/3/5CCM@ZIF-8@SA. (**d**) Raman spectra of SA, ZIF-8@SA, 1/3/5CCM@ZIF-8@SA.

Figure 3. (**a**) TGA thermograms of SA, ZIF-8@SA,1/3/5CCM@ZIF-8@SA. (**b**) Linear absorption desorption isotherm of ZIF-8,1/3/5CCM@ZIF-8. (**c**) Pore size distribution of ZIF-8,1/3/5CCM@ZIF-8.

Table 1. Surface area, pore volume, and average pore diameter of ZIF-8 and 1/3/5CCM@ZIF-8.

	S_{BET} (m^2/g)	V_P (cm^3/g)	Pore Width (nm)
ZIF-8	968.3	0.461	3.302
1CCM@ZIF-8	951.9	0.446	2.907
3CCM@ZIF-8	949.1	0.388	2.897
5CCM@ZIF-8	876.2	0.315	2.813

The mechanical properties of composite hydrogels were also tested. As shown in Figure 4a, the tensile strength increased by adding CCM and ZIF-8. Compared with the SA hydrogel (7.66 MPa), the ultimate tensile strengths of the ZIF-8@SA and 5CCM@ZIF-8@SA were 10.67 MPa and 11.95 MPa, respectively. This might be attributed to the formation of supplementary chemical cross-linking (i.e., coordination bond, hydrogen bond) between ZIF-8 and SA [42,43]. Bending and twisting tests were conducted on 5CCM@ZIF-8@SA samples to verify the flexibility. As shown in Figure 4b, the 5CCM@ZIF-8@SA sample exhibits significant distortion and maintains the original shape after twisting. When the hydrogel adhered to the finger joints and with bending (Figure 4c), the sample did not break after bending at 120°, 90°, and 60°. These results indicate its good toughness and flexibility, which are promising for use as a wound dressing.

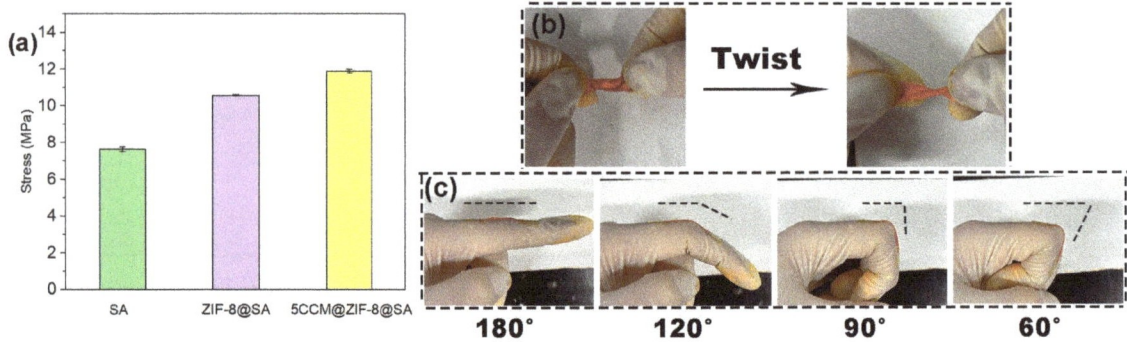

Figure 4. (a) Stress-strain image of SA, ZIF-8@SA, and 5CCM@ZIF-8@SA. (b) Twisting behavior of 5CCM@ZIF-8@SA. (c) Bending behavior of 5CCM@ZIF-8@SA.

2.2. Drug Release Behavior of CCM@ZIF-8@SA-Composite Hydrogel

As previously reported, encapsulated CCM is relatively stable against degradation [44] under neutral and alkaline conditions. As a result, the release of CCM in 1/3/5CCM@ZIF-8@SA in 0.01M PBS and Tween 20 (0.5 v/v%) (PH = 7.4) was investigated. From the quantitative analysis of turmeric quality of 1/3/5CCM@ZIF-8@SA samples, shown in Table 2, the encapsulation efficiency values of 1/3/5CCM@ZIF-8@SA CCM were 50%, 56.2%, and 76.7%, respectively. Based on the wavelength scans of CCM solutions with different concentrations of CCM and CCM standard curves, the relationship between CCM concentration and absorbance was obtained by linear fitting as y = 0.14617x + 0.02948, with a correlation coefficient R^2 = 0.99562 (Figure 5a,b). The cumulative percentage of CCM released from the three samples of 1CCM@ZIF-8@SA, 3CCM@ZIF-8@SA, and 5CCM@ZIF-8@SA could be calculated from this linear relationship, as shown in Figure 5c. The release of CCM increased rapidly in the first 4 h, after which the rising release trend gradually slowed and finally approached a steady state. This is because CCM is encapsulated inside the tiny pores of the ZIF-8 framework. At the initial stage of release, the CCM on the surface and near the pores diffused into the solution (related to the fast swelling capacity of the sample surface) [45], whereas most of the CCM inside remained encapsulated inside the particles [46]. Therefore, the release of CCM inside the granules would be slow, and a long-term release behavior could be achieved. A similar release pattern was observed for three samples. Compared to the relatively rapid release of 1CCM@ZIF-8@SA (about 16.4% in 72 h), CCM release from 3CCM@ZIF-8@SA and 5CCM@ZIF-8@SA samples was controlled and significantly delayed (about 22.1% and 25.9% in 72 h), and most importantly, the duration of the sustained release of CCM for 5CCM@ZIF-8@SA could reach 72 h or above with a slower releasing rate. Due to its eutectic surface, 1CCM@ZIF-8@SA released fast, possibly because it would dissolve rapidly in large amounts upon touching the PBS containing Tween 20 (0.5 v/v%) solution upon contact. In contrast, most CCM molecules entered the interior of ZIF-8 for 5CCM@ZIF-8@SA and 3CCM@ZIF-8@SA in addition to the surficial few molecules. In particular, CCM in the 5CCM@ZIF-8@SA group fully integrated with the pores in ZIF-8 to achieve the optimal coating rate of the drug, delaying the release rate and prolonging the release time. Based on the SEM images, this conclusion is consistent with the observed morphologies and structures.

Table 2. Sample quality and CCM loading volume.

Materials	SA	ZIF-8@SA	1CCM@ZIF-8@SA	3CCM@ZIF-8@SA	5CCM@ZIF-8@SA
Quality	0.0396 (±0.002) g	0.0462 (±0.004) g	0.0465 (±0.002) g	0.0471 (±0.006) g	0.0485 (±0.009) g
CCM loading	0	0	0.0003 g	0.0009 g	0.0013 g
CCM release (72 h)	0	0	0.0492 µg	0.199 µg	0.332 µg

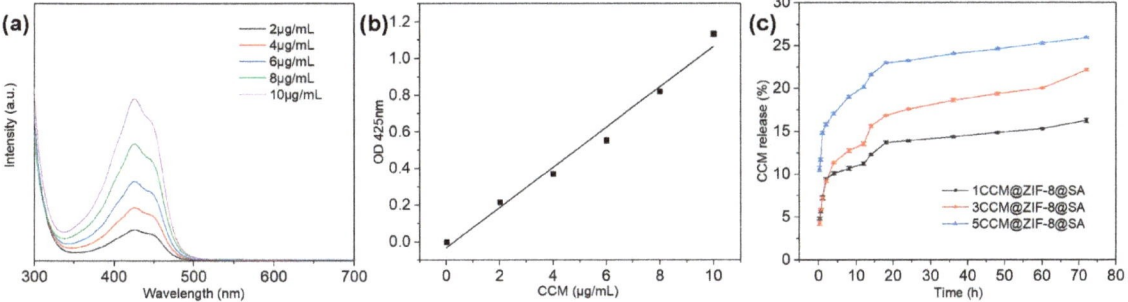

Figure 5. (**a**) Wavelength scans of ethanol solutions of different CCM concentrations, (**b**) Standard curve fitted to the actual concentration of CCM., (**c**) CCM released from CCM@ZIF-8@SA samples.

2.3. Antibacterial Activities of CCM@ZIF-8@SA-Composite Hydrogels

To assess the antibacterial activities of drug-loaded complex hydrogels, we selected Gram-positive *Staphylococcus aureus* (*S. aureus*, BNCC186335) and Gram-negative *Escherichia coli* (*E. coli*, BNCC336902) for antibacterial assays. Antibacterial performance is assessed by colony-forming unit (CFU) assay. After co-incubating the complex hydrogel with bacterial culture medium at 37.5 °C for 24 h, quantitative data (Figure 6a,c) and CFU assay (Figure 6b,d) showed that both the ZIF-8@SA and 1/3/5CCM@ZIF-8@SA samples showed good antimicrobial properties against both *S. aureus* and *E. coli*. Among them, ZIF-8@SA showed an antibacterial efficiency of 28% against *S. aureus* and 58% against *E. coli*. The antibacterial efficiencies against *S. aureus* increased to 30%, 42%, and 65% for 1/3/5CCM@ZIF-8@SA, respectively, 2–4 times higher than pure SA (18%). A similar trend was also found in killing *E. coli*, with values of 70%, 80%, and 86% for three samples. ZIF-8@SA performed better than bare SA, possibly because the release of Zn^{2+} from ZIF-8 could damage the bacteria. This may arise from the disruption of bacterial membrane permeability by Zn^{2+}, which inhibits glycolysis, glucosyltransferase production, and polysaccharide synthesis in bacteria [47,48]. The antibacterial effects of the 1/3/5CCM@ZIF-8@SA hydrogels were thus the synergistic effect of CCM and Zn^{2+}. Based on the coating results of the two types of bacteria and the quantitative analysis of OD values, the OD values followed a decreasing order of SA > ZIF-8@SA > 1CCM@ZIF-8@SA > 3CCM@ZIF-8@SA > 5CCM@ZIF-8@SA both for E. coli and S. aureus, with 5CCM@ZIF-8@SA performing the best. The excellent antibacterial properties of 1/3/5CCM@ZIF-8@SA highlight its practicality as a wound dressing.

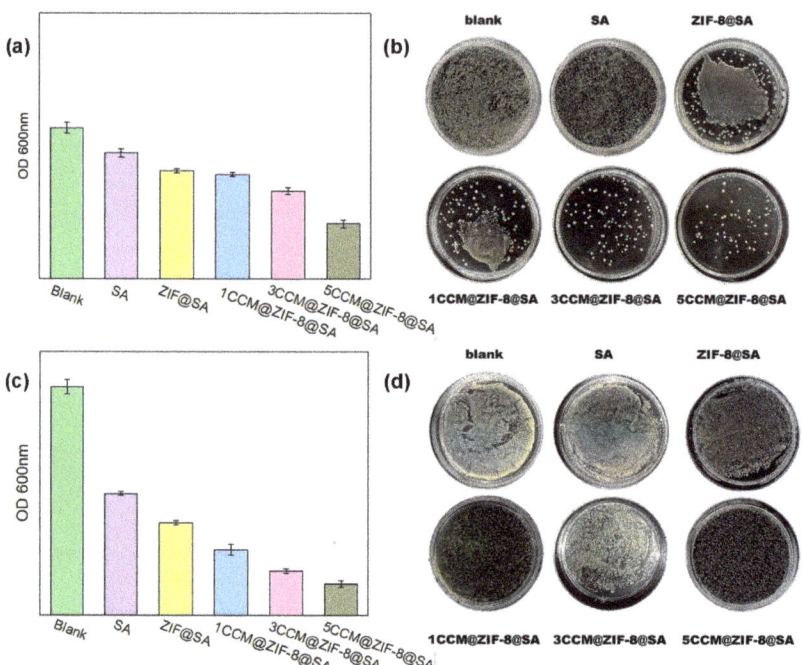

Figure 6. Quantitative measurement of E. coli (**a**), S. aureus (**c**) survival after 24 h of incubation, bacterial inhibition ability of the SA, ZIF-8@SA, and 1/3/5CCM@ZIF-8@SA against the clinically established bacterial pathogens [E. coli (**b**) and S. aureus (**d**)].

The minimum bacteriostatic concentration (MIC) and minimum bactericidal concentration (MBC) were used to determine the access to the bacteriostatic ability of bacteriostatic substances that act on E. coli and S. aureus with SA, ZIF8@SA, and 1/3/5CCM@ZIF-8@SA. As shown in Figure 7a, the MIC values of pure SA against E. coli and S. aureus were 3.3 mg/mL and 5.3 mg/mL, indicating that the basal SA did not have good antibacterial properties. The MIC values of ZIF-8@SA against E. coli and S. aureus were 0.83 mg/mL and 1.67 mg/mL, which indicated that after loading ZIF-8, the hydrogel had a particular antibacterial ability. In the antibacterial experiment of 1/3/5 ZIF-8@CCM@SA samples, with the increase in CCM loading, the MIC value of E. coli gradually decreased to 0.67, 0.33, and 0.25 mg/mL. In addition, the MIC value of S. aureus gradually decreased to 0.83, 0.41, and 0.33 mg/mL. Similarly, the MBC results (Figure 7b) showed the same trend. The MBC values followed a decreasing order of SA > ZIF-8@SA > 1CCM@ZIF-8@SA > 3CCM@ZIF-8@SA > 5CCM@ZIF-8@SA both for E. coli and S. aureus, with 5CCM@ZIF-8@SA performing the best. Therefore, ZIF-8 could improve the stability and utilization efficiency of CCM and, finally, increase the bacteriostatic ability of CCM @ ZIF-8 @ SA.

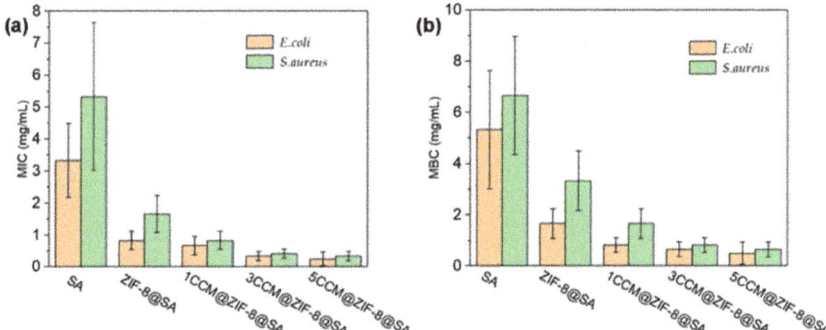

Figure 7. (**a**) MIC values and (**b**) MBC values of SA, ZIF8@SA, and 1/3/5CCM@ZIF-8@SA against *E. coli* and *S. aureus*.

3. Materials and Methods

3.1. Materials

2-mIM, CCM, SA, zinc chloride, and methanol (MeOH, for synthesis) were purchased from Beijing Bailinway Technology Co., Ltd., (Beijing, China), and phosphate buffer (0.01 M) was purchased from Beijing Lanyi Chemical Products Co., Ltd., (Beijing, China). Gram-positive *S. aureus* (BNCC186335) and Gram-negative *E. coli* (BNCC336902) were purchased from Beinong Yuhe Technology Development Co., Ltd., (Beijing, China).

3.2. Synthesis

- (i) Preparation of SA

SA hydrogels were prepared using an ionic cross-linking method. First, sodium alginate (300 mg) was dissolved in water (10 mL) and thoroughly mixed in a surface dish to obtain aqueous sodium of solution alginate. A Petri dish containing 1 g of dissolved sodium alginate solution was weighed on an analytical balance before being transferred to the refrigerator frozen section for 24 h to solidify. $ZnCl_2$ (545 mg) was dissolved in H_2O (20 mL) and magnetically stirred to homogenize the solution. The freeze-fixed aqueous sodium alginate solution was mixed dropwise with 6 mL of the $ZnCl_2$ solution to create the SA hydrogel. The reaction was allowed to run for approximately 2 h. Finally, the residual $ZnCl_2$ solution on the hydrogel surface was washed with deionized water.

- (ii) Preparation of ZIF-8@SA

ZIF-8@SA hydrogels were synthesized by co-precipitation at room temperature. 2-mIM (330 mg) was dissolved in a MeOH solution (10 mL), and the mixture was magnetically agitated (as described above) until it became clear. The solution above was then added to a 35 mm Petri plate with the SA hydrogel created in step (i), and the reaction was allowed to run for 4 h. After the reaction, the hydrogel was oven-dried at 60 °C for 8 h. Finally, a ZIF-8@SA hydrogel with a white solid attached to the surface was obtained.

- (iii) Preparation of 1/3/5CCM@ZIF-8@SA

Similarly, CCM@ZIF-8@SA hydrogels were synthesized using a room-temperature co-precipitation method. 2-mIM (330 mg) and CCM (1.0, 3.0, and 5.0 mg) were dissolved in MeOH solution (10 mL), naming solution A/B/C, and magnetically stirred until the solution became clear. Then, the SA hydrogel prepared in step (i) was soaked in the above solution (in a 35 mm petri dish), and the reaction was allowed to continue for 4 h. The resulting hydrogel was dried in an oven at 60 °C for 8 h. Three types of hydrogels were obtained, all with CCM solids attached to their surfaces, and were named 1CCM@ZIF-8@SA, 3CCM@ZIF-8@SA, and 5CCM@ZIF-8@SA, respectively.

3.3. Characterizations

The analytical balance (OHAUS brand) used for weighing was purchased from Cangzhou Zhenqian Instrument Company (Cangzhou, China). The magnetic stirrer (IKA C-MAG HS 7) used for solution stirring was purchased from Beijing Chenshi Technology Trade Company (Beijing, China). The oven (DHG-9148A) used for sample drying was purchased from Shanghai Jinghong Laboratory Equipment Company (Shanghai, China). The optical images of the sample were recorded by the phone's rear camera. The topography and microstructure were observed by the Regulus-8100 cryo-field emission scanning electron microscope with an acceleration voltage of 15 kV. Prior to the experiments, hydrogel samples were prepared. Due to the poor conductivity of the material, aside from using conductive glue to fix the water gel sample onto the experimental sample stage, pre-sputter gold treatment was also required. The XRD patterns were obtained on a Rigaku TTR3 X-ray diffractometer with a CuKα target (λ = 1.5 Å). The dried hydrogel samples were ground into powder before the experiment, and an appropriate amount of hydrogel powder was placed on a sample holder for scanning. The scanning range of 2θ was 5° to 40°, and the scanning speed was 10°/min. Raman scattering spectra were obtained on a micro-focus Raman spectrometer (inVia-Reflex) with an excitation wavelength of 785 nm. Similarly, the ground hydrogel powder was placed on a glass slide for the experiment. The scanning range of Raman spectra was 0–1800 cm^{-1}, and the scanning step of 2 cm^{-1} was used to detect the crystal structure of the samples. The Cary 7000 UV-Vis diffuse reflectance spectrophotometer was used to test the hydrogel samples' optical response range and intensity changes with a 200–700 nm scanning wavelength. Fourier transform infrared spectroscopy (Excalibur 3100) further detected the samples' functional groups in the range of 500–4000 cm^{-1}. The wet-state composite hydrogel was stretched at a 5 mm/min speed on a universal tensile testing machine (TA.HD PLUS), and the tensile strength was calculated by dividing the maximum stress by the area. In the drug release test, the release concentration of CCM was measured using a UV-Vis spectrophotometer (UV-2600).

3.4. Drug Release Tests

- (i) The standard curve of CCM:

CCM (100 mg) was dissolved in 10 mL PBS and Tween 20 (0.5 v/v%) to prepare as the stock solution. Then, 1 mL was extracted from the above solution and diluted, and finally obtain a gradient of 10, 8, 6, 4, and 2 μg/mL of CCM solution. The UV-Vis absorption of these solutions was measured using a spectrophotometer in the wavelength range of 300–700 nm. The absorbance intensities at 425 nm were recorded for each concentration of the CCM solution.

- (ii) CCM encapsulation efficiency

The total amount of CCM in 1/3/5CCM@ZIF-8@SA was obtained using the following method:
Calculated from the following equation:
Encapsulation efficiency (%) = $\frac{Loading}{Total\ amount\ of\ CCM} \times 100\%$
Loading: the mass of 1/3/5CCM@ZIF-8@SA–the mass of ZIF-8@SA;
Total amount of CCM: the mass of curcumin in 6 mL solution A/B/C.

- (iii) The drug release test of CCM@ZIF-8@SA-composite hydrogels:

The as-prepared CCM@ZIF-8@SA-composite hydrogel was soaked in 15 mL of PBS containing 0.5 v/v% Tween 20 (pH = 7.4). At specific time intervals (20 min, 40 min, 1, 2, 4, 8, 12, 14, 24, 36, 60, and 72 h), 3 mL of the soaking solution was analyzed. The absorbance intensities at 425 nm were measured for each sample. The concentration of CCM in each sample was determined by comparing the absorbance of the samples at 425 nm with the standard curve. Each experiment was performed three times in parallel, and the average results from the three experiments were used for data processing and analysis.

The 72 h release of CCM was calculated according to the following equation:

$$\text{Release (\%)} = \frac{\text{Released amount of CCM}}{\text{Total amount of CCM}} \times 100\%$$

3.5. Antimicrobial Properties

- (i) Sample processing:

The as-prepared hydrogels were placed in a centrifuge tube and irradiated under an ultraviolet lamp for 30 min.

- (ii) Medium configuring:

Briefly, 4 g of Luria–Bertani (LB) broth was added to 200 mL of deionized water and stirred to obtain a liquid medium, and 9.6 g of agar medium was added to 300 mL of deionized water and stirred thoroughly to obtain a solid medium. Then two different media were placed into the LX-B50L autoclave for sterilization (120 °C, 30 min). After the sterilization process was complete, the media were allowed to cool down. At last, the liquid medium was sealed with a sealing film and placed on the ultra-clean stage for future use. The solid medium was poured into 90 mm Petri dishes at a temperature of 70 °C. Approximately 10 mL of the medium was poured into each dish, and once it solidified, it was sealed with a sealing film and placed upside down on the ultra-clean stage for future use.

- (iii) Bacterial resuscitation:

The strains used in this experiment are Gram-positive *S. aureus* (BNCC186335) and Gram-negative *E. coli* (BNCC336902). Before the experiment, 50 mL centrifuge tubes, respiratory membranes, and other experimental consumables were placed on a laminar flow hood and exposed to UV light for 30 min for sterilization. During the experiment, a small number of bacteria was taken with a heated inoculation loop and transferred into a 30 mL liquid medium. The culture was then sealed with a breathable membrane and incubated on a shaker at 37 °C for 18 h.

- (iv) Bacterial proliferation:

Bacterial proliferation was analyzed by measuring the optical density (OD) value and the coating method. After 18 h of bacterial recovery, the bacterial solution was removed from the shaker and diluted thrice with a liquid culture medium. Then, 3 mL of the diluted bacterial solution was taken and placed in a cuvette, and the OD value of the bacteria was measured using a UV-vis spectrophotometer. The bacterial concentration was then diluted to 10^6 CFU·mL^{-1} based on the OD value (each OD value = 10^8 CFU·mL^{-1}). The experimental group added 5 mL of bacterial liquid and a hydrogel sample to a 15 mL centrifuge tube. In contrast, in the control group, only 5 mL of bacterial liquid was added to a 15 mL centrifuge tube. Finally, the centrifuge tubes were sealed with a breathable membrane and placed in a constant temperature (37 °C) incubator for 24 h of cultivation.

- (v) Determination of bacterial bacteriostatic rate:

After co-culturing for 24 h, the OD value at 600 nm was measured using a UV spectrophotometer. Next, the bacterial solution after co-culturing was diluted 10^3, 10^4, 10^5, and 10^6 times. Then 100 µL of different concentrations of the bacterial liquid was taken and coated onto the pre-prepared solid culture medium using spreaders. The culture plates were sealed with sealing film and transferred to a 37 °C incubator for cultivation. After 24 h of cultivation, the culture plates were taken out, and the number of bacterial colonies was observed and recorded using photography. The relative antibacterial rate was calculated according to the following formula:

Relative bacteriostatic rate (%) = $\frac{A-B}{A} \times 100\%$

A—the OD values of the colonies in the blank group;

B—the OD values of the colonies in the experimental group.

4. Conclusions

In summary, a flexible CCM-loaded Zn-MOF hydrogel was facilely prepared by combining the drug with organic ligands through Zn^{2+} crosslinking and Zn^{2+} centered coordination. The results showed that CCM-loaded ZIF-8 crystals could be homogeneously synthesized onto SA hydrogel, with CCM becoming part of the ZIF-8 ligand. This would strengthen the encapsulation of CCM inside ZIF-8 frameworks and simultaneously enhance the crosslinking by the supplementary intermolecular forces. Consequently, flexible and tough composite hydrogels can be obtained to adapt to the mechanical behaviors of human skin. More importantly, this hierarchical structure could achieve long-term drug release and obtain exceptional antimicrobial properties against *E. coli* and *S. aureus*. In conclusion, such composite hydrogel promises a wound-healing system that can guide the future design of practical chronic wound dressings.

Author Contributions: J.L. conducted the experiment and processed the data; Q.F. drew some illustrations; Y.Y. conceived the work; Y.C., J.L. and Y.Y. wrote the paper; M.I.H. review and editing; Y.C. and L.-N.W. visualization; and all the authors contributed to the general discussion. All authors have read and agreed to the published version of the manuscript.

Funding: This work was financially supported by the National Key R&D Program of China (No. 2021YFB3802200); the Scientific and Technological Innovation Foundation of Shunde Graduate School, USTB (BK22BE012); and the International Exchange and Growth Program for Young Teachers, USTB (QNXM20210017).

Institutional Review Board Statement: Not applicable.

Informed Consent Statement: Not applicable.

Data Availability Statement: The data that support the findings of this study are available from the corresponding author upon reasonable request.

Conflicts of Interest: The authors declare that they do not have any commercial or associative interest that represents a conflict of interest in connection with the work submitted.

References

1. Liu, J.Z.; Dong, J.; Zhang, T.; Peng, Q. Graphene-based nanomaterials and their potentials in advanced drug delivery and cancer therapy. *J. Control. Release* **2018**, *286*, 64–73. [CrossRef] [PubMed]
2. Naghdi, T.; Golmohammadi, H.; Vosough, M.; Atashi, M.; Saeedi, I.; Maghsoudi, M.T. Lab-on-nanopaper: An optical sensing bioplatform based on curcumin embedded in bacterial nanocellulose as an albumin assay kit. *Anal. Chim. Acta* **2019**, *1070*, 104–111. [CrossRef] [PubMed]
3. Fan, Y.T.; Yi, J.; Zhang, Y.Z.; Yokoyama, W. Fabrication of curcumin-loaded bovine serum albumin (BSA)-dextran nanoparticles and the cellular antioxidant activity. *Food Chem.* **2018**, *239*, 1210–1218. [CrossRef] [PubMed]
4. Ghosh, S.; Banerjee, S.; Sil, P.C. The beneficial role of curcumin on inflammation; diabetes and neurodegenerative disease: A recent update. *Food Chem. Toxicol.* **2015**, *83*, 111–124. [CrossRef] [PubMed]
5. Zheng, B.J.; McClements, D.J. Formulation of More Efficacious Curcumin Delivery Systems Using Colloid Science: Enhanced Solubility, Stability, and Bioavailability. *Molecules* **2020**, *25*, 2791. [CrossRef]
6. Wang, X.Y.; Fan, Y.L.; Yan, J.J.; Yang, M. Engineering polyphenol-based polymeric nanoparticles for drug delivery and bioimaging. *Chem. Eng. J.* **2022**, *439*, 135661. [CrossRef]
7. Almawash, S. Solid lipid nanoparticles, an effective carrier for classical antifungal drugs. *Saudi Pharm. J.* **2023**, *31*, 1167–1180. [CrossRef]
8. Zhu, H.J.; Li, B.F.; Chan, C.Y.; Ling, B.L.Q.; Tor, J.; Oh, X.Y.; Jiang, W.B.; Ye, E.Y.; Li, Z.B.; Loh, X.J. Advances in Single-component inorganic nanostructures for photoacoustic imaging guided photothermal therapy. *Adv. Drug Deliv. Rev.* **2023**, *192*, 114644. [CrossRef]
9. Pourmadadi, M.; Abbasi, P.; Eshaghi, M.M.; Bakhshi, A.; Manicum, A.L.E.; Rahdar, A.; Pandey, S.; Jadoun, S.; Diez-Pascual, A.M. Curcumin delivery and co-delivery based on nanomaterials as an effective approach for cancer therapy. *J. Drug Deliv. Sci. Technol.* **2022**, *78*, 103982. [CrossRef]
10. Ban, E.; Kim, A. Coacervates: Recent developments as nanostructure delivery platforms for therapeutic biomolecules. *Int. J. Pharm.* **2022**, *624*, 122058. [CrossRef]
11. Hirschbiegel, C.M.; Zhang, X.; Huang, R.; Cicek, Y.A.; Fedeli, S.; Rotello, V.M. Inorganic nanoparticles as scaffolds for bioorthogonal catalysts. *Adv. Drug Deliv. Rev.* **2023**, *195*, 114730. [CrossRef]

12. Jang, E.H.; Kim, G.L.; Park, M.G.; Shim, M.K.; Kim, J.-H. Hypoxia-responsive, organic-inorganic hybrid mesoporous silica nanoparticles for triggered drug release. *J. Drug Deliv. Sci. Technol.* **2020**, *56*, 101543. [CrossRef]
13. Mohan, B.; Kamboj, A.; Virender; Singh, K.; Priyanka; Singh, G.; Pombeiro, A.J.L.; Ren, P. Metal-organic frameworks (MOFs) materials for pesticides, heavy metals, and drugs removal: Environmental safety. *Sep. Purif. Technol.* **2023**, *310*, 123175. [CrossRef]
14. Acharya, A.P.; Sezginel, K.B.; Gideon, H.P.; Greene, A.C.; Lawson, H.D.; Inamdar, S.; Tang, Y.; Fraser, A.J.; Patel, K.V.; Liu, C.; et al. In silico identification and synthesis of a multi-drug loaded MOF for treating tuberculosis. *J. Control. Release* **2022**, *352*, 242–255. [CrossRef]
15. Gupta, A.; Keddie, D.J.; Kannappan, V.; Gibson, H.; Khalil, I.R.; Kowalczuk, M.; Martin, C.; Shuai, X.; Radecka, I. Production and characterisation of bacterial cellulose hydrogels loaded with curcumin encapsulated in cyclodextrins as wound dressings. *Eur. Polym. J.* **2019**, *118*, 437–450. [CrossRef]
16. Norahan, M.H.; Pedroza-Gonz, S.C.; Sanchez-Salazar, M.G.; Alvarez, M.M.; Santiago, G.T.D. Structural and biological engineering of 3D hydrogels for wound healing. *Bioact. Mater.* **2023**, *24*, 197–235. [CrossRef]
17. Liu, K.; Chen, Y.Y.; Zha, X.Q.; Li, Q.M.; Pan, L.H.; Luo, J.P. Research progress on polysaccharide/protein hydrogels: Preparation method, functional property and application as delivery systems for bioactive ingredients. *Food Res. Int.* **2021**, *147*, 110542. [CrossRef]
18. Li, H.B.; Cheng, F.; Wei, X.J.; Yi, X.T.; Tang, S.Z.; Wang, Z.Y.; Zhang, Y.S.; He, J.M.; Huang, Y.D. Injectable, self-healing, antibacterial, and hemostatic N,O-carboxymethyl chitosan/oxidized chondroitin sulfate composite hydrogel for wound dressing. *Mater. Sci. Eng. C* **2021**, *118*, 111324. [CrossRef]
19. Zhang, X.J.; Lin, G.; Kumar, S.R.; Mark, J.E. Hydrogels prepared from polysiloxane chains by end linking them with trifunctional silanes containing hydrophilic groups. *Polymer* **2009**, *50*, 5414–5421. [CrossRef]
20. Badsha, M.A.H.; Khan, M.; Wu, B.L.; Kumar, A.; Lo, I.M.C. Role of surface functional groups of hydrogels in metal adsorption: From experimental to mechanism. *J. Hazard. Mater.* **2021**, *408*, 124463. [CrossRef]
21. El-Bindary, A.A.; Toson, E.A.; Shoueir, K.R.; Aljohani, H.A.; Abo-Ser, M.M. Metal–organic frameworks as efficient materials for drug delivery: Synthesis, characterization, antioxidant, anticancer, antibacterial and molecular docking investigation. *Appl. Organomet. Chem.* **2020**, *34*, e5905. [CrossRef]
22. Soltani, S.; Akhbari, K. Facile and single-step entrapment of chloramphenicol in ZIF-8 and evaluation of its performance in killing infectious bacteria with high loading content and controlled release of the drug. *Crystengcomm* **2022**, *24*, 1934–1941. [CrossRef]
23. Jabbar, A.; Rehman, K.; Jabri, T.; Kanwal, T.; Perveen, S.; Rashid, M.A.; Kazi, M.; Ahmad Khan, S.; Saifullah, S.; Shah, M.R. Improving curcumin bactericidal potential against multi-drug resistant bacteria via its loading in polydopamine coated zinc-based metal-organic frameworks. *Drug Deliv.* **2023**, *30*, 2159587. [CrossRef] [PubMed]
24. Ren, X.; Chang, L.; Hu, Y.; Zhao, X.; Xu, S.; Liang, Z.; Mei, X.; Chen, Z. Au@MOFs used as peroxidase-like catalytic nanozyme for bacterial infected wound healing through bacterial membranes disruption and protein leakage promotion. *Mater. Des.* **2023**, *229*, 111890. [CrossRef]
25. Tian, M.; Zhou, L.; Fan, C.; Wang, L.; Lin, X.; Wen, Y.; Su, L.; Dong, H. Bimetal-organic framework/GOx-based hydrogel dressings with antibacterial and inflammatory modulation for wound healing. *Acta Biomater.* **2023**, *158*, 252–265. [CrossRef]
26. Chen, Y.; Li, D.; Zhong, Y.; Lu, Z.; Wang, D. NIR regulated upconversion nanoparticles@metal-organic framework composite hydrogel dressing with catalase-like performance and enhanced antibacterial efficacy for accelerating wound healing. *Int. J. Biol. Macromol.* **2023**, *235*, 123683. [CrossRef]
27. Fan, Y.; Lüchow, M.; Zhang, Y.; Lin, J.; Fortuin, L.; Mohanty, S.; Brauner, A.; Malkoch, M. Nanogel Encapsulated Hydrogels As Advanced Wound Dressings for the Controlled Delivery of Antibiotics. *Adv. Funct. Mater.* **2020**, *31*, 2006453. [CrossRef]
28. Cai, D.; Chen, S.; Wu, B.; Chen, J.; Tao, D.; Li, Z.; Dong, Q.; Zou, Y.; Chen, Y.; Bi, C.; et al. Construction of multifunctional porcine acellular dermal matrix hydrogel blended with vancomycin for hemorrhage control, antibacterial action, and tissue repair in infected trauma wounds. *Mater. Today Bio* **2021**, *12*, 100127. [CrossRef]
29. Huang, K.; Liu, W.; Wei, W.; Zhao, Y.; Zhuang, P.; Wang, X.; Wang, Y.; Hu, Y.; Dai, H. Photothermal Hydrogel Encapsulating Intelligently Bacteria-Capturing Bio-MOF for Infectious Wound Healing. *ACS Nano* **2022**, *16*, 19491–19508. [CrossRef]
30. Zhang, W.; Wang, B.; Xiang, G.; Jiang, T.; Zhao, X. Photodynamic Alginate Zn-MOF Thermosensitive Hydrogel for Accelerated Healing of Infected Wounds. *ACS Appl. Mater. Interfaces* **2023**, *15*, 22830–22842. [CrossRef]
31. Zhao, X.; Chang, L.; Hu, Y.; Xu, S.; Liang, Z.; Ren, X.; Mei, X.; Chen, Z. Preparation of Photocatalytic and Antibacterial MOF Nanozyme Used for Infected Diabetic Wound Healing. *ACS Appl. Mater. Interfaces* **2022**, *14*, 18194–18208. [CrossRef]
32. Gutiérrez, M.; Ferrer, M.L.; Mateo, C.R.; Monte, F.D. Freeze-drying of aqueous solutions of deep eutectic solvents: A suitable approach to deep eutectic suspensions of self-assembled structures. *Langmuir* **2009**, *25*, 5509–5515. [CrossRef]
33. Su, H.; Sun, F.; Jia, J.; He, H.; Wang, A.; Zhu, G. A highly porous medical metal-organic framework constructed from bioactive curcumin. *Chem. Commun.* **2015**, *51*, 5774–5777. [CrossRef]
34. Wu, C.S.; Xiong, Z.H.; Li, C.; Zhang, J.M. Zeolitic imidazolate metal organic framework ZIF-8 with ultra-high adsorption capacity bound tetracycline in aqueous solution. *RSC Adv.* **2015**, *5*, 82127–82137. [CrossRef]
35. Devarayapalli, K.C.; Vattikuti, S.V.P.; Yoo, K.S.; Nagajyothi, P.C.; Shim, J. Rapid microwave-assisted construction of ZIF-8 derived ZnO and ZnO@Ta$_2$O$_5$ nanocomposite as an efficient electrode for methanol and urea electro-oxidation. *J. Electroanal. Chem.* **2020**, *878*, 114634. [CrossRef]

36. Geng, C.; Liu, X.; Ma, J.; Ban, H.; Bian, H.; Huang, G. High strength, controlled release of curcumin-loaded ZIF-8/chitosan/zein film with excellence gas barrier and antibacterial activity for litchi preservation. *Carbohydr. Polym.* **2023**, *306*, 120612. [CrossRef]
37. Moussawi, R.N.; Patra, D. Modification of nanostructured ZnO surfaces with curcumin: Fluorescence-based sensing for arsenic and improving arsenic removal by ZnO. *RSC Adv.* **2016**, *6*, 17256–17268. [CrossRef]
38. Kurkcuoglu, G.S.; Kavlak, I.; Kinik, B.; Sahin, O. Experimental and theoretical studies on the molecular structures and vibrational spectra of cyanide complexes with 1,2-dimethylimidazole: M(dmi)(2)Ni(mu-CN)(4) (M = Cu, Zn or Cd). *J. Mol. Struct.* **2020**, *1199*, 126892. [CrossRef]
39. Kumari, G.; Jayaramulu, K.; Maji, T.K.; Narayana, C. Temperature Induced Structural Transformations and Gas Adsorption in the Zeolitic Imidazolate Framework ZIF-8: A Raman Study. *J. Phys. Chem. A* **2013**, *117*, 11006–11012. [CrossRef]
40. Radhakrishnan, D.; Narayana, C. Guest dependent Brillouin and Raman scattering studies of zeolitic imidazolate framework-8 (ZIF-8) under external pressure. *J. Chem. Phys.* **2016**, *144*, 134704. [CrossRef]
41. Zheng, M.; Liu, S.; Guan, X.G.; Xie, Z.G. One-Step Synthesis of Nanoscale Zeolitic Imidazolate Frameworks with High Curcumin Loading for Treatment of Cervical Cancer. *ACS Appl. Mater. Interfaces* **2015**, *7*, 22181–22187. [CrossRef] [PubMed]
42. Wang, H.; Lu, Z.; Wang, F.; Li, Y.; Ou, Z.; Jiang, J. A novel strategy to reinforce double network hydrogels with enhanced mechanical strength and swelling ratio by nano cement hydrates. *Polymer* **2023**, *269*, 125725. [CrossRef]
43. Xu, P.; Shang, Z.; Yao, M.; Li, X. Mechanistic insight into improving strength and stability of hydrogels via nano-silica. *J. Mol. Liq.* **2022**, *357*, 119094. [CrossRef]
44. Tiwari, A.; Singh, A.; Garg, N.; Randhawa, J.K. Curcumin encapsulated zeolitic imidazolate frameworks as stimuli responsive drug delivery system and their interaction with biomimetic environment. *Sci. Rep.* **2017**, *7*, 12598. [CrossRef]
45. Nikpour, S.; Ansari-Asl, Z.; Sedaghat, T.; Hoveizi, E. Curcumin-loaded Fe-MOF/PDMS porous scaffold: Fabrication, characterization, and biocompatibility assessment. *J. Ind. Eng. Chem.* **2022**, *110*, 188–197. [CrossRef]
46. Kang, L.; Liang, Q.; Abdul, Q.; Rashid, A.; Ren, X.; Ma, H. Preparation technology and preservation mechanism of gamma-CD-MOFs biaological packaging film loaded with curcumin. *Food Chem.* **2023**, *420*, 136142. [CrossRef]
47. Phan, T.N.; Buckner, T.; Sheng, J.; Baldeck, J.D.; Marquis, R.E. Physiologic actions of zinc related to inhibition of acid and alkali production by oral streptococci in suspensions and biofilms. *Oral. Microbiol. Immunol.* **2010**, *19*, 31–38. [CrossRef]
48. Liu, Z.; Tan, L.; Liu, X.; Liang, Y.; Zheng, Y.; Yeung, K.W.K.; Cui, Z.; Zhu, S.; Li, Z.; Wu, S. Zn^{2+}-assisted photothermal therapy for rapid bacteria-killing using biodegradable humic acid encapsulated MOFs. *Colloids Surf. B Biointerfaces* **2020**, *188*, 110781. [CrossRef]

Disclaimer/Publisher's Note: The statements, opinions and data contained in all publications are solely those of the individual author(s) and contributor(s) and not of MDPI and/or the editor(s). MDPI and/or the editor(s) disclaim responsibility for any injury to people or property resulting from any ideas, methods, instructions or products referred to in the content.

Article

3D Hierarchical Porous and N-Doped Carbonized Microspheres Derived from Chitin for Remarkable Adsorption of Congo Red in Aqueous Solution

Taimei Cai [1], Huijie Chen [2], Lihua Yao [1] and Hailong Peng [2,*]

1. School of Life Science, Jiangxi Science and Technology Normal University, Nanchang 330013, China
2. School of Chemistry and Chemical Engineering, Nanchang University, Nanchang 330031, China
* Correspondence: penghailong@ncu.edu.cn

Abstract: A novel adsorbent of N-doped carbonized microspheres were developed from chitin (N-doped CM-chitin) for adsorption of Congo red (CR). The N-doped CM-chitin showed spherical shape and consisted of carbon nanofibers with 3D hierarchical architecture. There were many micro/nano-pores existing in N-doped CM-chitin with high surface area (455.703 m^2 g^{-1}). The N element was uniformly distributed on the carbon nanofibers and formed with oxidize-N graphitic-N, pyrrolic-N, and pyridinic-N. The N-doped CM-chitin showed excellent adsorption capability for CR and the maximum adsorption amount was approximate 954.47 mg g^{-1}. The π-π/n-π interaction, hydrogen-bond interactions, and pore filling adsorption might be the adsorption mechanisms. The adsorption of N-doped CM-chitin was considered as a spontaneous endothermic adsorption process, and which well conformed to the pseudo-second-order kinetic and Langmuir isotherm model. The N-doped CM-chitin exhibited an effective adsorption performance for dynamic CR water with good reusability. Therefore, this work provides new insights into the fabrication of a novel N-doped adsorbent from low-cost and waste biomasses.

Keywords: chitin; N-doped carbonized microspheres; congo red; remarkable adsorption

1. Introduction

Water pollution has become a severe issue in environment due to the rapid expansion in industrialization and urbanization [1]. Among water pollutants, the highly stable toxic dyes in aquatic ecosystems are one of the major pollutants present in the wastewater, which mostly resulted from the leather, textile, food and paint industries [2]. As we all know, dyes contain harmful aromatic structure that make them highly lethal and even carcinogenic to body. Furthermore, dyes are easily broken into benzidine in the aqueous environment and then exhibited high level of toxicity for human body [3,4]. Congo red (CR, Figure S1A,B) is a typical azo anion reactive dye with good solubility in aqueous solution. CR is mainly applied to textiles, printing and dyeing, paper, rubber and plastics industries [5,6]. Nearly 70% of the 10,000 dyes used in textile manufacturing alone are azo dyes [7]. CR displayed severe toxicities of eye stimulator, nausea, vomiting, diarrhea, bold clotting, and drowsiness [6,8]. CR also adversely affect the photosynthetic processes of aquatic creatures because their colored nature reduces the light penetration [9,10]. Therefore, highly efficient removal of CR methods from aquatic ecosystem are necessary.

Generally, CR removal approaches can be divided into chemical, physical, and biological techniques [11,12]. The physical dye sequestration method has been considered as the most efficient method due to the stability of dyes to light, non-biodegradable, and resistant to aerobic digestion treatment [9,13]. Among these physical approaches, adsorption can be regarded as the best technique for CR pollutant removal because of the advantages of low cost, simple operation process, high efficiency, easy recovery, and repeatability [14,15]. Recently, great range of materials can be used as adsorbents for CR removal, such as metal oxides [16,17], Metal–organic frameworks (MOFs) [18], biochars [19], graphenes [20], activated carbons [21], and Zeolitic imidazolate frameworks (ZIFs) [22]. However, these adsorbents generally involved some drawbacks of high-cost and organic solvents during the complicated manufacturing processes, which could lead to unacceptably second pollution for environment and then seriously limit their applications [23,24]. Therefore, fabrication of environmentally friendly, economical, and superior performance adsorbent is necessary. Up to now, lots of renewable biomass resources have been used to development of high-performance adsorbent for pollutant adsorption and removal due to the advantages of low cost, easy of obtainment, nontoxicity to environment, and sustainable development [25]. Thus, functional carbon-based nanomaterials derived from renewable biomass of cellulose, starch, chitosan and chitin are become more and more popular [26–29].

At present, chitin is mainly produced from discarded shrimp and crab shells during processing of aquatic products. Shrimp shell as shrimp or edible waste is not fully utilized, but the content of chitin in shrimp shell accounts for 15–20%, and the content of chitin in dried shrimp shell is as high as 20–30% [30]. As a natural macromolecular polysaccharide, chitin is received increasing attention because of its properties of biodegradability, biocompatibility, and nontoxicity [31], which widely exists in arthropods (crustaceans, insects, and arachnids), single celled organisms, invertebrates and fish [32,33]. Additionally, chitin contain numerous amino and hydroxyl groups (Figure S1C), and which can be direct converted into nitrogen-doped carbon materials (Figure S1D) via direct carbonization method [34]. Chitin dissolve in NaOH/urea aqueous system at low temperature shows worm-like chains, which can quickly self-aggregate in parallel via hydrogen bonding and hydrophobic interactions to form nanofibers at elevated temperature [32]. The corresponding nanofibrous microspheres were constructed via sol-gel transition method from the chitin nanofibers solution. After carbonization, N-doped CM-chitin formed with hierarchical porosity, a stable 3D interconnected structure, and a high specific surface area [34,35]. Meanwhile, N-doped CM-chitin has the structure of disorder carbon phase with pyridine-N-oxide, graphitic-N, pyrrolic-N, and pyridinic-N [35]. Owing to these advantages, N-doped CM-chitin has been used as adsorbent to removal contaminants from water, such as neonicotinoid residues [35], volatile organic compounds [36], and methyl orange [37]. Up till now, N-doped CM-chitin was just used to statically removal contaminants in stationary vessel. Therefore, N-doped CM-chitin act as the adsorbent for dynamical separation and removal of contaminants might be more meaningful for practical application. Meanwhile, to the best of our knowledge, N-doped CM-chitin was used to adsorption and removal of CR is rarely reported.

Inspired by the previous reports and issues, in this study, an adsorbent of N-doped CM-chitin was fabricated from chitin, which exhibited properties of hierarchical porosity, a stable 3D interconnected structure, and a high specific surface area. The adsorption performance toward CR was explored for the first time, and the kinetic, thermodynamic, and equilibrium were all evaluated for the adsorption mechanism. Meanwhile, the N-doped CM-chitin was also used as adsorbent for CR removal in dynamic water. Therefore, this study may provide new insights into recycling biowaste resources to produce novel adsorbent for contaminants removal from real water samples.

2. Results and Discussion
2.1. Fabrication of N-Doped CM-Chitin

Chitin, which obtained from sea food wastes, can be easily dissolved in NaOH/urea aqueous solution via a freezing-thawing method [32]. The chitin chains were formed in NaOH/urea solution and can self-aggregate in parallel to form nanofibers via hydrogen bonding and hydrophobic interactions. The corresponding fibrous microspheres derived from chitin (FM-chitin) were developed using a "bottom-up" fabrication method in isooctane phase with surfactants Tween-85 and Span 85 at 0 °C with stirring, and the obtained FM-chitin were induced by heating at 60 °C. The FM-chitin can be directly carbonized to formation of N-doped carbonized microspheres (N-doped CM-chitin), which showed advantages of high surface aera, hierarchical porous structure, different doping N elements, and environmentally friendly processes. Because of these advantages, the N-doped CM-chitin may be a potential adsorbent for pollutants removal in the wastewater treatment field.

2.2. Characterizations

The morphology of FM-chitin and N-doped CM-chitin was shown in Figure 1, which shows that FM-chitin exhibited a spherical shape with 3D interconnected porous structure and nano/micro-pores on the surface (Figure 1A–C), which may be due to the phase separation induced by the occupying H_2O during the sol-gel process and the solvent-rich regions contributed to the pore formation [38]. Obviously, there was no essential change in the morphology for N-doped CM-chitin after carbonization (Figure 1D,E), the chitin nanofibers converted into carbon nanofibers with high density after carbonization (Figure 1F), which may be attributed to the shrinkage of nanofibers during the calcining process. Fortunately, the 3D hierarchical porous structure was almost preserved, which was good for adsorption performance. However, the relatively denser surface was observed for N-doped CM-chitin after CR adsorption. The morphology of CR was also investigated (Figure S2), it can be observed that CR showed a good crystal plate morphology. After adsorption (Figure 1G–I), CR crystal cannot be found on the surface of N-doped CM-chitin, suggesting that the native crystalline CR had changed into amorphous CR. It was confirmed that CR has adsorbed onto the fibrous carbon surface of N-doped CM-chitin via the pores and channels. The elements of N-doped CM-chitin were also determined using EDX mapping, showing that the C, O and N elements were uniformly distributed over the carbon nanofibers of N-doped CM-chitin (Figure 1J).

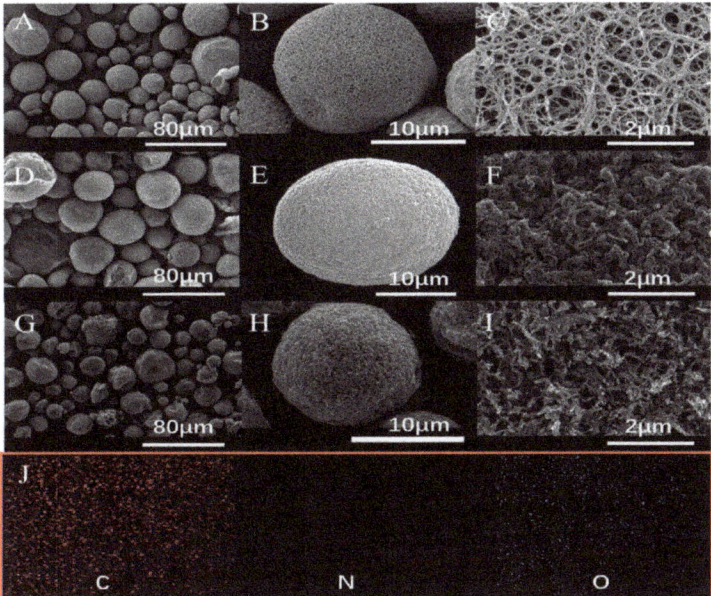

Figure 1. SEM images of FM-chitin (**A–C**), N-doped CM-chitin before (**D–F**) and after (**G–I**) adsorption of CR, and elemental mappings (**J**) of C, N, and O of N-doped CM-chitin.

Figure S3A shows the XPS survey of N-doped CM-chitin before and after adsorption of CR, which were carried out to further investigate the elemental contents and bonding states. Three peaks attributed to C1s, N1s and O1s were observed in the full survey spectra of N-doped CM-chitin. A new binding energy peak at 166.6 eV appeared after adsorption of CR, which was owing to the S2p element from CR molecule (Figure S3B). The high-resolution XPS spectra of N-doped CM-chitin before and after adsorption of CR was also measured. The C1s peaks could be deconvolved into the forms of C = O, C-O/C-N, and C-C/C-H at 288.2, 286.3, and 284.8 eV (Figure S3C), respectively. The C-C/C-H became the primary bonding state to form hydrophobic carbon skeleton structure of N-doped CM-chitin. However, C1s exhibited a new peak at the binding energy of 283.50 eV after adsorption of CR (Figure S3D), representing the C = C groups from the phenyl of CR molecule. The O1s spectrum of N-doped CM-chitin before adsorption can be resolved into three peaks at 535.8, 533.3 and 531.7 eV, which was attributed to N-O, O-H, C = O/C-O-C, respectively. After adsorption of CR, a new peak at 531.1 eV of S-O/S = O can be found (Figure S3F). The S-O/S = O bond interaction between surface oxygen functional groups of N-doped CM-chitin and (Φ-N_{Azo} str) groups of CR was beneficial to improve the adsorption capability. The N1 s spectrum can be deconvoluted into the forms of oxidize-N (N^+O), graphitic-N (N-Q), pyrrolic-N (N-5), and pyridinic-N (N-6), the corresponding peaks was at 403.6, 401.0, 399.1 and 398.2 eV (Figure S3G), respectively. After adsorption, three new peaks at 406.7, 402.9, and 399.9 eV of N = N, C-N, and N-H were appeared (Figure S3H), which may be due to the interaction π–π/n-π and H bonding interactions between N-doped CM-chitin and CR. In conclusion, CR molecules were also adsorbed on the carbon nanofibers surface of N-doped CM-chitin.

Figure 2A shows the FT-IR spectra of N-doped CM-chitin before and after adsorption of CR. There are some new absorption peaks at 588, 713, 754, and 1172 cm^{-1} also appeared for the N-doped CM-chitin after adsorption of CR. The peak at 588 cm^{-1} was attributed to the shift of out-of-plane flexural stretch vibration peaks of C-H of arene. The peaks of 713 and 754 cm^{-1} were the shift of pure benzylamine bands, and the 1172 cm^{-1} peak was the shift of S-O single bond or S = O double bond, which may be due to the characteristic peaks of sulfo group in CR molecule.

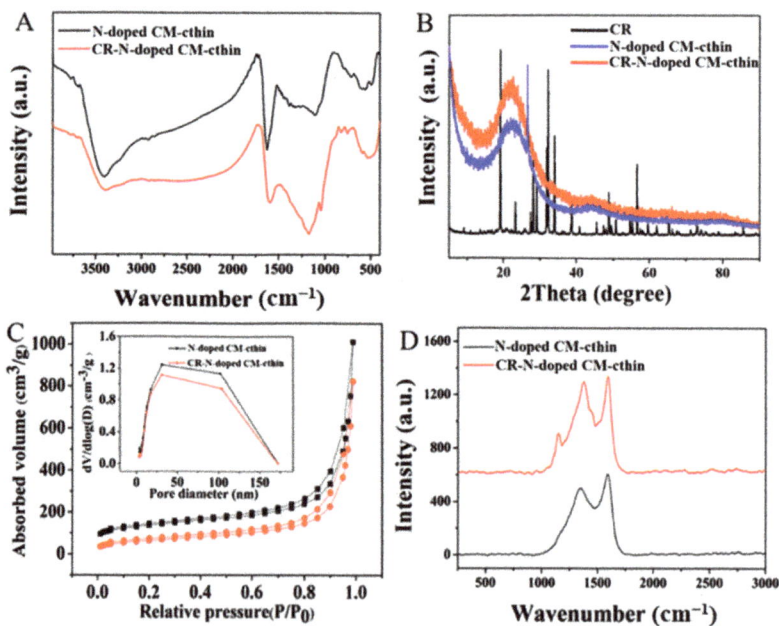

Figure 2. FT-IR spectra (**A**), XRD curves (**B**), Nitrogen adsorption-desorption isotherms (**C**), and Raman spectra (**D**) of N-doped CM-chitin before and after adsorption of CR.

The XRD results of the N-doped CM-chitin and CR were shown in Figure 2B. The N-doped CM-chitin exhibited the similar XRD peaks before and after adsorption of CR. The peaks of 24.30° and 43.21° was corresponded to the (002) and (101) reflection, respectively, which indicated that N-doped CM-chitin have amorphous and graphite crystalline structures [39]. The pure CR exhibited a series of sharp peaks from 10 to 60°, which suggested that CR exhibited the crystalline nature. However, these sharp peaks of CR almost disappeared after adsorption into the N-doped CM-chitin, which indicated that the native crystalline CR had changed into amorphous CR. The XRD results are consistent with SEM, and CR crystals do not appear on the surface of N-doped CM-chitin (Figure S2A,B).

The surface areas of N-doped CM-chitin before and after adsorption were determined using N_2 adsorption and desorption method (Figure 2C). As can be seen, the N_2 adsorption and desorption isotherms of N-doped CM-chitin exhibited type II: the N_2 uptakes increase with the increase in relative pressures, especially at high relative pressures, and there is minor hysteresis between the adsorption and desorption isotherms. The specific surface, micropore volume, and total pore volume of N-doped CM-chitin was 455.703 $m^2\ g^{-1}$, 0.098 $cm^3\ g^{-1}$, and 1.565 $cm^3\ g^{-1}$, respectively. However, these porosity parameters of N-doped CM-chitin after adsorption decreased and the value was only 230.919 $m^2\ g^{-1}$, 0.014 $cm^3\ g^{-1}$, and 1.274 $cm^3\ g^{-1}$, respectively. These parameters decreased due to the adsorption of CR molecule on the nanofibrous carbon surface, which is in good agreement with the results of XPS. Meanwhile, the pore size distribution results were determined and shown in Figure 2C (inset), suggesting that N-doped CM-chitin have a meso/macropore dominant hierarchical structure with an extended continuous pore size distribution. This interconnected hierarchical porous nanofibrous architecture would provide more favorable adsorption for CR.

Raman spectra of N-doped CM-chitin before and after adsorption of CR displays in Figure 2D. The peaks at 1357 cm^{-1} and 1591 cm^{-1} were observed for N-doped CM-chitin before adsorption, as well as 1375 cm^{-1} and 1591 cm^{-1} after adsorption, which represent the D and G band of carbon structure, respectively [35]. The D band was attributed to the presence of sp3 carbon atoms, and the G band was related to the in-plane vibration of sp2 carbon atoms [40,41]. The intensity ratio of G band to D band (I_G:I_D) presented the graphitization degree of carbon. The I_G:I_D for N-doped CM-chitin was 0.843 and 0.910 before and after adsorption of CR, respectively, indicating that the crystallinity decrease, and the ratio of disorder structures become larger. Meanwhile, a new peak at 1153 cm^{-1} was appeared, which might be due to the azobenzene stretching of CR [42].

2.3. Optimization of Adsorption Conditions

The effect of N-doped CM-chitin amounts (10, 20, 30, 40, and 50 mg) on the adsorption capacity and removal efficiency for CR (200 mg L^{-1}, 50 mL) was evaluated (Figure 3A). It can be observed that the removal efficiency increased from 92.64% to 99.79% with increasing the consumption of adsorbent in the same volume of solution. The removal efficiency has less change with adsorbent increasing from 20 mg to 50 mg, and the results also showed that the equilibrium absorption capacity (926.39 mg g^{-1}) occurred in adsorbent dose of 10 mg (Figure 3A). The equilibrium absorption capacity decreased from 926.39 mg g^{-1} to 199.57 mg g^{-1} with amount of N-doped CM-chitin increasing from 10 mg to 50 mg, which may be the adsorbents aggregation and overlapping of adsorption sites during the adsorption process [43]. Therefore, 10 mg of the adsorbent dose was used in the following experiments.

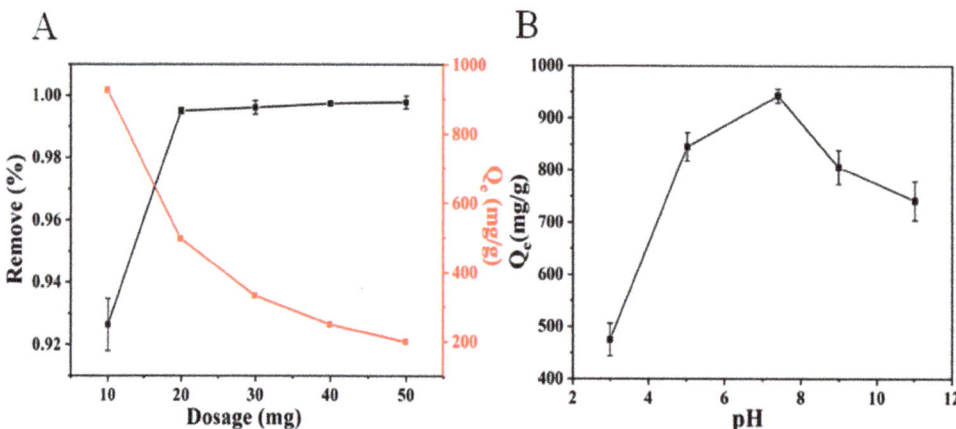

Figure 3. Effect of N-doped CM-chitin dosages on the adsorption capability toward CR (200 mg L^{-1}) (**A**), and effect of pH on the N-doped CM-chitin adsorption capability toward CR (200 mg L^{-1}) (**B**).

The pH value of solution is one of the vital factors in the adsorption capacity due to affecting the surface charge properties of CR and N-doped CM-chitin, and then the electrostatic interaction can be affected under different pH solutions. The CR solution was initially configured with pH of 7.4 and which was set up using 0.1 mol L^{-1} HCl and 0.1 mol L^{-1} NaOH solution. The adsorption capability of N-doped CM-chitin for CR were investigated in different pH solutions with ranging from 3.0 to 11.0. As shown in Figure 3B, the adsorption capacity of CR increased with the pH increasing from 3.0 to 7.4, and the adsorption capability was 474.85 and 942.24 mg g^{-1} at pH value of 3.0 and 7.4, respectively. However, the adsorption capability decreased from 942.24 to 741.31 mg g^{-1} with pH value increasing from 7.4 to 11.0. CR molecule existing in dissociated form as anionic dye ions with negatively charged (dye-SO$_3^-$). Under acidic solution, N-doped CM-chitin protonated to form positively charge sites [44], which resulting in the electrostatic interaction between CR and N-doped CM-chitin. However, the adsorption capability of N-doped CM-chitin not only depended on electrostatic interaction, but also other interactions of hydrogen bonding and π-π interactions. In this study, the N-doped CM-chitin exhibited highest adsorption capability at pH 7.4. Therefore, the pH 7.4 considered as the optimum pH value during the N-doped CM-chitin adsorption processes.

The effect of the initial CR concentration (100, 150, 200, 300, and 400 mg L^{-1}) on the adsorption capability was also evaluated at pH 7.4 under different temperatures (290 K, 300 K, and 310 K). As shown in Figure 4A, the adsorption capability of CR on N-doped CM-chitin increased from 494.56 to 1003.94 mg g^{-1} (290 K), 490.80 to 1230.74 mg g^1 (300 K), and 498.75 to 1076.89 mg g^{-1} (310 K) with the concentration of CR increasing from 100 to 400 mg L^{-1}. This might be due to two reasons: the increased in initial concentration caused the increase in driving force of the concentration gradient between the bulk solution and N-doped CM-chitin, which could enhance the diffusion of CR into N-doped CM-chitin. Additionally, more collision times occurred between CR and N-doped CM-chitin when increasing CR concentration [45]. In the process of increasing CR concentration, the adsorption capacity of the process increased, the change of adsorption capacity is small and then 200 mg L^{-1} was adopted in subsequent experiments.

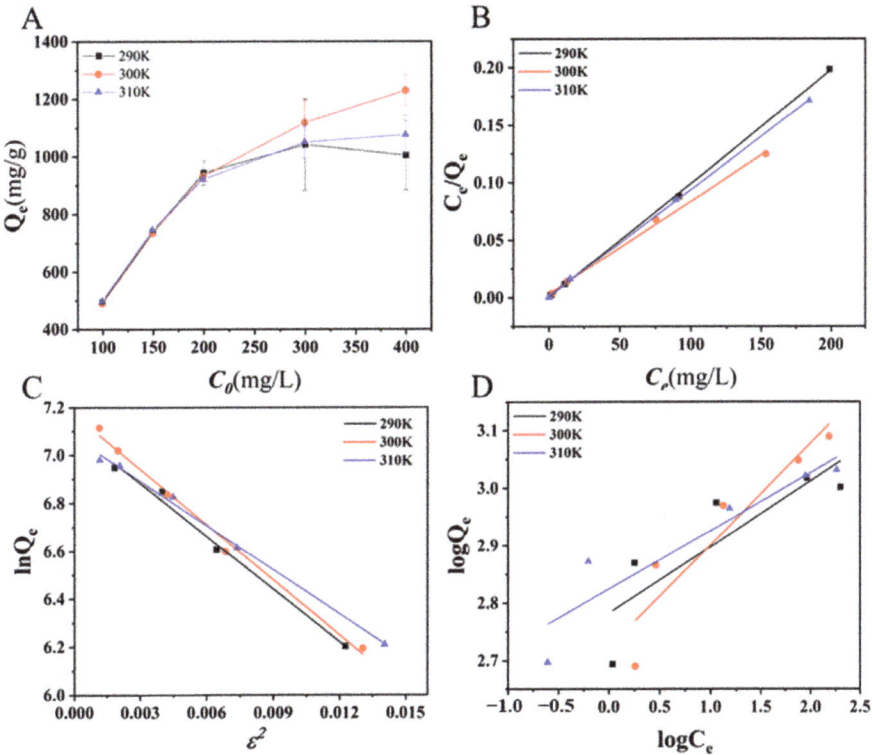

Figure 4. Effect of temperature on the N-doped CM-chitin adsorption capacity (**A**) and adsorption isotherms of CR on N-doped CM-chitin through Langmuir (**B**), Dubinin–Radushkevich (**C**), and Freundlich models (**D**).

2.4. Adsorption Isotherms, Kinetics, and Thermodynamics

Figure 4A shows the isothermal adsorption results of CR by N-doped CM-chitin at different temperatures of 290, 300, and 310 K in pH 7.4 solution. To further investigate the adsorption process, three models of Langmuir (Equations (S1) and (S2)), Dubinin-Radushkevieh (D-R) (Equations (S3)–(S5)), and Feundlich (Equation (S6)) were used to fit the isothermal data. The fitting results were described in Figure 4B–D and listed in Table S1. The Langmuir model was better than D-R and Freundlich isotherm models to describe the adsorption of CR into N-doped CM-chitin due to the highest R^2 values at all different temperatures. Meanwhile, the R_L value was all lower than 1 at different temperatures. The Freundlich constant n were larger than 1 and the E was lower than 8 kJ mol^{-1}. Obviously, N-doped CM-chitin exhibited a monolayer and favorable physical adsorption process.

Figure 5A shows the kinetic curve of CR adsorption by N-doped CM-chitin at different time. The adsorption capability increased quickly within 90 min, which was due to the hierarchical porous, high surface area, and N-doping structure of N-doped CM-chitin. The adsorption capability attenuated gradually from 90 min to 180 min, and which has no obvious increased after 180 min. After 90 min, the diffusion resistance of CR increased with an increasing amount of CR on the microspheres surface, resulting in a lower adsorption rate. Additionally, the remained CR concentration gradually decreased and the adsorption driving force also gradually decreased, which further decreased the adsorption rate of N-doped CM-chitin for CR after 90 min. Therefore, the kinetic curve of CR adsorption by N-doped CM-chitin was fitted using the pseudo-first order kinetic model (Equations (S7) and (S8) and Figure 5B), pseudo-second order kinetic model

(Equations (S9) and (S10) and Figure 5C), and diffusion model (Equation (S11) and Figure 5D) and corresponding parameter results were listed in Table S2. Obviously, $Q_{e,cal}$ (1000 mg g^{-1}) obtained from the pseudo-second order kinetic model has good agreement with the experimental result 954.47 mg g^{-1} with the highest R^2 (0.9996), which showed that the adsorption is subject to chemisorption such as hydrogen bond and π-π interactions. The adsorption processes can be expressed by two linear models using the intra-particle diffusion model, showing that the adsorption rate was not only mainly depended on intra-particle diffusion process, but also affected by surface layer adsorption or boundary layer diffusion. As shown in Figure 5D, CR diffuses rapidly on the fibrous carbon surface via the pore channels at the first stage, and instantaneous adsorption occurs on the outer surface of the fibrous carbon of N-doped CM-chitin. The second stage almost horizontal suggesting a dramatic decrease in adsorption rate.

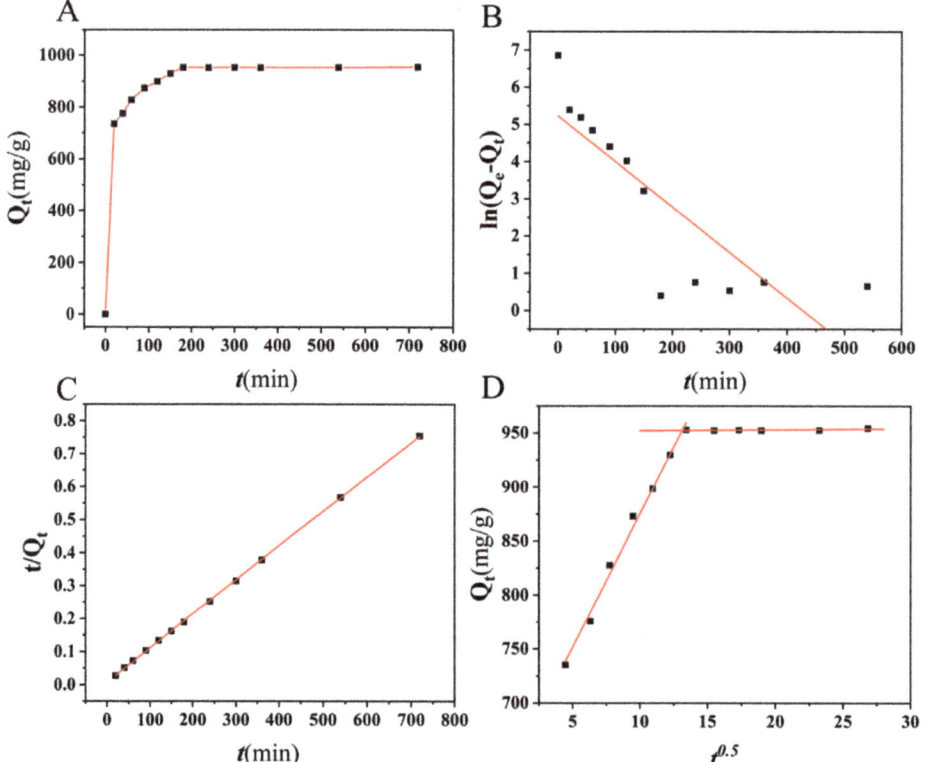

Figure 5. Adsorption kinetics of N-doped CM-chitin toward CR (**A**), Pseudo-first-order kinetic model (**B**), Pseudo-second-order kinetic model (**C**), and Weber-Morris kinetic model (**D**).

For investigation the effect of temperature on the adsorption of CR on N-doped CM-chitin, the adsorption experiments were carried out at 290, 300, and 310 K. The energy of activations (ΔH) and enthalpy change (ΔS) was calculated from the Van 't Hoff equation (Equation (S12) and Figure S4), the Gibbs free energy change (ΔG) was measured from Equation (S13), and the thermodynamic results were listed in Table S3. From the results, ΔH > 0 and negative ΔG, indicating that the adsorption process is endothermic nature and spontaneous. ΔS > 0 showed an entropy increase irregularities at the solid-liquid interface on the carbon nanofibers surface of N-doped CM-chitin.

2.5. Effect of Salt Ions

Previous reports suggest that oxyanions, monovalent and divalent cations present in the waters of an ecosystem could significantly influence the sorption of organic compounds [46]. Figure 6 shows the effect of added cations (Na^+, K^+, Mg^{2+} and Ca^{2+}) and anions (Cl^-, NO_3^-, SO_4^{2-} and CO_3^{2-}) on the adsorption of CR onto N-doped CM-chitin. In the experiment, the ion concentration increased from 10 to 50 mmol L. As shown Figure 6, it was found that cation has a greater promoting effect on N-doped CM-chitin adsorption of CR than anion. For cations, $Mg^{2+} \approx Ca^{2+} > K^+ > Na^+$ promotes adsorption. The order of the radius of cations is $Na^+ > K^+ > Ca^{2+} > Mg^{2+}$, and the larger the radius of cations, the less the polymerization capacity of dyes, and which caused the less promotion of adsorption. For anions, the inhibition of adsorption is $CO_3^{2-} > SO_4^{2-} > Cl^- > NO_3^-$. With the same valence state, the smaller the hydration radius of ions, the stronger the affinity and the stronger the inhibition of adsorption. Therefore, adsorbent affinity for Cl^- is greater than NO_3^-.

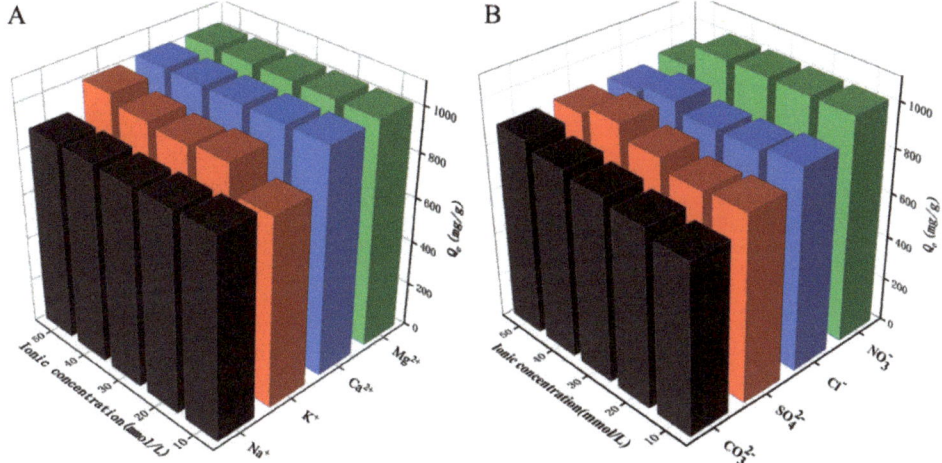

Figure 6. Effect of cations on the adsorption of CR (200 mg L^{-1}) onto N-doped CM-chitin (**A**), and effect of coexisting anions on the adsorption of CR (200 mg L^{-1}) onto N-doped CM-chitin (**B**).

2.6. Possible Adsorption Mechanism

As shown Figure 7A, the possible adsorption mechanism of N-doped CM-chitin was proposed according to the abovementioned characterization results as followings: (1) the π-π and n-π interaction can be formed between the π electrons in N-doped CM-chitin and the benzene/naphthalene ring structure of CR; (2) the hydrogen-bond interactions also can be formed between these carbonyl, hydroxyl and hydroxylamine groups of N-doped CM-chitin and CR; (3) the electrostatic interaction could be happened between the negative charges of CR and positive charges of N-doped CM-chitin; (4) the pore filling adsorption may be also play vital role during the adsorption process [35]. The adsorption capability of N-doped CM-chitin were compared with the other sorbent used in CR adsorption as listed in Table S4, which showed that the adsorption capability of N-doped CM-chitin was better than those of other sorbents in Table S4. The excellent adsorption capability of N-doped CM-chitin was attributed to the multiple adsorption mechanisms, porous structure, and high surface area.

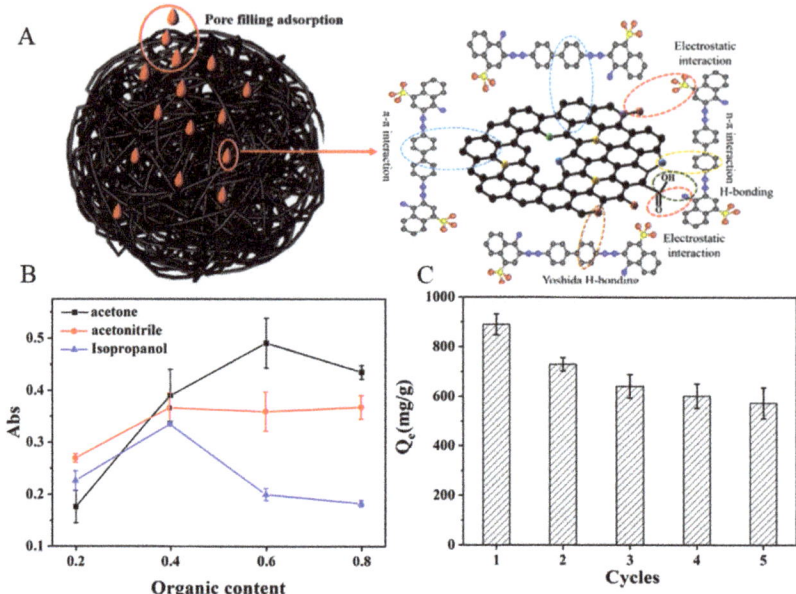

Figure 7. Possible adsorption mechanism of N-doped CM-chitin toward to CR (**A**), desorption rates of different elutant (**B**), and reusability of N-doped CM-chitin (10 mg) for adsorpiton of CR (200 mg L^{-1}) at 298.15 K (**C**).

2.7. Regeneration and Reuse

The regeneration of N-doped CM-chitin is very important for real application. Thus, five consecutive adsorption-desorption experiments were carried out to evaluate the regeneration performance. In this study, HCl, NaOH, NaCl and acetone solution were used as eluent for N-doped CM-chitin desorption. During the experiments, it was found that the desorption performance of CR by acid, base and salt solution was almost no desorption capability (Figure S5). However, the organic solvent (acetone, acetonitrile, and isopropanol solution) had good effective for the desorption performance (Figure 7B). It can be observed that acetone solution exhibited best elution capability than that of acetonitrile and isopropanol. Therefore, acetone solution (60.00%) was used as the desorption eluent for the N-doped CM-chitin generation. As shown Figure 7C, the adsorption amount exhibited a downward trend with the adsorption times increasing, and the adsorption amount still maintained about 573.15 mg g^{-1} after recycling 5 times. The decreasing adsorption capability might be due to the losing of N-doped CM-chitin during the adsorption-desorption processes, which was consistent with the results of increased disorder of materials shown by Raman spectroscopy. The other reason may be because treatment with acetone solution cannot completely remove CR in the N-doped CM-chitin at the previous adsorption step.

2.8. Dynamic Removal of N-Doped CM-Chitin toward CR

The highly adsorption efficient and the facile application of N-doped CM-chitin are important in the adsorption field. Therefore, dynamic removal experiments were carried out with an N-doped CM-chitin syringe device. As is shown in Figure 8 and Video S1, the real adsorption and removal of CR using N-doped CM-chitin dynamic injector system was evaluated. When CR solution was syringed through the N-doped CM-chitin packing, the CR solution become colorless (Figure 8A). Moreover, the penetration curve of Congo red solution to N-doped CM-chitin is shown in Figure 8B, the N-doped CM-chitin still remain excellent adsorption capability and the CR solution also changed into colorless. Above all, the N-doped CM-chitin exhibited effective adsorption and removal capability for dynamic CR solution, which is suitable for the practical application in water pollution treatment.

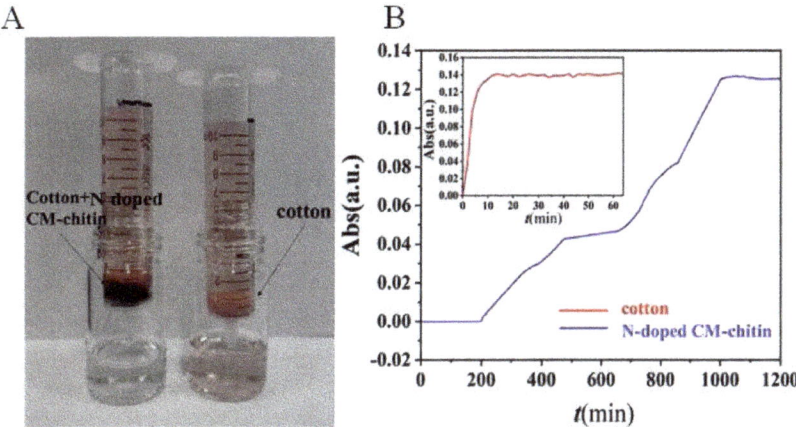

Figure 8. Optical photograph of the dynamic removal of CR (20 mg L^{-1}) by cotton + N-doped CM-chitin and cotton (**A**). Breakthrough curves of cotton and N-doped CM-chitin (**B**).

2.9. Comparison of Adsorption Capacity with Different Sorbents

The adsorption performance of N-doped CM-chitin for CR was compared with other adsorbents and the results were listed in Table S4. It can be seen from Table S4 that the adsorption capacity of N-doped CM-chitin was higher than those of other sorbents. The excellent adsorption capability of N-doped CM-chitin was attributed to the multiple adsorption mechanisms, porous structure, and high surface area. Therefore, N-doped CM-chitin can be applied as a potential adsorbent for practical application in water pollution removal.

3. Materials and Methods

3.1. Materials

Chitin was supplied by the Golden-shell Biochemical Co., Ltd. (Zhejiang, China) and was purified according to previous report [32,40]. Span 85, isooctane and Tween 85 were purchased from Aladdin (Shanghai, China). CR (>75.00%) was bought from the Solarbio Life Science (Beijing, China). All chemicals used were of analytical grade. Ultrapure water used during the whole processes was manufactured by the Hyperpure water system (Millopore, Bdford, MA, USA).

3.2. Fabrication of N-Doped CM-Chitin

Briefly, 94 g of aqueous solution containing NaOH (10.34 g), and urea (3.76 g) was prepared and then stirred continuously for 5 min. After that, 6 g of purified chitin was added into the NaOH/urea solution with stirring for another 10 min, which was placed into a refrigerator at −30 °C for 4 h and then thawed at room temperature. After three cycles of freezing-thawing processes, a transparent and homogeneous chitin aqueous phase was obtained. The oil phase (200 g of isooctane and 8.8 g of Span 85) was added into flask at 0 °C under stirring for 30 min. After that, the chitin aqueous phase was dropped into the oil phase within 5 min and further stirred for 60 min. The emulsification containing 20 g isooctane and 4.8 g Tween85 was added into the flask and stirred for another 60 min. The emulsified chitin solution was transferred to a water bath (60 °C) within 5 min to form the fibers rapidly and woven into microspheres. Subsequently, the fibrous microspheres derived from chitin (FM-chitin) were separated and washed with deionized water and ethanol to completely remove removal of isooctane, Tween-85 and Span-85. The FM-chitin were subjected to solvent-exchange with tertbutanol for 12 h and followed by freeze-dried

(−68 °C) for 24 h. Finally, the dried FM-chitin were carbonized under N_2 atmosphere in a tube furnace heating from room temperature to 650 °C at a rate of 3 °C/min and the equilibrating for 2 h, and then the nitrogen-doped carbonized microspheres derived from chitin (N-doped CM-chitin) were obtained after cooling to room temperature.

3.3. Characterizations

The morphology of different samples was observed with scanning electron microscope (SEM) (JSM-6701F, JEOL, Tokyo, Japan). The phase structure of CR before and after adsorption was determined using X-ray diffractometry (XRD, D8 ADVANCE, Bruker, Germany). X-ray photoelectron spectroscopy (XPS) measurements were performed using 2000 a XPS system (ESCALAB250Xi, ThermoFisher Scientific, Waltham, MA, USA). Fourier transform infrared (FT-IR) spectra was recorded on an infrared spectrometer (Nicolet5700, Thermo Nicolet, Waltham, MA, USA). The nitrogen adsorption-desorption isotherms were performed using the Brunauer-Emmett-Teller (BET) (Autosorb-2, Quantachrome, Boynton Beach, FL, USA). Raman spectroscopy was performed using a Confocal Raman microscope (Renishaw, UK).

3.4. Adsorption Experiments

Adsorption of N-doped CM-chitin toward CR were performed in the triangle flask was shaken for a rate about 120 rpm under dark condition. The experiment will be repeated three times in parallel to ensure the accuracy of the data. For the effect of the pH value, 10 mg of N-doped CM-chitin was added into 50 mL CR solution with different initial pH (3.0, 5.0, 7.4, 9.0, and 11.0) value for 12 h. The kinetic experiments of N-doped CM-chitin (10 mg) were conducted in CR solution (50 mL, 200 mg L^{-1}) and 3 mL solution was withdraw at specific time intervals. For the adsorption isotherms, N-doped CM-chitin (10 mg) was added into CR solutions (50 mL) with concentration ranging from 100 to 400 mg L^{-1}. Meanwhile, the dosage effect of N-doped CM-chitin varied from 10 mg to 50 mg for the adsorption capability was also investigated in CR solution (50 mL, 200 mg L^{-1}). The CR concentration was measured using UV-vis spectroscopy at 498 nm. The removal efficiency and equilibrium adsorption capacity of CR by N-doped CM-chitin was calculated using the following formulas [16,45]:

$$R(\%) = \frac{C_0 - C_e}{C_0} \times 100\% \qquad (1)$$

$$q_e = \frac{(C_0 - C_e)}{M} \times V \qquad (2)$$

where R is the removal rate of CR, C_0 is initial concertation of CR (mg L^{-1}), C_e is the equilibrium concentration (mg L^{-1}), q_e (mg L^{-1}) is the equilibrium adsorption capacity of N-doped CM-chitin, V is the volume of solution (L), C_e is the equilibrium concentration (mg L^{-1}), and M is the quantity of N-doped CM-chitin (mg).

3.5. Desorption, Regeneration, and Re-Usability of N-Doped CM-Chitin

N-doped CM-chitin (10 mg) was added into CR solution (50 mL, 200 mg L^{-1}) and shaken with 120 rpm at 25 °C for 12 h to reach adsorption equilibrium. After filtration, CR on the surface of N-doped CM-chitin was washed using water. After that, N-doped CM-chitin were added into eluents (acetone, isopropanol, and acetonitrile) with desired concentrations and then shaken at 120 rpm at 25 °C for 12 h. The CR concentrations in the regenerated liquid were measured using UV at 498 nm and the desorption efficiency (DE) was then calculated by the following equation [47]:

$$DE\% = \frac{(A_0 - A_d)}{A_0} \times 100\% \qquad (3)$$

where A_0 is the initial absorbance of CR solution (200 mg L^{-1}), A_d is the residual absorbance of UV after desorption. For further investigation the regeneration and re-usability, the re-adsorption experiments were carried out in CR solution (50 mL, 200 mg L^{-1}) for 5 times abide by the procedure described in adsorption section.

3.6. Dynamic Removal of N-Doped CM-Chitin toward CR

Dynamic removal experiments were carried out with syringe device. N-doped CM-chitin were packed into an injector (10 mL), which acted as an adsorption column to form a static bed (50 mg N-doped CM-chitin and 0.2 g cotton). A certain amount of cotton was placed at the bottom of the injector to prevent the outflow of N-doped CM-chitin. The CR solution (10 mL, 20 mg L^{-1}) was added into the injector from the top and flowed naturally from the fixed bed. Samples were taken every 3 min. The absorbance was measured with UV-vis spectra and the penetration curve was drawn [48,49].

4. Conclusions

In conclusion, a novel adsorbent (N-doped CM-chitin) with 3D porous framework architecture was developed from chitin, and which was used to removal of CR from water for the first time. Different N forms of oxidize-N, graphitic-N, pyrrolic-N, and pyridinic-N distributed in the carbon nanofibers of N-doped CM-chitin. Meanwhile, N-doped CM-chitin had micro/nano-pores and high specific surface area. The N-doped CM-chitin showed excellent removal capability with maximum adsorption amount of 954.47 mg g^{-1} for CR (200 mg L^{-1}). The significant adsorption capability might be due to the hydrophobic interaction, electrostatic interaction, π-π/n-π interaction and hydrogen-bond interaction between N-doped CM-chitin and CR. The CR adsorption into the N-doped CM-chitin was considered as a spontaneous endothermic process, and well conformed to the pseudo-second-order kinetic and Langmuir isotherm model. Interestingly, the N-doped CM-chitin also exhibited effective removal capability for the dynamic CR water with long-time stability. Therefore, this study provides new insight into fabrication of novel N-doped adsorbent from low-cost and waste biomass resource, and which has great potential for practical application in water pollution removal.

Supplementary Materials: The supporting information can be downloaded at: https://www.mdpi.com/article/10.3390/ijms24010684/s1. References [50–64] are cited in Supplementary Materials.

Author Contributions: Methodology, T.C.; Validation, H.C.; Formal analysis, T.C.; Data curation, H.C.; Writing—original draft, T.C. and H.C.; Supervision, H.P.; Funding acquisition, L.Y. All authors have read and agreed to the published version of the manuscript.

Funding: This work was financially supported by the Jiangxi "Double Thousand Plan" Cultivation Program for Distinguished Talents in Scientific and Technological Innovation (JXSQ2019201011), and Natural Science Foundation of Jiangxi Province (20224BAB203022, 20224ACB203016, and 20212BAB203034).

Institutional Review Board Statement: Not applicable.

Informed Consent Statement: Not applicable.

Data Availability Statement: Not applicable.

Conflicts of Interest: The authors declare no conflict of interest.

References

1. Azhdari, R.; Mousavi, S.M.; Hashemi, S.A.; Bahrani, S.; Ramakrishna, S. Decorated graphene with aluminum fumarate metal organic framework as a superior non-toxic agent for efficient removal of Congo Red dye from wastewater. *J. Environ. Eng.* **2019**, *7*, 103437. [CrossRef]
2. Gupta, V.K.; Kumar, R.; Nayak, A.; Saleh, T.A.; Barakat, M.A. Adsorption removal of dyes from aqueous solution onto carbon nanotubes: A review. *Adv. Colloid Interface Sci.* **2013**, *193–194*, 24–34. [CrossRef] [PubMed]
3. Sathishkumar, K.; AlSalhi, M.S.; Sanganyado, E.; Devanesan, S.; Arulprakash, A.; Rajasekar, A. Sequential electrochemical oxidation and bio-treatment of the azo dye congo red and textile effluent. *J. Photoch. Photobio. B* **2019**, *200*, 111655. [CrossRef] [PubMed]

4. Shaban, M.; Abukhadra, M.R.; Shahien, M.G.; Ibrahim, S.S. Novel bentonite/zeolite-NaP composite efficiently removes methylene blue and Congo red dyes. *Environ. Chem. Lett.* **2018**, *16*, 275–280. [CrossRef]
5. Chowdhyry, A.; Kumari, S.; Khan, A.A.; Hussain, S. Synthesis of mixed phase crystalline CoN12S4 nanomaterial and selective mechanism for adsorption of Congo red from aqueous solution. *J. Environ. Chem. Eng.* **2021**, *9*, 106554. [CrossRef]
6. Sharma, G.; AlGarni, T.S.; Kumar, P.S.; Bhogal, S.; Kumar, A.; Sharma, S.; Stadler, F.J. Utilization of Ag_2O–Al_2O_3–ZrO_2 decorated onto rGO as adsorbent for the removal of Congo red from aqueous solution. *Environ. Res.* **2021**, *197*, 111179. [CrossRef]
7. Manzoor, J.; Sharma, M. Impact of textile dyes on human health and environment. In *Impact of Textile Dyes on Public Health and the Environment*; IGI Global: Hershey, PA, USA, 2020; pp. 162–169.
8. Sathiyavimal, S.; Vasantharaj, S.; Shanmugavel, M.; Manikandan, E.; Nguyen-Tri, P.; Brindhadevi, K.; Pugazhendhi, A. Facile synthesis and characterization of hydroxyapatite from fish bones: Photocatalytic degradation of industrial dyes (crystal violet and Congo red). *Prog. Org. Coat.* **2020**, *148*, 105890. [CrossRef]
9. Raval, N.P.; Shah, P.U.; Ladha, D.G.; Wadhwani, P.M.; Shah, N.K. Comparative study of chitin and chitosan beads for the adsorption of hazardous anionic azo dye Congo Red from wastewater. *Desalin. Water Treat.* **2016**, *57*, 9247–9262. [CrossRef]
10. Shetti, N.P.; Malode, S.J.; Malladi, R.S.; Nargund, S.L.; Shukla, S.S.; Aminabhavi, T.M. Electrochemical detection and degradation of textile dye Congo red at graphene oxide modified electrode. *Microchem. J.* **2019**, *146*, 387–392. [CrossRef]
11. Buthiyappan, A.; Aziz, A.R.A.; Daud, W.M.A.W. Recent advances and prospects of catalytic advanced oxidation process in treating textile effluents. *Rev. Chem. Eng.* **2016**, *32*, 1–47. [CrossRef]
12. Najafi, M.; Bastami, T.R.; Binesh, N.; Ayati, A.; Emamverdi, S. Sono-sorption versus adsorption for the removal of congo red from aqueous solution using NiFeLDH/Au nanocoposite: Kinetics, thermodynamics, isotherm studies, and optimization of process parameters. *J. Ind. Eng. Chem.* **2022**, *116*, 489–503. [CrossRef]
13. Sharma, A.; Siddiqui, Z.M.; Dhar, S.; Mehta, P.; Pathania, D. Adsorptive removal of congo red dye (CR) from aqueous solution by Cornulaca monacantha stem and biomass-based activated carbon: Isotherm, kinetics and thermodynamics. *Sep. Sci. Technol.* **2019**, *54*, 916–929. [CrossRef]
14. Amran, F.; Zaini, M.A.A. Sodium hydroxide-activated Casuarina empty fruit: Isotherm, kinetics and thermodynamics of methylene blue and congo red adsorption. *Environ. Technol. Innov.* **2021**, *23*, 101727. [CrossRef]
15. Zhao, X.R.; Wang, W.; Zhang, Y.J.; Wu, S.Z.; Li, F.; Liu, J.P. Synthesis and characterization of gadolinium doped cobalt ferrite nanoparticles with enhanced adsorption capability for Congo Red. *Chem. Eng. J.* **2014**, *250*, 164–174. [CrossRef]
16. Yamaguchi, T.; Xie, J.; Oh, J.M. Synthesis of a mesoporous Mg–Al–mixed metal oxide with P123 template for effective removal of Congo red via aggregation-driven adsorption. *J. Solid State Chem.* **2021**, *293*, 121758.
17. Achour, Y.; Bahsis, L.; Ablouh, E.; Yazid, H.; Laanari, M.R.; Haddad, M.E. Insight into adsorption mechanism of Congo red dye onto Bombax Buonopozense bark activated-carbon using central composite design and DFT studies. *Surf. Interfaces* **2021**, *23*, 100977. [CrossRef]
18. Singh, S.; Perween, S.; Ranjan, A. Dramatic enhancement in adsorption of congo red dye in polymer-nanoparticles composite of polyaniline-zinc titanate. *J. Environ. Chem. Eng.* **2021**, *9*, 105149. [CrossRef]
19. Nodehi, R.; Shayesteh, H.; Kelishami, A.R. Enhanced adsorption of congo red using cationic surfactant functionalized zeolite particles. *Microchem. J.* **2020**, *153*, 104281. [CrossRef]
20. Jagusiak, A.; Goclon, J.; Panczyk, T. Adsorption of Evans blue and Congo red on carbon nanotubes and its influence on the fracture parameters of defective and functionalized carbon nanotubes studied suing computational methods. *Appl. Surf. Sci.* **2021**, *539*, 148236. [CrossRef]
21. Li, Z.C.; Hanafy, H.; Zhang, L.; Sellaoui, L.; Netto, M.S.; Oliveira, M.L.S.; Seliem, M.K.; Dotto, G.L.; Bonilla-Petriciolet, A.; Li, Q. Adsorption of congo red and methylene blue dyes on an ashitaba waste and a walnut shell-based activated carbon from aqueous solutions: Experiments, characterization and physical interpretations. *Chem. Eng. J.* **2020**, *388*, 124263. [CrossRef]
22. Ahma, R.; Ansari, K. Comparative study for adsorption of congo red and methylene blue dye on chitosan modified hybrid nanocomposite. *Process Biochem.* **2021**, *108*, 90–102. [CrossRef]
23. Chatterjee, S.; Lee, D.S.; Lee, M.W.; Woo, S.H. Enhanced adsorption of congo red from aqueous solutions by chitosan hydrogel beads impregnated with cetyl trimethyl ammonium bromide. *Bioresour. Technol.* **2009**, *100*, 2803–2809. [CrossRef] [PubMed]
24. Vimonses, V.; Lei, S.; Jin, B.; Chow, C.W.K.; Saint, C. Kinetic study and equilibrium isotherm analysis of Congo Red adsorption by clay materials. *Chem. Eng. J.* **2009**, *148*, 354–364. [CrossRef]
25. Ummartyotin, S.; Pechyen, C. Strategies for development and implementation of bio-based materials as effective renewable resources of energy: A comprehensive review on adsorbent technology. *Renew. Sust. Energ. Rev.* **2016**, *62*, 654–664. [CrossRef]
26. Abdel-Rahman, L.H.; Abu-Dief, A.M.; Sayed, M.A.A.E.; Zikry, M.M. Nano sized moringa oleifera an effective strategy for Pb(II) inos removal from aqueous solution. *Chem. Mater. Res.* **2016**, *8*, 8–22.
27. Gholami, M.; Zare-Hoseinabadi, A.; Mohammadi, M.; Taghizadeh, S.; Behbahani, A.B.; Amani, A.M.; Solghar, R.A. Preparation of ZnXFe3-XO4@ chitosan Nanoparticles as an Adsorbent for Methyl Orange and Phenol. *J. Environ. Treat. Tech.* **2019**, *7*, 245–249.
28. Nasrollahzadeh, M.; Sajjadi, M.; Iravani, S.; Varma, R.S. Starch, cellulose, pectin, gum, alginate, chitin and chitosan derived nanomaterials for sustainable water treatment: A review. *Carbohydr. Polym.* **2021**, *251*, 116986. [CrossRef]
29. Hammi, N.; Chen, S.; Dumeignil, F.; Royer, S.; Kadib, A.E. Chitosan as s sustainable precursor for nitrogencontaining carbon nanomaterials: Synthsis and uses. *Materals Today Sustain.* **2020**, *10*, 100053. [CrossRef]

30. Kaya, M.; Baran, T.; Mentes, A.; Asaroglu, M.; Sezen, G.; Tozak, K.O. Extraction and characterization of α-chitin and chitosan from six different aquatic invertebrates. *Food Biophys.* **2014**, *9*, 145–157. [CrossRef]
31. Zamora, A.Z.; Mena, G.J.; Gonzalez, E.G. Removal of congo red from the aqueous phase by chitin and chitosan from waste shrimp. Desalination and Water Treatment. *Desalination Water Treat.* **2016**, *57*, 14674–14685. [CrossRef]
32. Zheng, S.; Cui, Y.; Zhang, J.W.; Gu, Y.X.; Shi, X.W.; Peng, C.; Wang, D.H. Ntrogen doped microporous carbon nanospheres derived from chitin nanogels as attractive materials for supercapacitors. *RSC Adv.* **2019**, *9*, 10976. [CrossRef] [PubMed]
33. Khan, A.; Goepel, M.; Colmenares, J.C.; Glaser, R. Chitosan-based N-doped carbon materials for electrocatalytic and photocatalytica applications. *ACS Sustain. Chem. Eng.* **2020**, *8*, 4708–4727. [CrossRef]
34. Zheng, S.; Zhang, J.W.; Deng, H.B.; Du, Y.M.; Shi, X.W. Chitin derived nitrogen-doped porous carbons with ultrahigh specific surface area and tailored hierarchical porosity for high performance supercapacitors. *J. Bioresour. Bioprod.* **2021**, *6*, 142–151. [CrossRef]
35. Ahmed, M.J.; Hameed, B.H.; Hummadi, E.H. Review on recent progress in chiosan/chitin-carbonaceous material composites for the adsrortpion of water pollutants. *Carbohydr. Polym.* **2020**, *247*, 116690. [CrossRef] [PubMed]
36. Nie, L.; Duan, B.; Lu, A.; Zhang, L.N. Pd/TiO_2@ carbon microspheres derived from chitin for highly efficient photocatalytic degradation of volatile organic compounds. *ACS Sustain. Chem. Eng.* **2018**, *7*, 1658–1666. [CrossRef]
37. Zhou, Y.S.; Cai, T.M.; Liu, S.; Liu, Y.Y.; Chen, H.J.; Li, Z.T.; Peng, H.L. N-doped magnetic three-dimensional carbon microspheres@ TiO_2 with a porous architecture for enhanced degradation of tetracycline and methyl orange via adsorption/photocatalysis synergy. *Chem. Eng. J.* **2021**, *411*, 128615. [CrossRef]
38. Duan, B.; Liu, F.; He, M.; Zhang, L.N. Ag–Fe_3O_4 nanocomposites@chitin microspheres constructed by in situ one-pot synthesis for rapid hydrogenation catalysis. *Green Chem.* **2014**, *16*, 2835–2845. [CrossRef]
39. Duan, B.; Gao, X.; Yao, X.; Fang, Y.; Huang, L.; Zhou, J.; Zhang, L.N. Unique elastic N-doped carbon nanofibrous microspheres with hierarchical porosity derived from renewable chitin for high rate supercapacitors. *Nano Energy* **2016**, *27*, 482–491. [CrossRef]
40. Cui, W.G.; Lai, Q.B.; Zhang, L.; Wang, F.M. Quantitative measurements of sp3 content in DLC films with Raman spectroscopy. *Surf. Coat. Tech.* **2010**, *205*, 1995–1999. [CrossRef]
41. Ferrari, A.C.; Robertson, J. Interpretation of Raman spectra of disordered and amorphous carbon. *Phys. Rev. B* **2000**, *61*, 14095. [CrossRef]
42. Yamamoto, Y.S.; Hasegawa, K.; Hasegawa, Y.; Takahashi, N.; Kitahama, Y.; Fukuoka, S.; Itoh, T. Direct conversion of silver complexes to nanoscale hexagonal columns on a copper alloy for plasmonic applications. *Phys. Chem. Chem. Phys.* **2013**, *15*, 14611–14615. [CrossRef] [PubMed]
43. Zhang, X.M.; Liu, J.Y.; Kelly, S.J.; Huang, X.J.; Liu, J.H. Biomimetic snowflake-shaped magnetic micro-/nanostructures for highly efficient adsorption of heavy metal ions and organic pollutants from aqueous solution. *J. Mater. Chem. A* **2014**, *2*, 11759–11767. [CrossRef]
44. Hu, D.Y.; Wang, P.; Li, J.; Wang, L.J. Functionalization of microcrystalline cellulose with n; n-dimethyldodecylamine for the removal of Congo red dye from an aqueous solution. *Bioresources* **2014**, *9*, 5951–5962. [CrossRef]
45. Kammah, M.E.; Elkhatib, E.; Gouveia, S.; Cameselle, C.; Aboukila, E. Cost-effective ecofriendly nanoparticles for rapid and efficient indigo carmine dye removal from wastewater: Adsorption equilibrium, kinetics and mechanism. *Environ. Technol. Innov.* **2022**, *28*, 102595. [CrossRef]
46. Sun, X.F.; Wang, S.G.; Liu, X.W.; Gong, W.X.; Bao, N.; Ma, Y. The effects of pH and ionic strength on fulvic acid uptake by chitosan hydrogel beads. *Colloid. Surface. A* **2008**, *324*, 28–34. [CrossRef]
47. Dichiara, A.B.; Smith, J.B.; Rogers, R.E. Enhanced adsorption of carbon nanocomposites exhausted with 2,4-dichlorophenoxyacetic acid after regeneration by thermal oxidation and microwave irradiation. *Environ. Sci. Nano* **2014**, *1*, 113–116. [CrossRef]
48. Good man, S.M.; Bura, R.; Dichiara, A.B. Facile impregnation of graphene into porous wood filters for the dynamic removal and recovery of dyes from aqueous soluitons. *ACS Appl. Nano Mater.* **2018**, *1*, 5682–5690. [CrossRef]
49. Dichiara, A.B.; Harlander, S.F.; Rogers, R.E. Fixed bed adsorption of diquat dibromide from aqueous solution using carbon nanotubes. *RSC Adv.* **2015**, *5*, 61508–61512. [CrossRef]
50. Hou, F.; Wang, D.; Ma, X.; Fan, L.; Ding, T.; Ye, X.; Liu, D. Enhanced adsorption of Congo red using chitin suspension after sonoenzymolysis. *Ultrason. Sonochem.* **2021**, *70*, 105327. [CrossRef]
51. Zolgharnein, J.; Asanjarani, N.; Mousavi, S. Optimization and Characterization of Tl(I) Adsorption onto Modified Ulmus carpinifolia Tree Leaves. *Clean Soil Air Water* **2011**, *39*, 250–258. [CrossRef]
52. Zhao, D.; Yu, Y.; Chen, J. Zirconium/polyvinyl alcohol modified flat-sheet polyvinyldene fluoride membrane for decontamination of arsenic: Material design and optimization, study of mechanisms, and application prospects. *Chemosphere* **2016**, *155*, 630–639. [CrossRef] [PubMed]
53. Ho, Y.; McKay, G. Pseudo-second order model for sorption processes. *Process Biochem.* **1999**, *34*, 451–465. [CrossRef]
54. Miraboutalebi, S.; Nikouzad, S.; Peydayesh, M.; Allahgholi, N.; Vafajoo, L.; McKay, G. Methylene blue adsorption via maize silk powder: Kinetic, equilibrium, thermodynamic studies and residual error analysis. *Process Saf. Environ. Prot.* **2017**, *106*, 191–202. [CrossRef]
55. Lim, L.; Priyantha, N.; Tennakoon, D.; Chieng, H.; Dahri, M.; Suklueng, M. Breadnut peel as a highly effective low-cost biosorbent for methylene blue: Equilibrium, thermodynamic and kinetic studies. *Arab. J. Chem.* **2017**, *10*, S3216–S3228. [CrossRef]

56. Lian, L.L.; Guo, L.P.; Guo, C.J. Adsorption of Congo red from aqueous solutions onto Ca-bentonite. *J. Hazard. Mater.* **2009**, *161*, 126–131. [CrossRef]
57. Mohebali, S.; Bastani, D.; Shayesteh, H. Equilibrium, kinetic and thermodynamic studies of a low-cost biosorbent for the removal of Congo red dye: Acid and CTAB-acid modified celery (Apium graveolens). *J. Mol. Struct.* **2019**, *1176*, 181–193. [CrossRef]
58. Lei, C.S.; Pi, M.; Jiang, C.J.; Cheng, B.; Yu, J.G. Synthesis of hierarchical porous zinc oxide (ZnO) microspheres with highly efficient adsorption of Congo red. *J. Colloid. Interf. Sci.* **2017**, *490*, 242–251. [CrossRef]
59. Hu, H.M.; Deng, C.H.; Sun, M.; Zhang, K.H.; Wang, M.; Xu, J.Y.; Le, H.R. Facile template-free synthesis of hierarchically porous NiO hollow architectures with high-efficiency adsorptive removal of Congo red. *J. Porous. Mat.* **2019**, *26*, 1743–1753. [CrossRef]
60. Liu, J.Y.; Yu, H.J.; Wang, L. Superior absorption capacity of tremella like ferrocene based metal-organic framework in removal of organic dye from water. *J. Hazard. Mater.* **2020**, *392*, 122274. [CrossRef]
61. Liu, J.; Li, J.; Wang, G.; Yang, W.N.; Yang, J.; Liu, Y. Bioinspired zeolitic imidazolate framework (ZIF-8) magnetic micromotors for highly efficient removal of organic pollutants from water. *J. Colloid Interface Sci.* **2019**, *555*, 234–244. [CrossRef]
62. Lei, C.S.; Zhu, X.F.; Zhu, B.C.; Jiang, C.J.; Le, Y.; Yu, J.G. Superb adsorption capacity of hierarchical calcined Ni/Mg/Al layered double hydroxides for Congo red and Cr (VI) ions. *J. Hazard. Mater.* **2017**, *321*, 801–811. [CrossRef] [PubMed]
63. Li, J.; Fan, Q.H.; Wu, Y.J.; Wang, X.; Chen, C.L.; Tang, Z.Y.; Wang, X.K. Magnetic polydopamine decorated with Mg–Al LDH nanoflakes as a novel bio-based adsorbent for simultaneous removal of potentially toxic metals and anionic dyes. *J. Mater. Chem. A* **2016**, *4*, 1737–1746. [CrossRef]
64. Yu, Z.C.; Hu, C.S.; Dichiara, A.B.; Jiang, W.H.; Gu, J. Cellulose Nanofibril/carbon nanomaterial hybrid aerogels for adsorption removal of cationic and anionic organic dyes. *Nanomaterials* **2020**, *10*, 169. [CrossRef] [PubMed]

Disclaimer/Publisher's Note: The statements, opinions and data contained in all publications are solely those of the individual author(s) and contributor(s) and not of MDPI and/or the editor(s). MDPI and/or the editor(s) disclaim responsibility for any injury to people or property resulting from any ideas, methods, instructions or products referred to in the content.

Surface Functionalisation of Self-Assembled Quantum Dot Microlasers with a DNA Aptamer

Bethan K. Charlton, Dillon H. Downie, Isaac Noman, Pedro Urbano Alves, Charlotte J. Eling and Nicolas Laurand *

Technology & Innovation Centre, Institute of Photonics, University of Strathclyde, 99 George Street, Glasgow G1 1RD, UK; bethan.charlton@strath.ac.uk (B.K.C.); dillon.downie@strath.ac.uk (D.H.D.); isaac.noman@strath.ac.uk (I.N.); pedro.alves@strath.ac.uk (P.U.A.); charlotte.eling@strath.ac.uk (C.J.E.)
* Correspondence: nicolas.laurand@strath.ac.uk

Abstract: The surface functionalisation of self-assembled colloidal quantum dot supraparticle lasers with a thrombin binding aptamer (TBA-15) has been demonstrated. The self-assembly of CdSSe/ZnS alloyed core/shell microsphere-shape CQD supraparticles emitting at 630 nm was carried out using an oil-in-water emulsion technique, yielding microspheres with an oleic acid surface and an average diameter of 7.3 ± 5.3 µm. Surface modification of the microspheres was achieved through a ligand exchange with mercaptopropionic acid and the subsequent attachment of TBA-15 using EDC/NHS coupling, confirmed by zeta potential and Fourier transform IR spectroscopy. Lasing functionality between 627 nm and 635 nm was retained post-functionalisation, with oleic acid- and TBA-coated microspheres exhibiting laser oscillation with thresholds as low as 4.10 ± 0.37 mJ·cm^{-2} and 7.23 ± 0.78 mJ·cm^{-2}, respectively.

Keywords: colloidal quantum dots; microsphere lasers; whispering gallery mode laser; surface functionalisation; self-assembled supraparticles; aptamers

1. Introduction

Colloidal quantum dots (CQDs) are an important class of material with excellent properties for use as laser gain media, such as a high density of states, low optical gain temperature dependence, high quantum yield, and favourable photostability [1,2]. CQDs can be forced to self-assemble using an oil-in-water emulsion method to produce spherical supraparticles (SPs), with a relatively high refractive index of n_{eff} = 1.7 [3], which demonstrate whispering gallery mode (WGM) lasing under photoexcitation [4–7]. The emulsion technique for SP synthesis is facile and scalable, and resulting SPs work as microresonators, or cavities, with favourable Q factors [3,6] and low lasing thresholds [5]. WGMs are generated within SPs when light cycles around the surface of these cavities through total internal reflection enabled by the higher refractive index of an SP relative to the surrounding environment. Self-assembled SPs can be more easily tailored for multiple functionalities, for example, multi-wavelength emission for spectral coding, while their high refractive index conferred by the dense CQD assembly has the potential for stronger light confinement and, in turn, lasing in smaller structures. WGM lasers, or more generally active WGM resonators, have also gained huge interest for use in biological sensing applications due to the evanescent tail produced by WGM modes that makes the light emission of the resonators extremely sensitive to changes in the refractive index of the surrounding medium and changes to the surface chemistry [8]. Using WGM lasers for sensing is also favoured over fluorescent sensors as a result of the high signal-to-noise ratio and the reduction in emission linewidth when operating in the stimulated emission regime [9].

However, to produce functional SP laser biosensors that are water soluble and able to detect specific analytes in solution, the surface functionalisation of SPs is necessary. To our knowledge, there are currently no methods of developing functional, water soluble SP laser biosensors that can detect specific analytes in solution. As a result of the colloidal synthesis of the nanocrystals, CQDs are typically coated with hydrophobic ligands, such as oleic acid (OA), making CQDs insoluble in water and therefore unsuitable for biological applications [10]. Even though engineering of the surface chemistry of CQDs is established and has been studied extensively [7,11–18], ligand exchange directly on SPs has only recently been demonstrated [7,19,20].

Aptamers are engineered sections of DNA or RNA that can bind to specific proteins with high affinity and selectivity. For sensing applications, aptamers are ideal due to their small size, ease of production stability, and ease of modification [21]. Quantum dot aptasensors are a popular area of research, and many different types of sensors have been investigated [22–30]. WGM resonators have also been functionalised with aptamers and demonstrated very good sensitivity; however, these are often passive resonators made of silica or glass with stringent alignment tolerances for light injection into WGMs [22]. Thrombin is a popular target for many aptamer-functionalised sensors as it is involved in many pathological diseases, such as thrombosis and atherosclerosis. It functions as a coagulation factor and its presence in blood can indicate the presence of a blood clot. Current methods of thrombin detection are limited by the availability of suitable antibodies and can be time consuming [21,22].

This work addresses the above challenge by focusing on the surface functionalisation of such SPs. We report the synthesis and surface modification of SPs through the ligand exchange of oleic acid with mercaptopropionic acid (MPA) and subsequent EDC/NHS coupling with the thrombin binding aptamer TBA-15 to produce SP microlasers which could function as biosensors for the protein thrombin.

2. Results and Discussion

2.1. Synthesis and Characterisation

The synthesis of microsphere SPs consisting entirely of CdSSe/ZnS alloyed core-shell quantum dots was achieved using an oil-in-water emulsion method adapted from the literature [3]. Chloroform was chosen for its relatively low boiling point to allow the solvent to evaporate in a reasonable time at ambient temperatures, which has been shown to be the optimum drying condition for SP fabrication [31]. Environmental conditions are therefore important factors to consider during the fabrication of these microspheres and could cause large inconsistencies in the time required to produce solid SPs, which demanded constant monitoring of the emulsion to determine when the self-assembly process was finished. Surfactant choice is also important for producing SPs with favourable properties and stability [32], which is crucial for being able to alter the surface chemistry of SPs post-fabrication. The literature procedure was simplified by removing the microfluidic chip and generating emulsion droplets through vortexing the solution of CQDs and surfactant. This simplification results in broader size distributions in microspheres; however, there is still some control over the sizes of resulting SPs by manipulating the concentration of the CQD solution and surfactant used in the self-assembly process [33].

Scheme 1 depicts the surface functionalisation procedure conducted in this work. The CQDs begin coated in oleic acid, a molecule with a long and unreactive carbon chain that makes the nanocrystals insoluble in water. Although this hydrophobicity is required for the SP self-assembly method, retaining this surface chemistry is not suitable for biological and sensing applications due to the inert nature of long carbon chains. A direct ligand exchange was then carried out to replace the oleic acid on the surface of the SPs with mercaptopropionic acid (MPA). This exchange is possible due to the smaller, more thermodynamically favourable chain length of MPA compared with oleic acid [13]. The resulting SPs should have carboxylic acid groups at their outer surface [12], making the SPs water soluble and able to be functionalised using a wide range of reactions. Ligand exchange was confirmed by zeta potential measurements showing a change in surface charge from -19.7 ± 6.73 mV to -31.7 ± 5.12 mV, as shown in Table 1, before and after ligand exchange respectively. This decrease in charge was caused by the deprotonation of carboxylic acid groups at the SP surface in an acidic environment, resulting in a more negative zeta potential value at the SP surface [34,35]. Having the carboxylic acid functionality on the outer surface of the SPs allowed for further modification of the surface. EDC/NHS coupling was subsequently used to attach the aptamer TBA-15 to the SPs through the carboxylates of the SP surface and the amine modifier attached to the 5' end of the TBA-15 DNA chain [36]. TBA-15 was the aptamer of choice as it has the shortest DNA chain and is readily available. Consideration of the size of the molecules used for functionalisation is important because the addition of a target analyte must be within the evanescent field of the WGMs generated in the SP to be able to produce a change in lasing wavelength [37]. Functionalisation with TBA-15 was also verified using zeta potential measurements and Fourier transform infrared spectroscopy (FTIR). Zeta potential measurements before and after EDC/NHS coupling demonstrated a significant reduction in SP surface charge from -31.7 ± 5.12 mV to -12.5 ± 4.43 mV, as depicted in Table 1, which suggests successful coupling of the TBA-15 to the MPA-SP surface [38,39]. The FTIR spectra taken after each functionalisation step are shown in Figure 1, with the insets showing the areas of interest when analysing the spectra. In the spectrum for oleic acid-capped SPs (OA-SPs), the broad peak at 3120 cm^{-1} and the sharp, strong peak at 1545 cm^{-1} correspond to O–H stretching and C=O stretching for bound carboxyl groups, respectively, proving the presence of the carboxylic acid group of oleic acid [40]. The peaks at 2935 and 2865 cm^{-1} demonstrate the presence of C–H stretching from the alkene and alkane groups present in oleic acid, and peaks at 1463 and 1410 cm^{-1} also correspond to either alkane C–H bending from methyl end groups or O–H bending from the carboxylic acid and the CH$_2$ groups of the carbon chain, respectively. After ligand exchange with MPA, the 18-carbon chain of oleic acid was switched with a 3-carbon chain with a thiol at one end that binds to the surface of the SPs and a carboxylic acid present at the opposite end of the chain. This replacement is reflected in the MPA-SP FTIR spectrum with the almost complete disappearance of the peaks at 2935, 2865, 1463, and 1410 cm^{-1} [41]. The FTIR spectrum taken after EDC/NHS coupling contains a peak at 3145 cm^{-1} corresponding to N–H stretching, peaks at 1637 and 1560 cm^{-1} corresponding to C=O and C–N stretching (i.e., the amide I and II absorption bands), respectively, and the peaks at 1318 and 1285 cm^{-1} correspond to the amide III band. These three indicators suggest the presence of an amide group and therefore the successful EDC/NHS coupling of TBA-15 to the surface of the SPs [42–44]. TBA-15 is composed of a DNA sequence consisting only of guanine and thymine bases which contain N–H bonds, possibly contributing to the strong and broad nature of the peak at 3145 cm^{-1}. Alkane C–H bending peaks at 1407 and 1450 cm^{-1} are more pronounced than those in the spectrum for the MPA-capped SPs which would be expected due to an increase in the number of those bonds present after the addition of TBA-15 to the SP surface.

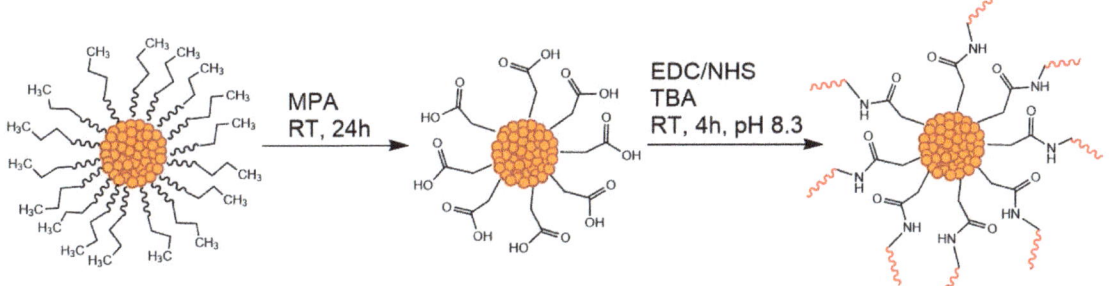

Scheme 1. Reaction scheme illustrating the functionalisation procedure used to bind the aptamer TBA-15 to the surface of an SP. The first step is the ligand exchange of oleic acid with 3-mercaptopropionic acid (MPA), and the second step is an EDC/NHS coupling carried out to attach TBA-15 to the carboxylic acid groups present on the MPA-SPs.

Table 1. Zeta potential measurements taken of SPs suspended in water after each step of the assembly and functionalisation procedure, taken at pH 6. ΔpH is the difference in the sample pH after the measurements were carried out.

Sample	Zeta Potential (mV)	σ (mV)	ΔpH
OA-SPs	−19.7	6.73	−0.42
MPA-SPs	−31.7	5.12	0.07
TBA-SPs	−12.5	4.43	−0.49

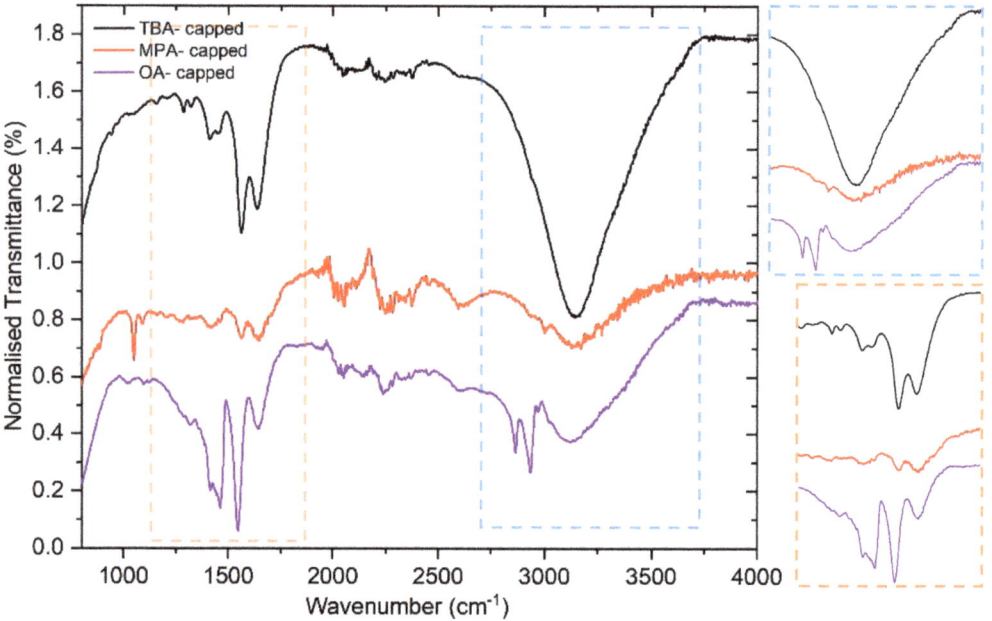

Figure 1. FTIR spectra of SPs capped with oleic acid (purple), MPA (red), and the aptamer TBA-15 (black). Inset: Zoom in of the FTIR spectra between 2600 cm^{-1} to 4000 cm^{-1} (blue) and 1200 cm^{-1} to 1850 cm^{-1} (orange).

Comparing SP sizes at each synthetic step can afford insights into the stability of these SPs to further functionalisation. After self-assembly, the average diameter of the SPs was found to be 7.3 ± 5.3 μm, then 7.2 ± 5.1 μm after ligand exchange, and finally 2.9 ± 1.2 μm after surface functionalisation. This indicates that there is no significant impact on the average SP size after ligand exchange whereas EDC/NHS coupling has a significant reducing effect on SP size. Although the average size values before and after ligand exchange would seem to indicate that the SPs are stable for further functionalisation procedures, in fact, there is an increase in the number of malformed, damaged, or collapsed SPs that can be observed under an optical microscope after ligand exchange. The SEM images shown in Figure 2 highlight the nature of damage that could be seen, particularly after EDC/NHS coupling. To quantify this, the percentage of damaged and collapsed SPs within each sample was estimated using images taken with a camera attached to an optical microscope. For the OA-SPs, the percentage of damaged spheres was estimated to be 26%, which increased to 37% after ligand exchange and then further increased to 52% for the TBA-capped SPs (TBA-SPs). These numbers demonstrate that these SPs are only partially stable to multistep post-assembly modification and functionalisation procedures; therefore, streamlining the post-assembly processing required for other applications is essential. The conditions for the optimal EDC/NHS coupling reaction rate ideally require a buffer solution and a pH of around 8.5 [36], though it has also been found that coupling can proceed at pH 7 at a slower rate [45]. Although the OA-SPs were stable in different solvents and buffer solutions, the MPA-SPs were not stable in buffer solutions and would dissolve back into quantum dots. The MPA-SPs would also dissolve at pH 8.5, hence the coupling being carried out between pH 6.5 and 7 in water with the addition of 1–2 µL of base every 20 min to keep the pH stable. The stability of SPs when exposed to further procedures could be improved using a surface coating to add an extra barrier between the SPs and the surrounding environment, making it harder for the spheres to break down [46]. For example, encasing SPs in materials such as silica in a sol–gel process resulted in WGM lasers that were able to exhibit lasing at temperatures as high as 450 K for time periods as long as 40 min [47]. Other possibilities include the modification of SPs with short polyethylene glycol (PEG)-containing functional groups that can be used to achieve further functionalisation using techniques such as click chemistry [15,16].

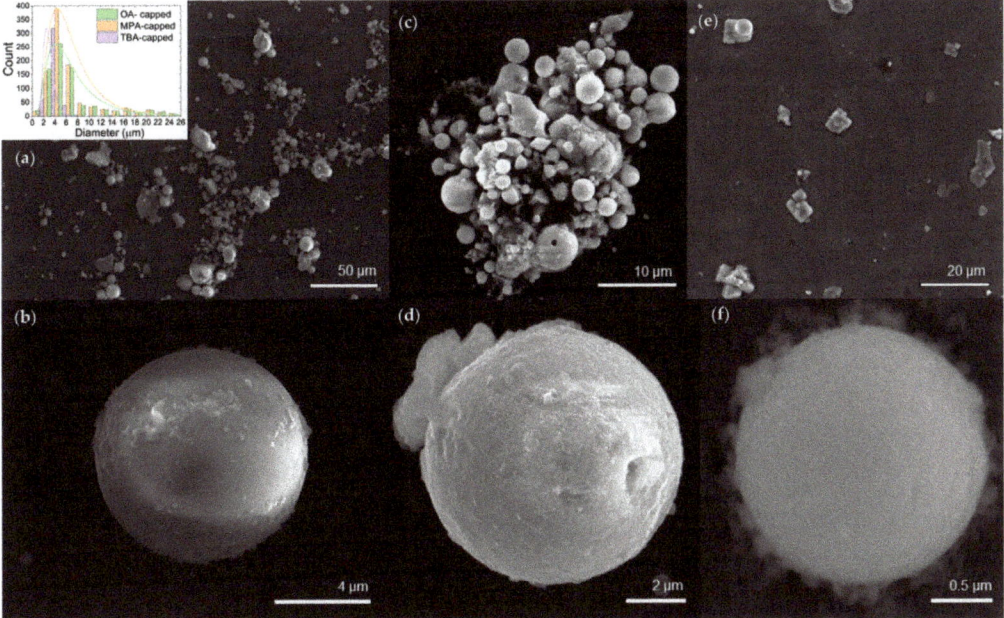

Figure 2. SEM micrographs of SP samples. (**a**) An OA-SP sample with an inset of the size distribution of SP samples measured by optical microscopy and (**b**) a close-up of a single OA-SP. (**c**) Image of a cluster of MPA-SPs and (**d**) a single MPA-SP. (**e**,**f**) Images of a TBA-SP sample and a single TBA-SP, respectively.

2.2. Optical Measurements

The laser characteristics of the SPs were obtained by optically pumping the individual SPs using a 355 nm Nd:YAG laser, described in the Section 3. Each SP sample was diluted in water at a ratio of 1:50 SPs to water and subsequently drop cast on a glass slide which was attached to a translation stage within the setup to enable pumping of individual SPs. Figure 3 shows examples of the emission spectra and the laser transfer functions used to determine the threshold energy for individual OA-, MPA-, and TBA-SPs with diameters of 5.5 µm, 4.1 µm, and 4.5 µm, respectively. The SPs in this work demonstrate WGM lasing which is made possible by the high refractive index of the CQDs used to assemble the cavities [48]. For the laser transfer functions, Figure 3a, the spectral emission intensity at different pump energy values was integrated over the spectral range of the dominant lasing peak, for example, between 631 nm 635 nm for the OA-capped SP, 634 nm to 638 nm for the MPA-SP, and 627 nm to 633 nm for the TBA-SP. The lasing threshold was 4.1 ± 0.37 mJ·cm^{-2}, 4.6 ± 0.41 mJ·cm^{-2}, and 7.2 ± 0.78 mJ·cm^{-2} for the OA-SP, MPA-SP and TBA-SP, respectively. There is therefore no significant difference in the laser properties of the OA-SP and the MPA-SP, although the laser slope efficiency is slightly lower for the MPA-SP. The laser spectra are also similar with a dominating peak and a less intense, red-shifted peak attributed to the next orbital WGM order (Figure 3a,b). There is a significant increase in the threshold post surface functionalisation from 4.6 ± 0.41 mJ·cm^{-2} to 7.2 ± 0.78 mJ·cm^{-2} for the MPA-SP and TBA-SP. Such increase in threshold may be explained by the addition of molecules on the SP surface decreasing the refractive index contrast, hence the light confinement, and possibly increasing the surface roughness which in turn can affect the ability of the SP to lase. Any further molecules being attached to the surface could further affect the SP threshold which could be exploited for biosensing applications. The Q factor of the SPs measured below the threshold were similar, between 210 and 240, for OA-SPs, MPA-SPs, and TBA-SPs. Therefore, it is difficult to ascertain any effects on the Q factor from this study. The measurement above, while typical, represents one SP selected within a population and there are statistical variations sample to sample.

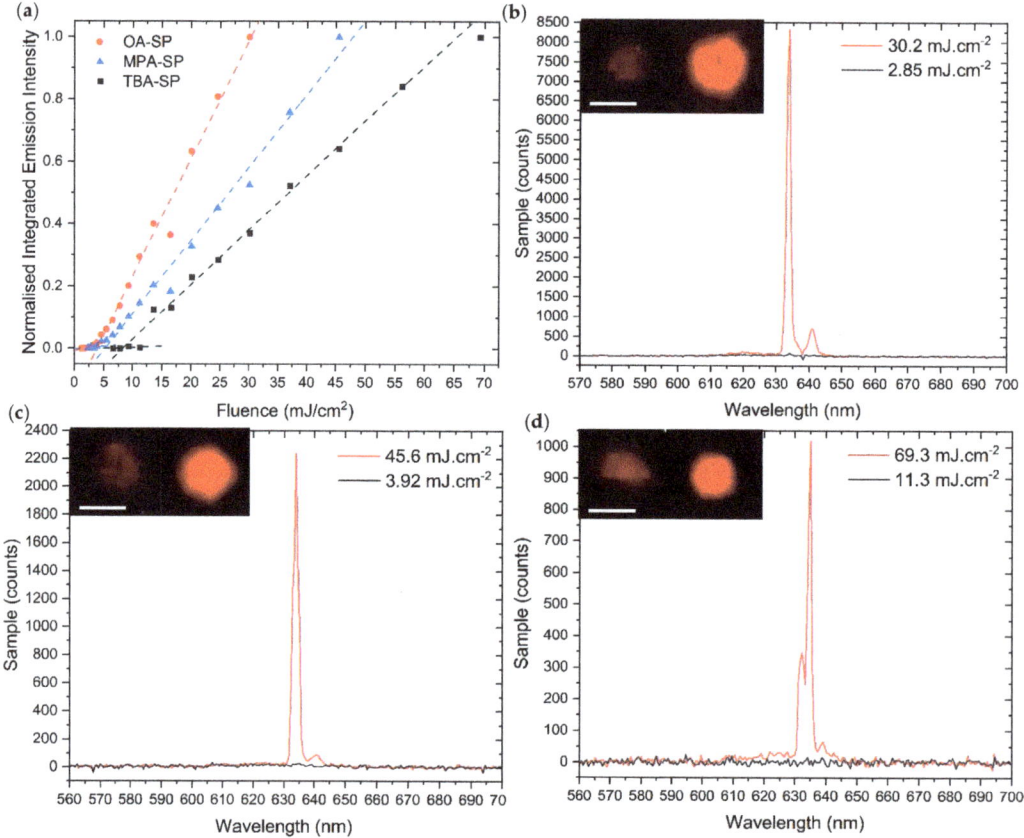

Figure 3. (**a**) Laser transfer function of the OA-, MPA- and TBA-capped SPs taken at the time the SPs were synthesised. The integrated emission intensity of each sphere is normalised for clarity. (**b**–**d**) are the PL spectra of single OA-, MPA- and TBA-capped SPs above (red) and below (black) the threshold, respectively. The inset of each PL spectrum is an image of the corresponding SP below (left) and above (right) the threshold. The scale bar is 5 microns.

To gain a better understanding on the factors affecting lasing threshold of the SPs and to establish the effect that surface functionalisation has on their lasing threshold, a one-way ANOVA (analysis of variance) was carried out using Wolfram Mathematica (Figure 4a). The ANOVA produced a high F ratio of 7.18 and a low p-value of 2.8×10^{-3} (Table 2). The F ratio is the ratio between the mean square values shown in Table 2, where a value above 1 demonstrates a difference in the means of the groups. The p-value tests the variance of each group, where a small p-value means that the standard deviations of the groups are different. Therefore, the combination of an F ratio much larger than 1 and the small p-value shown in Table 2 confirms that the mean threshold values of the SP groups are different. To elucidate which groups of SPs have a statistically significant difference in mean threshold values, a post hoc test (Tukey's test) was conducted. This test found that the mean value of the threshold was significantly different between only OA- and MPA-SPs at the 5% level. These results suggest that a ligand exchange from OA to MPA increases the lasing threshold on the SPs. No other statistically significant difference was found between the samples obtained in this study. The difference in sample sizes is due to the number of SPs per 10 µL sample capable of lasing after each stage of the functionalisation process. The size distribution of SPs in Figure 5a is shown on Figure 5b. While the OA-SPs were capable of lasing for a wide range of sizes, the data suggest that the subsequent ligand exchange and EDC/NHS coupling narrow the lasing capabilities of the SPs to smaller sizes. After synthesis, the average size of the SPs before and after ligand exchange with MPA remains similar (7.3 ± 5.3 µm and 7.2 ± 5.1 µm, respectively); therefore, a reduction in the sizes of the SPs available does not explain this observed reduction in the diameter of the MPA-SPs that demonstrate lasing. However, the average size of the SPs available would be a larger factor in the observed reduction in the number of TBA-SPs that can lase due to the significant reduction in the average size of the TBA-SPs to 2.9 ± 1.2 µm.

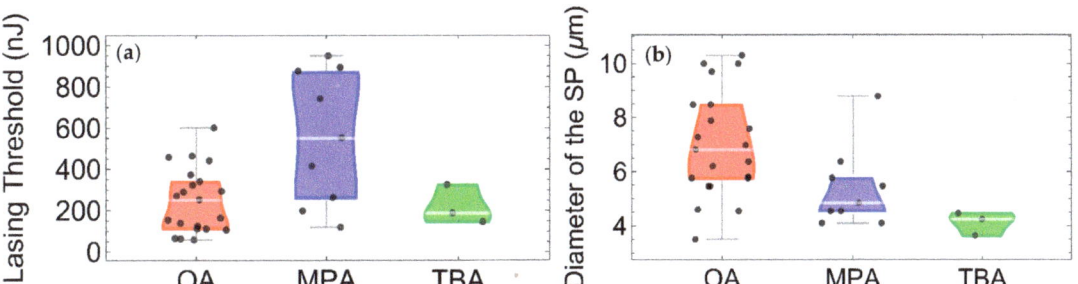

Figure 4. (**a**) Lasing threshold energy for each of the SPs studied. Data split according to the three ligands (red—OA, blue—MPA, and green—TBA) with the respective boxplots to demonstrate the locality, spread, and skewness groups of the data through their quartiles. (**b**) Diameter of each of the SPs studied. Data split according to the three ligands (red—OA, blue—MPA, and green—TBA) with the respective boxplots to demonstrate the locality, spread, and skewness groups of the data through their quartiles.

Table 2. One-way ANOVA results.

	Sum of Squares	Degrees of Freedom	Mean Square	F Ratio	p-Value
Between Groups	6.42×10^5	2	3.21×10^5	7.18	2.8×10^{-3}
Within Groups	1.34×10^6	30	4.47×10^4		
Total	1.98×10^6	32			

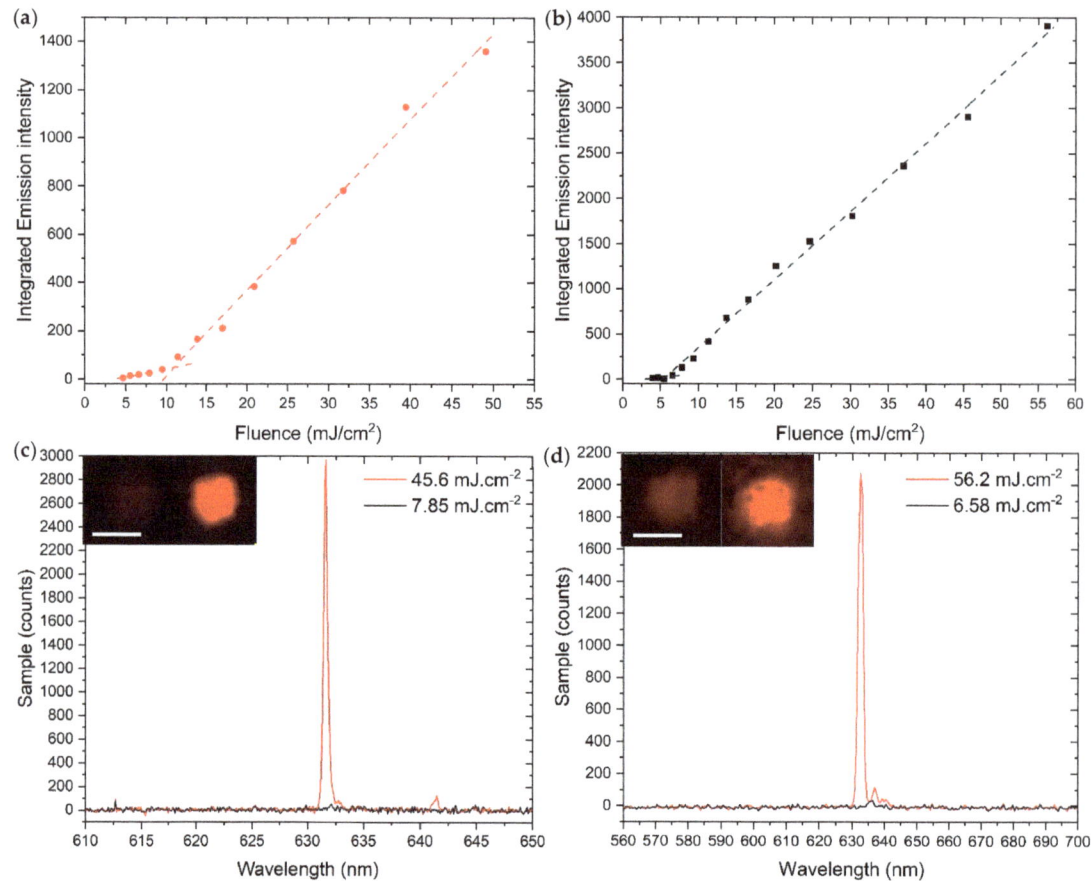

Figure 5. (**a**,**b**) Laser transfer function of the OA- and TBA-capped SPs taken after 35 days of being stored in water at 4 °C. (**c**,**d**) are the PL spectra of single OA- and TBA-capped SPs above (red) and below (black) the threshold, respectively. The inset of each PL spectrum is an image of the corresponding SP below (left) and above (right) the threshold. The scale bars are 5 microns.

The optical measurements depicted in Figure 5 were taken for SPs that had been stored in water at 4 °C for 35 days to gauge their stability over time. The samples used for the initial measurements where the SPs had been drop cast onto glass slides and stored dry were not able to lase compared to the SPs that had been stored in water. After this period, the threshold was 11.1 ± 1.0 mJ·cm^{-2} for an OA-capped sphere with a diameter of 5.8 µm whereas the TBA-SP measured after this storage period had a threshold of 5.63 ± 0.50 mJ·cm^{-2} with a diameter of 3.6 µm. Here, the emission intensity was integrated over 631 and 632.5 nm for the OA-SP and 630 to 634 nm for the TBA-SP. The comparison of these thresholds measured over a range of different SPs indicates that the thresholds measured after one month of storage are comparable to the values for fresh SPs, therefore showing that the SPs are stable after storage for a month at 4 °C. Owing to the sensitivity of WGM lasing to changes in the surrounding environment and the cavity, there are several factors which could contribute to the improved threshold observed for TBA-SPs after one month of storage. Therefore, this result requires more investigation in the future [49,50].

3. Materials and Methods

3.1. Materials

$CdSe_{1-x}S_x$/ZnS alloyed core/shell colloidal quantum dots were purchased from CD Bioparticles and the peak emission wavelength was 630 nm. Poly(vinyl alcohol), 3-mercaptopropionic acid (MPA), *N*-(3-Dimethylaminopropyl)-*N'*-ethyl carbodiimide (EDC), and *N*-Hydroxysuccinimide (NHS) were purchased from Merck. Thrombin binding aptamer with 15 bases (TBA-15) was purchased from Integrated DNA Technologies with the sequence 5'-NH-GGT TGG TGT GGT TGG -3' where NH is a 6-carbon amino linker.

3.2. Synthesis of $CdSe_{1-x}S_x$/ZnS Supraparticles

The procedure was adapted from the literature [3,5] and carried out under ambient conditions. A 20 mg/mL solution of $CdSe_{1-x}S_x$/ZnS alloyed core/shell CQDs was prepared in chloroform. Approximately 100 µL of this solution was added to 450 µL of a 2.5 w% solution of poly(vinyl alcohol) (PVA) in deionised water and stirred vigorously at room temperature for 4.5 h with constant monitoring of the emulsion after the first 2 h. The reaction mixture was centrifuged at 8000 rpm for 10 min, the supernatant was discarded, and the pellet was resuspended in DI water and stored at 4 °C. Microsphere size distributions were measured using ImageJ software using images taken from an optical microscope equipped with a Thorlabs™ camera. SEM images were captured using a JEOL JSM-IT100 operated at 20 kV.

3.3. Surface Functionalisation of SPs

The procedure was adapted from literature to make it suitable for SPs rather than CQDs [35,51]. Approximately 200 µL of oleic acid-capped SPs (OA-SPs) were re-dispersed in 500 µL of a 3:1 water/ethanol mix. Excess MPA (500µL) was added, and the solution was stirred at room temperature overnight. The reaction mixture was centrifuged at 8000 rpm for 10 min and the precipitated MPA-SPs were washed with ethanol twice to remove unbound MPA. The MPA-SPs were air dried and then stored at 4 °C.

The procedure adapted from the literature [52,53]. Approximately 200 µL of MPA-SPs were activated by the addition of 30 µL of 40 mM EDC and 30 µL of 15 mM NHS with NaOH (10 mM in H_2O) added to increase the pH to between 7.5 and 8 before leaving the mixture to stir for 1 h. Then, 20 µL of 10 µM TBA-15 was added and the solution was left stirring for 4 h with the pH of the reaction mixture kept at approximately pH 6.5–7.0 by the addition of 2 µL of 10 mM NaOH every 20 min. The SPs were characterised using a Malvern Zeta Potentiometer and FTIR which was obtained using a Nicolet iS5 FTIR Spectrometer.

3.4. Optical Characterisation

Approximately 10 µL samples of SPs after each functionalisation step were drop cast on glass slides and left to dry. Lasing measurements of self-assembled microlasers were obtained using a 355 nm, 5 ns pulsed Nd:YAG laser at a 10 Hz repetition rate with a beam spot area of ~2.6×10^{-5} cm^2. The beam intensity was altered using a variable wheel neutral density attenuator and focused on the sample using a 10× magnified objective lens. An Avantes AvaSpec-2048-4-DT spectrometer was used to acquire spectral data.

4. Conclusions

In conclusion, SPs of CdSeS/ZnS CQDs were successfully modified with the thrombin binding aptamer TBA-15 and are capable of lasing post-functionalisation, with lasing functionality retained after storage in water for over 1 month. Surface functionalisation is essential to realising the potential of these self-assembled SP lasers for a myriad of applications; therefore, this proof of concept paves the way for the use of SPs in biosensing applications. While surface functionalisation could reduce the toxicity of SPs, due to the nature of toxicity of the CQDs used, these SPs would only be used for in vitro sensing, for example, in a lab-on-a-chip sensing device. Using a facile and relatively mild procedure, the oleic acid originally coating the surface of the SPs can be exchanged with MPA to create a carboxylic acid-coated surface which can undergo EDC/NHS coupling to attach TBA-15 to the surface of the SPs in a three-step modification process. Zeta potential values of -21.5 ± 7.22 mV, $+24.1 \pm 4.99$ mV, and -22.4 ± 7.13 mV for OA-, MPA-, and TBA-SPs, respectively, confirmed the successful surface functionalisation, along with the FTIR spectra. The SPs before and after functionalisation demonstrated low lasing thresholds, with an OA- and TBA-SP exhibiting thresholds as low as 4.10 ± 0.37 mJ·cm^{-2} and 7.23 ± 0.78 mJ·cm^{-2}, respectively. The thresholds of an OA- and TBA-SP were measured after storage in water for over 1 month and remained similar to those measured for fresh SPs, with values of 11.1 ± 1.0 mJ·cm^{-2} and 5.63 ± 0.50 mJ·cm^{-2} for the OA- and TBA-SP, respectively. Obtaining SPs with this surface chemistry creates a platform that is incredibly versatile because EDC/NHS coupling is a very well-known reaction that can be used to attach any molecule containing an amine group to the surface of MPA-capped SPs [36]. The stability of SPs when exposed to further procedures could be improved using a surface coating to add an extra barrier between the SPs and the surrounding environment, making it harder for the spheres to break down [46]. For example, encasing SPs in materials such as silica enabled lasing at temperatures as high as 450 K for time periods as long as 40 min [47]. Other possibilities include the modification of SPs with short polyethylene glycol (PEG) chains containing functional groups that can be used to achieve further functionalisation using different techniques [15,16]. Adding an extra layer of protection to the SPs could also improve the stability of the spheres over time; however, these SPs have already been shown to be stable when stored in water. Work is currently underway to investigate the biosensing capabilities of these SPs.

Author Contributions: B.K.C., C.J.E. and N.L. conceived and designed the experiments.; B.K.C. performed the experiments, with C.J.E. and I.N. contributing data for the statistical analysis and D.H.D. and C.J.E. obtaining the SEM data. B.K.C. and P.U.A. analysed the data. B.K.C. wrote the original paper and C.J.E., P.U.A. and N.L. reviewed and edited the manuscript. All authors have read and agreed to the published version of the manuscript.

Funding: This research was funded by the Leverhulme Trust under the Research Leadership Award RL-2019-038.

Data Availability Statement: The dataset can be found at: https://doi.org/10.15129/d8b31125-23cb-4325-b136-38073dd6a4a6, 18 September 2023.

Acknowledgments: We would like to thank Paul R. Edwards for training and use of the SEM, Sian Sloan-Dennison for training and access to the zeta potentiometer, and Patrick Allen for access to the FTIR spectrometer.

Conflicts of Interest: The authors declare no conflict of interest.

References

1. Kim, J.Y.; Voznyy, O.; Zhitomirsky, D.; Sargent, E.H. 25th Anniversary Article: Colloidal Quantum Dot Materials and Devices: A Quarter-Century of Advances. *Adv. Mater.* **2013**, *25*, 4986–5010. [CrossRef]
2. Moreels, I.; Rainaó, G.; Gomes, R.; Hens, Z.; Stöferle, T.; Mahrt, R.F. Nearly Temperature-Independent Threshold for Amplified Spontaneous Emission in Colloidal CdSe/CdS Quantum Dot-in-Rods. *Adv. Mater.* **2012**, *24*, OP231–OP235. [CrossRef]
3. Montanarella, F.; Urbonas, D.; Chadwick, L.; Moerman, P.G.; Baesjou, P.J.; Mahrt, R.F.; Van Blaaderen, A.; Stöferle, T.; Vanmaekelbergh, D. Lasing Supraparticles Self-Assembled from Nanocrystals. *ACS Nano* **2018**, *12*, 12788–12794. [CrossRef]
4. Plunkett, A.; Eldridge, C.; Schneider, G.A.; Domènech, B. Controlling the Large-Scale Fabrication of Supraparticles. *J. Phys. Chem. B* **2020**, *124*, 11263–11272. [CrossRef]
5. Alves, P.U.; Laurand, N.; Dawson, M.D. Multicolor Laser Oscillation in a Single Self-Assembled Colloidal Quantum Dot Microsphere. In Proceedings of the 2020 IEEE Photonics Conference, IPC 2020, Virtual, 29 September–1 October 2020; pp. 5–6.
6. Alves, P.U.; Jevtics, D.; Strain, M.J.; Dawson, M.D.; Laurand, N. Enhancing Self-Assembled Colloidal Quantum Dot Microsphere Lasers. In Proceedings of the 2021 IEEE Photonics Conference (IPC), Virtual, 18–21 October 2021; IEEE: New York, NY, USA, 2021; pp. 1–2.
7. Kim, K.H.; Dannenberg, P.H.; Yan, H.; Cho, S.; Yun, S.H. Compact Quantum-Dot Microbeads with Sub-Nanometer Emission Linewidth. *Adv. Funct. Mater.* **2021**, *31*, 2103413. [CrossRef]

8. Beier, H.T.; Coté, G.L.; Meissner, K.E. Modeling Whispering Gallery Modes in Quantum Dot Embedded Polystyrene Microspheres. *J. Opt. Soc. Am. B* **2010**, *27*, 536. [CrossRef]
9. Francois, A.; Himmelhaus, M. Whispering Gallery Mode Biosensor Operated in the Stimulated Emission Regime. *Appl. Phys. Lett.* **2009**, *94*, 031101. [CrossRef]
10. Chinnathambi, S.; Shirahata, N. Recent Advances on Fluorescent Biomarkers of Near-Infrared Quantum Dots for In Vitro and In Vivo Imaging. *Sci. Technol. Adv. Mater.* **2019**, *20*, 337–355. [CrossRef]
11. Chan, W.C.W.; Nie, S. Quantum Dot Bioconjugates for Ultrasensitive Nonisotopic Detection. *Science* **1998**, *281*, 2016–2018. [CrossRef]
12. Aldana, J.; Lavelle, N.; Wang, Y.; Peng, X. Size-Dependent Dissociation PH of Thiolate Ligands from Cadmium Chalcogenide Nanocrystals. *J. Am. Chem. Soc.* **2005**, *127*, 2496–2504. [CrossRef]
13. Elimelech, O.; Aviv, O.; Oded, M.; Banin, U. A Tale of Tails: Thermodynamics of CdSe Nanocrystal Surface Ligand Exchange. *Nano Lett.* **2020**, *20*, 6396–6403. [CrossRef] [PubMed]
14. Kolb, H.C.; Finn, M.G.; Sharpless, K.B. Click Chemistry: Diverse Chemical Function from a Few Good Reactions. *Angew. Chem. Int. Ed.* **2001**, *40*, 2004–2021. [CrossRef]
15. Chen, Y.; Cordero, J.M.; Wang, H.; Franke, D.; Achorn, O.B.; Freyria, F.S.; Coropceanu, I.; Wei, H.; Chen, O.; Mooney, D.J.; et al. A Ligand System for the Flexible Functionalization of Quantum Dots via Click Chemistry. *Angew. Chem. Int. Ed.* **2018**, *57*, 4652–4656. [CrossRef] [PubMed]
16. Mao, G.; Ma, Y.; Wu, G.; Du, M.; Tian, S.; Huang, S.; Ji, X.; He, Z. Novel Method of Clickable Quantum Dot Construction for Bioorthogonal Labeling. *Anal. Chem.* **2021**, *93*, 777–783. [CrossRef]
17. Masteri-Farahani, M.; Mosleh, N. Modified CdS Quantum Dots as Selective Turn-on Fluorescent Nanosensor for Detection and Determination of Methamphetamine. *J. Mater. Sci. Mater. Electron.* **2019**, *30*, 21170–21176. [CrossRef]
18. Iyer, A.; Chandra, A.; Swaminathan, R. Hydrolytic Enzymes Conjugated to Quantum Dots Mostly Retain Whole Catalytic Activity. *Biochim. Biophys. Acta-Gen. Subj.* **2014**, *1840*, 2935–2943. [CrossRef]
19. Marino, E.; Bharti, H.; Xu, J.; Kagan, C.R.; Murray, C.B. Nanocrystal Superparticles with Whispering-Gallery Modes Tunable through Chemical and Optical Triggers. *Nano Lett.* **2022**, *22*, 4765–4773. [CrossRef]
20. Marino, E.; van Dongen, S.W.; Neuhaus, S.J.; Li, W.; Keller, A.W.; Kagan, C.R.; Kodger, T.E.; Murray, C.B. Monodisperse Nanocrystal Superparticles through a Source–Sink Emulsion System. *Chem. Mater.* **2022**, *34*, 2779–2789. [CrossRef]
21. Pasquardini, L.; Berneschi, S.; Barucci, A.; Cosi, F.; Dallapiccola, R.; Insinna, M.; Lunelli, L.; Conti, G.N.; Pederzolli, C.; Salvadori, S.; et al. Whispering Gallery Mode Aptasensors for Detection of Blood Proteins. *J. Biophotonics* **2013**, *6*, 178–187. [CrossRef]
22. Conti, G.N.; Berneschi, S.; Soria, S. Aptasensors Based on Whispering Gallery Mode Resonators. *Biosensors* **2016**, *6*, 28. [CrossRef]
23. Choi, J.H.; Chen, K.H.; Strano, M.S. Aptamer-Capped Nanocrystal Quantum Dots: A New Method for Label-Free Protein Detection. *J. Am. Chem. Soc.* **2006**, *128*, 15584–15585. [CrossRef] [PubMed]
24. Zhang, H.; Jiang, B.; Xiang, Y.; Zhang, Y.; Chai, Y.; Yuan, R. Aptamer/Quantum Dot-Based Simultaneous Electrochemical Detection of Multiple Small Molecules. *Anal. Chim. Acta* **2011**, *688*, 99–103. [CrossRef] [PubMed]
25. Li, Y.; Liu, L.; Fang, X.; Bao, J.; Han, M.; Dai, Z. Electrochemiluminescence Biosensor Based on CdSe Quantum Dots for the Detection of Thrombin. *Electrochim. Acta* **2012**, *65*, 1–6. [CrossRef]
26. Levy, M.; Cater, S.F.; Ellington, A.D. Quantum-Dot Aptamer Beacons for the Detection of Proteins. *ChemBioChem* **2005**, *6*, 2163–2166. [CrossRef]
27. Liu, J.; Jung, H.L.; Lu, Y. Quantum Dot Encoding of Aptamer-Linked Nanostructures for One-Pot Simultaneous Detection of Multiple Analytes. *Anal. Chem.* **2007**, *79*, 4120–4125. [CrossRef] [PubMed]
28. Wu, S.; Duan, N.; Ma, X.; Xia, Y.; Wang, H.; Wang, Z.; Zhang, Q. Multiplexed Fluorescence Resonance Energy Transfer Aptasensor between Upconversion Nanoparticles and Graphene Oxide for the Simultaneous Determination of Mycotoxins. *Anal. Chem.* **2012**, *84*, 6263–6270. [CrossRef]
29. Wang, C.; Lim, C.Y.; Choi, E.; Park, Y.; Park, J. Highly Sensitive User Friendly Thrombin Detection Using Emission Light Guidance from Quantum Dots-Aptamer Beacons in 3-Dimensional Photonic Crystal. *Sens. Actuators B Chem.* **2016**, *223*, 372–378. [CrossRef]
30. Liu, X.; Ren, J.; Su, L.; Gao, X.; Tang, Y.; Ma, T.; Zhu, L.; Li, J. Novel Hybrid Probe Based on Double Recognition of Aptamer-Molecularly Imprinted Polymer Grafted on Upconversion Nanoparticles for Enrofloxacin Sensing. *Biosens. Bioelectron.* **2017**, *87*, 203–208. [CrossRef]
31. Marino, E.; Marino, E.; Keller, A.W.; An, D.; Van Dongen, S.; Van Dongen, S.; Kodger, T.E.; MacArthur, K.E.; Heggen, M.; Kagan, C.R.; et al. Favoring the Growth of High-Quality, Three-Dimensional Supercrystals of Nanocrystals. *J. Phys. Chem. C* **2020**, *124*, 11256–11264. [CrossRef]
32. Wang, J.; Kang, E.; Sultan, U.; Merle, B.; Inayat, A.; Graczykowski, B.; Fytas, G.; Vogel, N. Influence of Surfactant-Mediated Interparticle Contacts on the Mechanical Stability of Supraparticles. *J. Phys. Chem. C* **2021**, *125*, 23445–23456. [CrossRef]
33. Wang, Y.; Ta, V.D.; Leck, K.S.; Tan, B.H.I.; Wang, Z.; He, T.; Ohl, C.D.; Demir, H.V.; Sun, H. Robust Whispering-Gallery-Mode Microbubble Lasers from Colloidal Quantum Dots. *Nano Lett.* **2017**, *17*, 2640–2646. [CrossRef]
34. Wei, Y.; Chen, L.; Zhao, S.; Liu, X.; Yang, Y.; Du, J.; Li, Q.; Yu, S. Green-Emissive Carbon Quantum Dots with High Fluorescence Quantum Yield: Preparation and Cell Imaging. *Front. Mater. Sci.* **2021**, *15*, 253–265. [CrossRef]

35. Vo, N.T.; Ngo, H.D.; Do Thi, N.P.; Nguyen Thi, K.P.; Duong, A.P.; Lam, V. Stability Investigation of Ligand-Exchanged CdSe/ZnS-Y (Y =3-Mercaptopropionic Acid or Mercaptosuccinic Acid) through Zeta Potential Measurements. *J. Nanomater.* **2016**, *2016*, 8564648. [CrossRef]
36. Marcel, J.E. Fischer Amine Coupling Through EDC/NHS: A Practical Approach. In *Surface Plasmon Resonance: Methods and Protocols*; Mol, N.J., Fischer, M.J.E., Eds.; Methods in Molecular Biology; Humana: Totowa, NJ, USA, 2010; Volume 627, pp. 55–73. ISBN 978-1-60761-669-6.
37. Reynolds, T.; François, A.; Riesen, N.; Turvey, M.E.; Nicholls, S.J.; Hoffmann, P.; Monro, T.M. Using Whispering Gallery Mode Micro Lasers for Biosensing within Undiluted Serum. In Proceedings of the SPIE BioPhotonics Australasia, Adelaide, Australia, 16–19 October 2016; Hutchinson, M.R., Goldys, E.M., Eds.; SPIE: Bellingham, WA, USA, 2016; Volume 10013, p. 100132X.
38. Tan, K.X.; Ujan, S.; Danquah, M.K.; Lau, S.Y. Design and Characterization of a Multi-Layered Polymeric Drug Delivery Vehicle. *Can. J. Chem. Eng.* **2019**, *97*, 1243–1252. [CrossRef]
39. Nguyen, V.T.; Lee, B.H.; Kim, S.H.; Gu, M.B. Aptamer-Aptamer Linkage Based Aptasensor for Highly Enhanced Detection of Small Molecules. *Biotechnol. J.* **2016**, *11*, 843–849. [CrossRef]
40. Singh, N.; Prajapati, S.; Prateek; Gupta, R.K. Investigation of Ag Doping and Ligand Engineering on Green Synthesized CdS Quantum Dots for Tuning Their Optical Properties. *Nanofabrication* **2022**, *7*, 89–103. [CrossRef]
41. Bel Haj Mohamed, N.; Haouari, M.; Zaaboub, Z.; Hassen, F.; Maaref, H.; Ben Ouada, H. Effect of Surface on the Optical Structure and Thermal Properties of Organically Capped CdS Nanoparticles. *J. Phys. Chem. Solids* **2014**, *75*, 936–944. [CrossRef]
42. Bharathi, M.V.; Roy, N.; Moharana, P.; Ghosh, K.; Paira, P. Green Synthesis of Highly Luminescent Biotin-Conjugated CdSe Quantum Dots for Bioimaging Applications. *New J. Chem.* **2020**, *44*, 16891–16899. [CrossRef]
43. Li, P.; Luo, C.; Chen, X.; Huang, C. An Off-on Fluorescence Aptasensor for Trace Thrombin Detection Based on FRET between CdS QDs and AuNPs. *RSC Adv.* **2022**, *12*, 35763–35769. [CrossRef] [PubMed]
44. Ji, Y.; Yang, X.; Ji, Z.; Zhu, L.; Ma, N.; Chen, D.; Jia, X.; Tang, J.; Cao, Y. DFT-Calculated IR Spectrum Amide I, II, and III Band Contributions of N-Methylacetamide Fine Components. *ACS Omega* **2020**, *5*, 8572–8578. [CrossRef] [PubMed]
45. Zhang, Q.; Li, R.X.; Chen, X.; He, X.X.; Han, A.L.; Fang, G.Z.; Liu, J.F.; Wang, S. Study of Efficiency of Coupling Peptides with Gold Nanoparticles. *Chin. J. Anal. Chem.* **2017**, *45*, 662–667. [CrossRef]
46. Eling, C.J.; Laurand, N.; Gunasekar, N.K.; Edwards, P.R.; Martin, R.W. Silica Coated Colloidal Semiconductor Quantum Dot Supracrystal Microlasers. In Proceedings of the 2022 IEEE 9th International Conference on Photonics, Virtual, 8–10 August 2022; pp. 2022–2023. [CrossRef]
47. Chang, H.; Zhong, Y.; Dong, H.; Wang, Z.; Xie, W.; Pan, A.; Zhang, L. Ultrastable Low-Cost Colloidal Quantum Dot Microlasers of Operative Temperature up to 450 K. *Light Sci. Appl.* **2021**, *10*, 60. [CrossRef] [PubMed]
48. Righini, G.C.; Soria, S. Biosensing by WGM Microspherical Resonators. *Sensors* **2016**, *16*, 905. [CrossRef]
49. Wu, Y.; Leung, P.T. Lasing Threshold for Whispering-Gallery-Mode Microsphere Lasers. *Phys. Rev. A-At. Mol. Opt. Phys.* **1999**, *60*, 630–633. [CrossRef]
50. Cai, L.; Pan, J.; Zhao, Y.; Wang, J.; Xiao, S. Whispering Gallery Mode Optical Microresonators: Structures and Sensing Applications. *Phys. Status Solidi Appl. Mater. Sci.* **2020**, *217*, 1900825. [CrossRef]
51. Rahman, S.A.; Ariffin, N.; Yusof, N.A.; Abdullah, J.; Mohammad, F.; Zubir, Z.A.; Aziz, N.M.A.N.A. Thiolate-Capped CdSe/ZnS Core-Shell Quantum Dots for the Sensitive Detection of Glucose. *Sensors* **2017**, *17*, 1537. [CrossRef] [PubMed]
52. Lao, Y.H.; Chi, C.W.; Friedrich, S.M.; Peck, K.; Wang, T.H.; Leong, K.W.; Chen, L.C. Signal-on Protein Detection via Dye Translocation between Aptamer and Quantum Dot. *ACS Appl. Mater. Interfaces* **2016**, *8*, 12048–12055. [CrossRef]
53. Zhu, L.; Hao, H.; Ding, C.; Gan, H.; Jiang, S.; Zhang, G.; Bi, J.; Yan, S.; Hou, H. A Novel Photoelectrochemical Aptamer Sensor Based on Cdte Quantum Dots Enhancement and Exonuclease I-Assisted Signal Amplification for Listeria Monocytogenes Detection. *Foods* **2021**, *10*, 2896. [CrossRef]

Disclaimer/Publisher's Note: The statements, opinions and data contained in all publications are solely those of the individual author(s) and contributor(s) and not of MDPI and/or the editor(s). MDPI and/or the editor(s) disclaim responsibility for any injury to people or property resulting from any ideas, methods, instructions or products referred to in the content.

Article

Metallic Nanowires Self-Assembled in Quasi-Circular Nanomolds Templated by DNA Origami

David Daniel Ruiz Arce [1], Shima Jazavandi Ghamsari [2], Artur Erbe [2,3,*] and Enrique C. Samano [1,*]

[1] Centro de Nanociencias y Nanotecnología, UNAM, Ensenada 22860, Mexico; ddruiza@ens.cnyn.unam.mx
[2] Helmholtz-Zentrum Dresden—Rossendorf, 01328 Dresden, Germany; s.jazavandi-ghamsari@hzdr.de
[3] Cluster of Excellence Center for Advancing Electronics Dresden (cfaed), TU Dresden, 01187 Dresden, Germany
* Correspondence: a.erbe@hzdr.de (A.E.); samano@ens.cnyn.unam.mx (E.C.S.)

Abstract: The self-assembly of conducting nanostructures is currently being investigated intensively in order to evaluate the feasibility of creating novel nanoelectronic devices and circuits using such pathways. In particular, methods based on so-called DNA Origami nanostructures have shown great potential in the formation of metallic nanowires. The main challenge of this method is the reproducible generation of very well-connected metallic nanostructures, which may be used as interconnects in future devices. Here, we use a novel design of nanowires with a quasi-circular cross-section as opposed to rectangular or uncontrolled cross-sections in earlier studies. We find indications that the reliability of the fabrication scheme is enhanced and the overall resistance of the wires is comparable to metallic nanostructures generated by electrochemistry or top-down methods. In addition, we observe that some of the nanowires are annealed when passing a current through them, which leads to a clear enhancement for the conductance. We envision that these nanowires provide further steps towards the successful generation of nanoelectronics using self-assembly.

Keywords: DNA nanotechnology; nanoelectronics; self-assembly; nanomaterials

1. Introduction

The enormous advances in the semiconductor industry have changed how people can interact with each other, receive healthcare, and work more efficiently, just to mention a few examples [1]. The production of smaller and cheaper electronic devices is foremost for future developments in electronics. The control of feature sizes below 10 nm in the fabrication of next-generation devices for sensing, computing, data storage, and communication systems is essential for reaching this target [2]. Developing novel, revolutionary materials and nano-manufacturing methods are required to achieve it. However, there is no doubt that the scientific and technological hurdles look almost insurmountable. An alternative way to overcome such hurdles is combining bottom-up and top-down techniques, which could be a roadmap to move into the right direction [3].

The convergence of recent advances in the DNA-based self-assembly approach and surface modification chemistry promises to bridge bottom-up and top-down schemes, enabling efficient construction of a new generation of devices [4–6]. Although non-origami DNA nanostructures have been reported [7], the cost-effective DNA Origami technique allows for the generation of 2D and 3D structures and for the controlled positioning of metal nanoparticles (NPs) [8]. Electronic devices consist of materials having a wide range of electrical conductivity values. Therefore, multilayer devices are fabricated using the deposition of insulating, semiconducting, and metallic overlayers, followed by patterning with photolithography and reactive ion etching in an ultra-clean room. On the contrary, a major breakthrough would be using simple, not-so-immaculate, but effective processes occurring in living species, namely, the self-assembly of organic molecules and macromolecules, as a first stage for manufacturing semiconductor devices [9].

A general scheme is devised in this work to fabricate gold nanowires (Au NWs) using custom-tailored 3D modules at the nanoscale. The modules are synthesized through the DNA Origami technique by hybridizing a scaffold and many short linear DNA single strands, ssDNA, complementary to the scaffold sequence [10,11].

The size and shape of the fundamental module, a hexagonal cross-section cylinder, is methodically conceived using a freeware named caDNAno. Each NW is based on a basic nanostructure, formed by assembling five modules with a biunique coupling using short extensions of complementary ssDNA of face-to-face neighboring modules, as observed at the top of Figure 1. To discriminate them, the five modules were labeled with different colors: purple—P, blue—B, green—G, yellow—Y, and orange—O. At each module, two Au NPs are previously attached at specific binding sites [12]. This approach allows for the casting of continuous Au NWs due to the precise positioning and controlled growth of the Au NPs by chemical reduction within the DNA basic nanostructure [13]. This two-step process is shown at the bottom of Figure 1. The shape of the seamless nanostructure reduces stress when the space inside the nanostructure is filled during this metalization procedure, and the resistivity of the resulting NWs is expected to decrease. The present study focuses on the feasibility of manufacturing high-quality metallic wires produced by controlling and confining the metal growth within three-dimensional mold nanostructures in contrast to our previous work using two-dimensional DNA Origami [14]. This demonstration of a first step towards nanofabrication using self-assembly will allow for the future development of nanoelectronic components once the generation of other metallic and semiconducting materials has been achieved using similar strategies. On top of that, self-recognition schemes, as presented here, can be used for biosensing [15], thus opening the way for further applications.

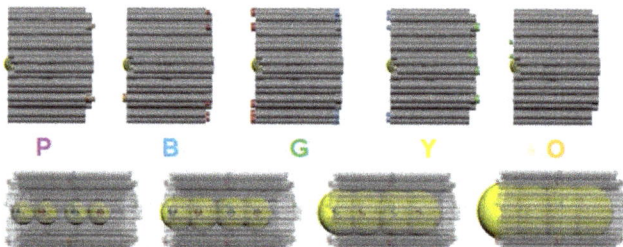

Figure 1. Coupling of five modules P—B—G—Y—O using extended staples to create the basic DNA nanostructure (**top**). Evolution of the metalization process by electroless metal deposition on Au NPs to achieve a gold nanowire, the DNA nanostructure is being used as a mold (**bottom**).

Au NWs are mainly used as microchip interconnects [16]. The NWs must be uniform and continuous to minimize their resistivity for this application. The initial stages of nucleation on Au NPs fixed to the DNA nanostructure and growth process are critical to satisfying these requirements. The first studies about the synthesis of DNA-templated NWs involved the coalescence of gold along DNA strands, either single or double [17]. The structure of the obtained NWs was a beads-along-a string morphology, and a major tour de force was the transformation from this framework to a neat metallic object having uniform geometry. The growth mechanism involves nucleation at binding sites on the DNA followed by the growth of spherical particles. Eventually, under favorable conditions, the final morphology of the overlaid metal is determined by two competitive processes: surface tension and adherence to the DNA template [17]. Although this effect has been observed in the fabrication of DNA-templated NWs based on the coalescence of gold around DNA linear strands immobilized on mica, their inferences can be extended if the templates are 3D DNA Origami nanostructures instead. Future studies will concentrate on improving the mechanical stability and cohesion of the actual DNA basic nanostructure, which serves as a template for the efficient assembly of the Au NPs, to enhance the electric conductance of the resulting NWs.

Gel electrophoresis, Atomic Force Microscopy (AFM), Transmission electron microscopy (TEM), and scanning electron microscopy (SEM) analytical techniques characterize the nanostructures. Randomly distributed Au NWs with a size below 200 nm were laid down on SiO_x/Si substrates and contacted with gold electrodes by electron beam lithography (EBL): A study of charge transport at room temperature of these wires shows ohmic behavior of the contacts. In addition, a reduction of the electrical resistance resembling current-induced annealing at atmospheric pressure is found. The conductance of the NWs shown here based on DNA nanostructures can possibly be explained by the so-called wind forces. The manufacturing of these Au NWs might provide a proof-of-concept for the feasibility of top-down nanofabrication combined with self-assembling systems, a long sought goal.

2. Results and Discussions

2.1. Visualization of the 3D Nanostructures

After annealing the scaffold with the staples and purifying the resulting modules, the structures were investigated using SEM, TEM, and AFM. The imaging of samples was first carried out utilizing tapping mode AFM in air. Figure 2a) shows a characteristic AFM image for the "blue" module. Most of the structures in the figure correspond to well-shaped single loose "blue" modules. However, the yield is low, with a few secondary structures, two to four modules overlapped by DNA-base stacking [18], already predicted by gel electrophoresis (not shown here). A cross-section analysis was also conducted, resulting in 36 nm and 32 nm for length and width, respectively. A similar visualization and cross-section analysis in AFM were also made for the other four modules, see Supplementary Information E Figures S8–S12. Table 1 summarizes the most relevant dimensions. These values agree with those predicted by the caDNAno design: length ∼33 nm and width ∼30 nm. The width measurement is an approximation because the module has a hexagonal cross-section. According to the Molecular Viewer model, based on the caDNAno design, the minimum and maximum width values are 17.6 nm and 42.8 nm, respectively. The physical contact between the specimen and tip apex when rastering during surface imaging is one of the characteristics of AFM as a topographic tool for sample analysis. The width of the hollow 3D module is distorted when the AFM tip scans a soft sample. Even in the tapping mode, the tip flattens the DNA module due to the normal stress on it; see the height values in the inserted cross-section in Figure 2a. This is why it is reasonable to compare the measured value with the width average value of the molecular model, ∼30 nm, in Figure 3 upper left.

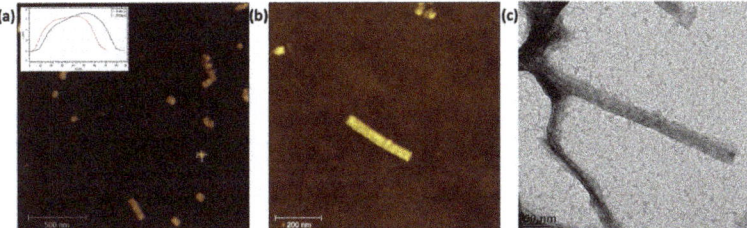

Figure 2. (a) AFM image, 2.5 μm × 2.5 μm image of the "blue" module deposited on SiO_x/Si. (b) AFM image, 0.6 μm × 0.6 μm image, of the basic nanostructure on SiO_x/Si. (c) TEM micrograph of the DNA basic nanostructure used as the mold.

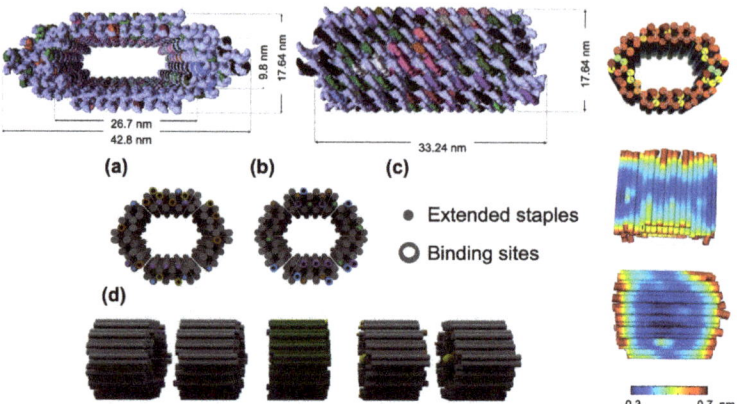

Figure 3. (**Upper left**) Molecular representation of the fundamental module as accomplished by means of Molecular Viewer from Autodesk. (**Upper right**) CanDo simulation for studying the structural stability of the primary module. (**Lower**) Schematics showing extended staples and binding sites for the unique coupling of the modules. (**a**) Back and (**b**) front view of a module. (**c**) The symbols are solid circles for extended staples and hollow ones for binding sites. (**d**) Assembly of the basic DNA nanostructure.

Table 1. Dimensions of modules as measured cross-section analysis of AFM images.

Module	Width	Length
Purple (P)	34(8) nm	33(8) nm
Blue (B)	36(8) nm	30(8) nm
Green (G)	32(8) nm	32(8) nm
Yellow (Y)	38(8) nm	30(8) nm
Orange (O)	40(8) nm	35(8) nm

All modules were bound together in a subsequent hybridization step to shape the basic nanostructure. Figure 2b shows a typical AFM image of an assembled DNA basic nanostructure with a ~190 nm length and ~30 nm width. As expected, no gaps were observed in the coupling of adjacent modules. The continuity of the mold, DNA basic nanostructure, was verified by TEM as well, a visualization technique with a high resolution. Figure 2c shows the image of two joined seamless molds with a 28 nm width.

As already mentioned, one of the advantages of the DNA Origami approach is locating and fixing inorganic material with a precision at the nanoscale. After functionalizing short ssDNA with Au NPs, two Au NPs with a 5 nm size and a separation of ~16 nm were inserted within each module, one in the middle and the other at one of the edges. By this means, Au NP–DNA Origami bioconjugates were generated by annealing the purified functionalized Au NPs, previously re-diluted in the proper buffer, with each module. The Au NPs insertion in each module was subsequently verified by AFM. According to the cross-section analysis in the AFM image insets in Figure 4, Au NPs were found inside some Au NP–module bioconjugates, as expected, but the yield is low. The measured distance between Au NPs is unreliable, as inferred from the images in Figure 4, due to the fact that the Au NPs are in the interior of the module and the functionalized ssDNA–Au NPs are flexible. The main function of the Au NPs is to become nucleation centers when gold coalesces on their surfaces during the metalization process.

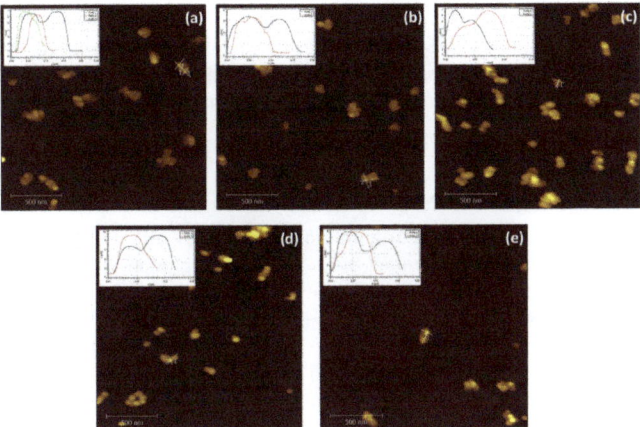

Figure 4. AFM micrographs of Au NP–module bioconjugates for the P, B, G, Y, and O modules. The scale bar is 500 nm.

After verifying the feasibility of Au NP–module bioconjugates, shown in Figure 4, and the synthesis of continuous basic nanostructures without Au NP, shown in Figure 2b, the mold formation for the Au NWs is followed. The Au NP–basic nanostructure bioconjugate, or mold, was shaped by the biunique assembly of the five Au NP–module bioconjugates. An AFM image of the basic nanostructure functionalized with Au NPs within is shown in Figure 5a. It is inferred from the cross-section analysis that the nanostructure has a length of ∼200 nm and width of 35 nm with 7 Au NPs inside. It is worth mentioning that 8 Au NPs, instead of 10 as designed, was the highest number of NPs which could be found in all images. The yield of NP attachment was thus observed to be low as well. This was probably due to steric effects within the nanostructure. Even though the inner cross-section for the basic nanostructure is minute, the literature reports the synthesis and characterization of Au NWs as having a comparable size to the one reported here [19]. Since the gaps between the NPs are filled by the subsequent metalization step, we can continue to form NWs even though some particles are missing inside the structures.

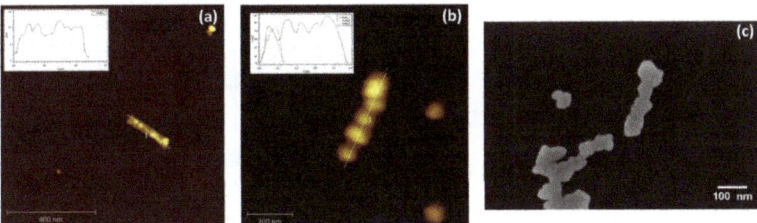

Figure 5. (**a**) AFM image of the basic nanostructure functionalized with Au NPs on SiO_x/Si. Seven NPs can be clearly identified, where one of the NPs is just off the straight line where the cross-section was made. AFM (**b**) and SEM (**c**) images of a Au NW on a SiO_x/Si substrate before being electrically contacted by EBL.

Finally, elongated Au NWs were created by the metalization of the Au NP–basic nanostructure bioconjugates and laid down on a SiO_x/Si substrate. The main purpose of the bioconjugate nanostructures is to be utilized as molds when the Au NWs are cast, which is the main object in this part of the fabrication process. Figure 5b shows the AFM cross-section analysis of a typical Au NW having dimensions of 360 nm and 56 nm in length and width, respectively. The structure of the extended Au NW is grainy rather than smooth, although it looks continuous. These dimensions and morphology of a similar NW were verified by SEM, Figure 5c, resulting in similar values for length and width. This type of structural appearance has been already observed in previous publications [20,21]. A successful metalization increases the chance to synthesize continuous and "almost-cylindrical" Au NWs due to an efficient reduction of gold on the Au NPs to enlarge their size, join them, and pack the available space inside the DNA container. The electron transport study in these Au NWs was then performed.

2.2. Electrical Characterization

After fabricating and successfully metalizing the NWs on a SiO_x/Si substrate, EBL was utilized to contact the NWs using alignment markers. Nine NWs were connected with two metallic contacts for defining source and drain electrodes in the electrical transport measurements; one example is shown in Figure 6a.

Initial measurements at room temperature and atmospheric pressure revealed a large variety of resistance values; characteristic I-V curves are shown in Figure 6b. Due to the limited voltage range, all curves are linear; therefore, a resistance value can be deduced for all NWs from the slope of each I-V-curve. One of the wires was conductive with a low resistance value of 78 Ω, while the other three have much higher resistances ranging from 1.5×10^6 Ω to 5×10^9 Ω, as shown in Table 2. The remaining five NWs were either short-circuited due to failure in the EBL process or no significant current was measured above noise level. SEM micrographs of the interconnects at every NW are shown in Supplementary Information F (Figure S13A,B). The high resistance values could be explained by the imperfect connection of the NWs to the Au-contacts, or because they might not be homogeneous and continuous, e.g., the existence of a gap, or gaps, somewhere along the wire. The most plausible argument of these two will be examined below.

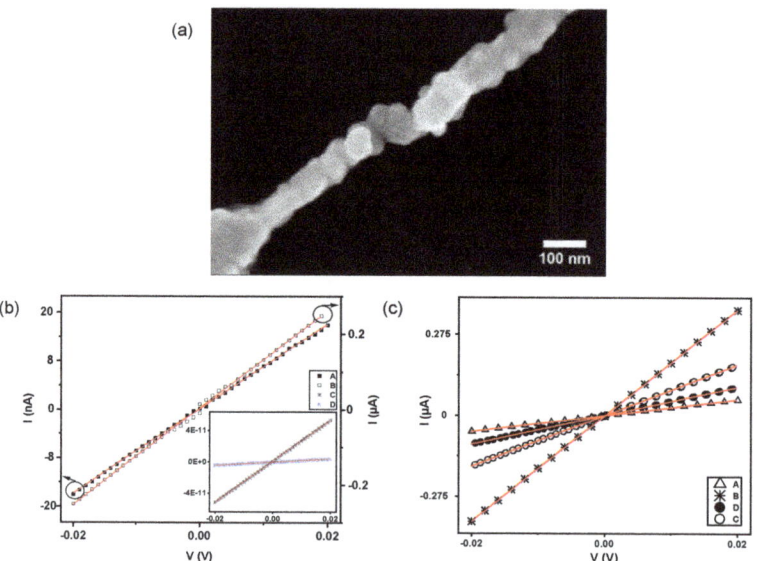

Figure 6. (**a**) A Au NW with two thermal contacts wired to the pads in the chip. (**b**) I-V measurements at room temperature (RT) and ambient pressure of four NWs using the "ambient setup". The left and right y-axis refers to the samples labeled as A and B, respectively. The inset corresponds to the I-V curves of the other two samples, labeled as C and D. (**c**) I-V measurements at room temperature (RT) in a system at a pressure of 1×10^{-6} mbar of the same devices displayed in (**b**) but using the "vacuum setup".

Table 2. Calculated resistances for the different NWs using a straight line fitting at room temperature and different pressure conditions. d and l are the diameters and lengths of the wires, respectively, as measured in the analysis of SEM images.

	$R_{RT,amb}$	$R_{RT,vac}$	d	l
A	1500(4) kΩ	365.0(1) Ω	120(10) nm	220(10) nm
B	78.0(4) Ω	56.0(2) Ω	142(10) nm	195(10) nm
C	410.0(2) MΩ	116.0(1) Ω	128(10) nm	183(10) nm
D	5.10(2) GΩ	206.0(4) Ω	116(10) nm	188(10) nm

The electrical measurements were reproduced in a different setup, termed the "vacuum setup" in the Methods section. In particular, Figure 6c shows the *I-V* curves of the same four previous NWs at room temperature from Figure 6b recorded using the "vacuum setup". The resistances were determined again from the linear fit to the measured *I-V* curves. Interestingly, the resistances were in the 56 Ω to 370 Ω range, as shown in Table 2. We now found a significant reduction in the resistance values for all wires, which were not damaged during the procedure (some of the wires were destroyed by electric shock). Even the NWs which showed tiny currents in the initial measurements were conductive in the same range as the other wires in the subsequent measurements. This decrease in resistance may be explained either by an alteration in contacts or by a modification of the morphology during the initial measurements. The resistance decrease is most prominent for the initially highly resistive wires, while it is only weak for the initially conductive wires because they are already in a low resistance state. To explain this extraordinary behavior in electron transport in metallic NWs, we compare our results to studies on similar systems in recent literature.

The use of 2D and 3D DNA Origami nanostructures to be harnessed as templates for fabricating Au NWs with dimensions and morphology determined by their geometry and shape has been reported recently [22,23]. Woolley et al. realized electric measurements at ambient conditions of C-shaped Au NWs, 130 nm long, laid down on 70 nm × 90 nm planar DNA Origami. They found resistance values in the range from 5.6×10^3 Ω to 11.7×10^6 Ω using four-probe contacts, which were defined by EBID; no post-deposition annealing was conducted on the NWs. [24]. Erbe et al. also used a self-assembled rectangular DNA Origami 90 nm × 70 nm) as a template to fabricate C-shape Au NWs with an overall length of 150 nm. The attachment probability of the Au NPs constituting the NWs was optimized for the desired patterns. To complement the previous studies, the electrical properties of these NWs were tested at temperatures ranging from 4.2 K to RT (293 K). Two C-shape NWs showed ohmic behavior at RT, and the resistances were found to be 2.3×10^6 Ω and 1.4×10^9 Ω. Despite their reduced dimensions and curved shape, these wires are as good in electrical quality as those previously manufactured, namely, longer straight wires [14]. Based on previous work, Woolley et al. studied the impact of low-temperature annealing on the morphology and conductance of DNA-patterned straight NWs at ambient conditions [25]. They found that the resistance of the wires decreased after being PMMA coated and annealed at 200 °C with four-point electrical measurements using EBID-defined contacts. The average NW resistance changed from 6.28×10^6 Ω (pre-annealing) to 1.90×10^5 Ω (post-annealing). For nanosized Au conductors, Rayleigh instabilities can cause the nanostructures to break up into small droplets even well below the melting temperature [26]. The authors claim that the diameter of the annealed structure is constrained by the PMMA coating, inhibiting the Rayleigh instability to some degree and imposing a mechanical constraint on the movement of gold. The presence of the thermally conductive EBID metal contacts, tungsten in this case, is not expected to influence the annealing of the Au NWs because both are expected to be at the substrate temperature [25].

The 3D nanostructures with an inner hollow have shown so far the best results in terms of electric conductance. Seidel et al. synthesized DNA nanostructures with a square cross-section, 40 nm long, 23 nm and 14 nm for outer and inner sides, respectively, using a modified version of the ssDNA M13mp18 genome having 7560 nt [27]. Later, they utilized this approach to couple DNA Origami mold monomers into a long linear structure to assemble conductive Au NWs [19]. The total length of each Au NW varied from 110 nm to 1000 nm with a diameter of 32(3) nm. Each mold monomer had a square cross-section with identical dimensions as in a previous report [27] but now had a 5 nm DNA-functionalized Au NP each to be used as a seed. The resistance values for approximately ten 200 nm-long NWs were widely spread from 90 Ω to 9×10^9 Ω [19]. The inhomogeneous growth of the Au NPs might be responsible for the discontinuity of the NWs and, therefore, the increase in resistance. They concluded that the charge transport of electrons in the DNA Origami template Au NWs was dominated by a single non-metallic interparticle interface that needs to be overcome by hopping at higher temperatures. An alternative method is using templates based on 3D DNA Origami nanostructures with a cross-section as close as possible to a circular shape to reduce non-metallic gaps at the interfaces. The results shown here in Figure 6 and Table 2 indicate how the conductance improves in Au NWs templated by 3D Origami if the cross-section is almost-circular, actually hexagonal in the design, possibly due to the reduction of internal strain of the metal confined to the mold during growth. Actually, the resistance values found here, see Table 2, are in the same order of magnitude to those reported in the literature, see Table 3 in [23], for similar 3D nanostructures.

An argument of plausibility will now be given to elucidate the huge variation in resistance of three out of four NWs after using two setups in different environments, compare Figure 6b,c. A simple explanation of these experimental results is provided in the spirit of Occam's razor. The high resistance values found in the devices utilizing the "ambient setup" could be explained by the possible existence of voids or vacancies within the NWs; i.e., they were probably neither completely continuous nor homogeneous after the metalization process. When passing current through such an NW, two forces act on the ions constituting the lattice of the wires: the direct electrostatic force on the electrons, and a possible contribution to the annealing of the wire by the so-called wind force. The wind force arises from the momentum transfer of electrons to metal ions, impurities, grain boundaries, vacancies, dislocations, and even phonon vibrations of ions from their equilibrium positions via scattering events when an electric field is applied [28]. These scattering processes thermally activate the migration of atoms by directionally decreasing their activation energies, E_a, and consequently result in a directed diffusion. The resulting material migration occurs in the direction of the electron flow. At low voltages, in a range from -20 mV to 20 mV, atoms or defects with the smallest activation energy barrier will become mobile in the contact region, move out of the contact due to diffusion, and redeposit in cooler regions [29]. Atoms located at dislocations, grain boundaries, surfaces, and vacancies have lower binding energies and, as a consequence, lower E_a values. An E_a value of 0.8 eV has been measured, even at low current densities, for passivated Au interconnects in which grain boundary diffusion acts as the dominant mechanism [30]. It should be noted that the electric fields arising in the structures are much larger than the fields required to induce the electrophoretic motion of isolated Au NPs in ds-DNA [31]. This further indicates that the removal of the DNA mold and the formation of continuous wires has been completely achieved in our NWs.

Due to difference in electrothermal properties, grain structure, the contribution of different atomic diffusion pathways, and the development of mechanical stresses, the wind-force-driven annealing in NWs differs from micro-scale interconnects in integrated circuits [32]. This process in NWs starts at the location of the highest divergence of atomic flux, which depends on the spatial profile of temperature, current density, and scattering cross-section. As mass is ejected from that location, the NWs narrow down over a length of approximately 10 nm [33]. Thinner and narrower Au wires are known to generate lower heat when the electron–phonon scattering length is on the order of 170 nm [34]. This whole process changes the electric resistance due to this force exerted on each metal ion core. This simple mechanism might fill the existing vacancies at the end of the experimental run in the "ambient setup", turning the Au NWs more conductive than before. As plating is typically used to produce conductive NWs, in general, annealing of plated NWs is highly relevant to the fabrication processes for NW electronics.

3. Materials and Methods

3.1. Module Design

The elongated Au NWs are cast using a 3D nanomold having a DNA module as the primary unit. This structure should withstand mechanical stress and keep its configuration during casting. To achieve these requirements, the modules used in this work are based on a variation of that created by the Peng Yin group [35], a rigid hollow 3D nanostructure synthesized by means of the DNA Origami method [10,36]. The fundamental module designed by caDNAno corresponds to a hollow tubule having a hexagonal cross-section formed by 80 helices. The module is attained by hybridization of a scaffold and several staples mixed in a buffer at the proper conditions. The Molecular Viewer software was used to obtain a quasi-realistic outline to scale by following bonds and bond angles based on the files provided by caDNAno [37], as shown in Figure 3. The design of the five modules shaping the nanomold can be simplified if the scaffold is precisely the same as the one used by the Peng Yin group, but the length and sequence of staples near the ends of the modules have been modified. By this means, the coupling of modules would be seamless because the extended staples simply bind with the complementary ones of those located close-by; no voids are left around the junction of modules (see Table S1 and the paragraph below the caption in Supplementary Information A for further information). This "peg-notch" assembly keeps the structural stability of the 3D nanomold.

Each module was ~33 nm long with ~30 nm and ~18 nm average outer and inner diameters, respectively, as observed in Figure 3 upper left. The mechanical and structural stability of each module was then studied using the CanDo computer code [38] based on the files generated by caDNAno. Flexibility maps of the DNA basic module assembly were determined and the thermally induced fluctuations were found to be in the 0.3 nm to 0.7 nm range, as shown in Figure 3 upper right. Similar results in CanDo were obtained for all five modules. The careful iterative use of caDNAno and CanDo design and analysis increases the chances of successfully synthesizing a DNA nanostructure. This whole method assures a high level of confidence in the modules' fabrication.

3.2. Basic Nanostructure Design (Five-Module Assembly)

The NWs' length is determined by an integer number of nanomolds, similar to links in a catena. For the sake of simplicity, the modules have been tagged with "colors" to distinguish them: P—B—G—Y—O. The main modification in the design of each module is harnessing the exceptional Watson–Crick base pairing between purines and pyrimidines, having the possibility to join adjacent modules in a particular manner. This is accomplished by extending certain staples, which have a minimum length of 8 nt, away from a module to be complementary to the scaffold of a neighboring module. As shown in Figure 3 lower, the front of the purple module (PF) links to the back of the blue one (BB); the front of the blue module (BF) connects to the back of the green one (GB); the front of the green module (GF) couples with the back of the yellow one (YB); and finally, the front of the yellow module (YF) joins to the back of the orange one (OB). In this case, the basic module [35] is the green one. The main goal of the nanostructure design is to prevent a union between undesired modules and avoid gaps around the faces of consecutive modules. This unique coupling between different modules allows for accomplishing a final basic nanostructure with controlled cross-section and length, resulting in a gold NW that is structurally stable and stiff.

3.3. Au NP–DNA Origami Bioconjugate Design

After functionalizing Au NPs with the suitable single strands, two NPs, 5 nm in size each, were positioned at the front and middle of each module, separated ~16 nm from center to center. This procedure was carried out before coupling the modules. Each Au NP was attached by four binding sites to the module. The sequence of the capturing strands for the Au NPs used here was devised by the Tim Liedl group in such a way that they point to opposite directions [39], i.e., the extended staples in each module for capturing Au NP1 and Au NP2 are directed along the $3' \rightarrow 5'$ and the $5' \rightarrow 3'$ directions, respectively, as displayed in Figure 7. The design aims to increase the binding probability for each NP. Similarly, the sequence of the functionalized Au NPs is complementary to those of the extended staples, not shown here. To ensure that each extended staple is pointing towards the module's interior, the staples were chosen using the Maya AutoDesk software to consider the intrinsic helicity of DNA [40].

3.4. Synthesis of Modules

The primary module is annealed by hybridization of a scaffold and several staples (Integrated DNA Technologies, Inc., Coralville, IA, USA) mixed in a buffer at the proper concentration and pH. The scaffold is a modification of the M13mp18 plasmid, 7249 nt, with a total length of 7249 + 311 nt, also termed as p7560 (Tilibit Nanosystem, Inc., Munchen, Germany). After optimizing the design for all modules and extended staples, the original plates and tubes containing the staples and scaffold were centrifuged to settle their content. Then, their actual concentrations were verified using a NanoDrop 2000 Spectrometer (Thermo Scientific, Inc., Waltham, MA, USA). These values resulted to be almost identical to the nominal concentrations for the p7560 single-strand scaffold, 100 nM, and staples, 200 µM. As a strategy for making easier and more efficient the synthesis process for the five modules, the standard and extended staples were split into "pools", namely, vials containing a small amount of solute in a standing buffer. In our case, 13 different "pools" were prepared to shape different regions of each module and capture Au NPs at the desired locations (see Supplementary Information A and B for further information).

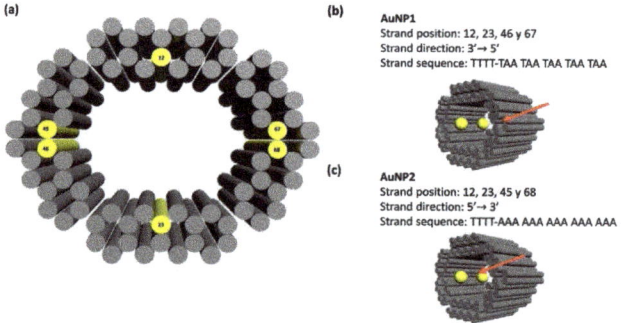

Figure 7. Au NPs location within the DNA Origami module with a "cylindrical" shape. (**a**) Selected helices (solid yellow circles) for extending some staples for attaching Au NPs. (**b**,**c**) Positions in the module and sequences of extended staples for capturing Au NP1 and Au NP2, respectively.

All pools were prepared by pipetting the suitable staples, blending them, and diluting them to a final concentration of 0.5 µM for each. A description of the necessary volumes of scaffold, buffer, and variety of pools for annealing all five modules is described in Supplementary Information B (Tables S2–S6). The mixture for each module was pipetted in a PCR microtube and annealed in a programmable thermal cycler; the temperature was raised to 80 °C and kept constant for 5 min, then the temperature was gradually decreased in steps: from 79 °C to 56 °C during 348 min, from 55 °C to 46 °C during 225 min, and from 45 °C to room temperature (25 °C) during 302 min. Afterward, the prepared modules were filtered by performing four buffer exchanges in a 1 × TAE/12.5 mM Mg^{2+} buffer (40 mM Tris-acetate, 1 mM EDTA, pH 8.2) with a Microcon DNA Fast Flow centrifugal filter (100 kDa MWCO, Millipore, Burlington, MA, USA) to remove the non-hybridized extra staples. The final concentration of each module was measured using a NanoDrop and resulted to be 10 nM.

3.5. Visualization of Modules and Nanomolds

Gel electrophoresis was performed to reveal the formation of each module and extract the modules for further preparation. An image of one of the resulting gels is shown in Supplementary Information C (Figure S1). The resulting nanostructures were analyzed by TEM. A Lacey carbon grid (Ted Pella, Inc., Redding, CA, USA) was employed to deposit 2.5 µL of a sample after annealing. The sample was deposited and adhered to the grid's network after waiting for 3 min, and then 2.5 µL of uranyl acetate at 2% was added for staining and incubated for 5 min. Afterward, a JEOL JEM 2000 FX TEM was used for imaging. Apart from TEM, all visualization and electrical characterization techniques require an insulating substrate. The modules, basic nanostructure, and Au NWs were deposited on 1 cm × 1 cm pieces of p-Si (100) wafer with a 300 nm SiO_2 electrical insulation layer, named SiO_x/Si substrate. Before dropping the DNA Origami, the wafer pieces were washed with ethanol and dried using a N_2 flow. They were pre-treated in an O_2 plasma (PICO, Diener Electronic-Plasma Surface Technology) at 7sccm O_2, power of 240 W for 5 min to render their surfaces hydrophilic. At that moment, each sample holder was rinsed with ethanol (20 s) and DI water (20 s) and dried in a N_2 stream. A volume of 20 µL of the sample was immediately dropped on the substrate surface at a final concentration of 1 nM, or 0.5 nM if 10 × TAE Mg^{2+} buffer was added, and incubated for 1 h. Afterward, the substrate was washed using HPLC-grade water and carefully dried in a N_2 stream. AFM and SEM then studied the morphology of all samples. The DNA assemblies were characterized by AFM operating in tapping mode using a Bruker MultiMode 3 and Bruker MultiMode 8 Scanning Probe Microscopes and aluminum reflex coated tips (Tap150Al-G from Nanoandmore, force constant 5 N m^{-1}, tip radius < 10 nm) for obtaining high-resolution images. The topographic AFM images were analyzed using the Gwyddion software (http://gwyddion.net/) and the one provided with the Bruker instruments (Billerica, MA, USA). To study the proper concentration of a divalent cation for stabilizing the DNA Origami modules and diminishing aggregation when specimens were laid down on the SiO_x/Si substrates, samples containing 125 mM, 200 mM and 350 mM Mg^{2+} diluted into 10 × TAE (40 mM Tris-acetate y 1 mM EDTA, pH 8.3) were prepared, and module deformation and aggregation was minimized when pipetted at 200 mM, see Supplementary Information D (Figures S2–S7). It was found that adding Mg^{2+} at 200 mM into the buffer was the proper amount. After proving that the nanomold can be constructed and analyzed by TEM and AFM, the casting of the NWs was carried out by a two-step process. Firstly, two Au NPs were placed inside each module. Secondly, an electroless chemical solution was used to reduce gold on the Au NPs surface to grow and merge them, filling the existing space within the stiff DNA mold.

3.6. Annealing of Au NP–Origami Bioconjugates

The Au NPs (Ted Pella, Inc.), 5 nm in size and nominal concentration of 5×10^{13} particles/mL, were conjugated to the specific oligonucleotide chains, complementary to those shown in Figure 7. The sequences of the capture strands for Au NP1 and Au NP2 are ATT ATT ATT ATT ATT TTTT-SH and TTT TTT TTT TTT TTT TTTT-SH and directed along the $5' \to 3'$ and the $3' \to 5'$ directions, respectively. The capture single strands had a thiol (SH) modification on either the $5'$ end or $3'$ end for binding to a Au NP, followed by a 4-base spacer (TTTT) for increased flexibility. The "salt aging" method was used to neutralize the inherent charge in Au NPs and thiolated single strands; both have a negative charge and still keep the colloidal stability of Au NPs. This method was followed for both Au NP1 and Au NP2 and has been previously reported [20]. Later, the disulfide–DNA sequence was added and mixed to the solution at a ratio of 1:5 Au NPs: ssDNA, left in a TekTrator® V type shaker at 120 rpm, and allowed to incubate for ≈48 h. After incubation, the Au NPs were backfilled with thiolated T5 sequences at a ratio of 1:60

to prevent Au NPs aggregation in the presence of high Mg^{2+} concentration buffers and left incubating for another 24 h. Excess DNA single strands were removed from the Au NPs–ssDNA conjugates by running the bioconjugate solutions for each Au NP on two 3 % agarose gels (1 × TAE Mg^{2+}, 12.5 mM) for 15 min at 100 V and 4 °C. The band containing Au NPs–ssDNA with the highest concentration (the most intense) was sliced and extracted from the gel, then finely chopped and centrifuged using a Freeze'N Squeeze kit (Bio-Rad Labs, Tokyo, Japan). Next, 1 × TAE buffer was added to the microtube containing the Au NP1 and Au NP2 bioconjugates until it reached the final concentration of 200 nM for each, measured using a NanoDrop. The final solvent was assumed to be 1 × TAE buffer with residues from the Freeze´N Squeeze product and the agarose gel. Now the Au NP–Module bioconjugates are ready to be hybridized. The Au NPs were attached by four binding sites, being the functionalized Au NPs complementary to those of the capturing strands in each module. The protocol for this part is almost identical to that developed by our group and found elsewhere [19]. A ratio of 6:1 of Au NP–ssDNA: Module was prepared to increase the likelihood of success. The original concentrations of Au NPs–ssDNA conjugates and each module resulted in 200 nM and 10 nM, respectively. Therefore, the following volumes: 0.6 µL of Au NP1–ssDNA, 0.6 µL of Au NP2–ssDNA, 2 µL of the module, and 16.8 µL of 1 × TAE were pipetted in five microtubes for each module. This means that the final concentrations of 1 nM for DNA Origami and 6 nM of Au NPs would be obtained in a volume of 20 µL in each vial. To ease and assure that the Au NP:ssDNA conjugates were transferred from the solution to the interior of each module, a sideral shaker with a heater (ThermoMixer Eppendorf) was used. The five vials containing each sample were secured on the shaker, warmed to 40 °C, then gradually cooled to 23 °C (room temperature) for 5 h at a frequency of 300 rpm. Later, the samples were imaged by AFM in the tapping mode in air. As mentioned, the Au NPs–Mold bioconjugates were obtained by hybridizing the five Au NPs–modules bioconjugates. Microtubes containing the same stoichiometry and volume, 5 µL, for each module bioconjugate, were pipetted and mixed in a buffer containing 5 mM Tris-HCl, 1 mM EDTA, 11 mM mM $MgCl_2$, pH 8.0, with 350 mM NaCl and incubated overnight. Thus, the final basic nanostructure, or mold, was formed and ready to be metalized.

3.7. Metalization of Au NP–Origami Bioconjugates

Subsequently, Au NPs within the mold were enhanced until they merged to shape quasi-cylindrical Au NWs. The growth of the Au NPs was controlled using the GoldEnhance™, EM Plus (Nanoprobes, Inc., Yaphank, NY, USA) metalization kit having four homogeneous mixtures. Same volumes, 10 µL, were taken from each solution, 1:1 ratio, and mixed with 40 µL of 1 × TAE buffer, 1:2 ratio, with 20 mM $MgCl_2$. This gold plating mixture, 80 µL, was separated into four equal volumes and dropped onto the Au NP–Origami conjugates, which were laid down on a clean SiO_x/Si substrate at equal intervals for a total of 10 min using a humidity chamber. Afterward, HPLC water was thoroughly sprayed onto the sample to inhibit the metalization procedure, and finally, N_2 was carefully blown to dry it. The Au NWs were ready to be contacted by EBL.

3.8. Fabrication of Contacts and Electric Characterization

A RAITH e-line Plus system was used to analyze the samples' morphology by SEM, fabricate the electric pads, and contact the NWs using alignment markers by lithography. EBL (e-line Plus) was used to manufacture the electrical contact pads and markers on p-Si (100) with a ≈280 nm thick SiO_x electrical insulation layer, previously plated with DNA-Origami-based gold wires. This random arrangement of NWs was firstly treated in an O_2 plasma (PICO, Diener Electronic-Plasma Surface Technology, Ebhausen, Germany) at a flow rate of 5 sccm, input power of 240 W for 30 min to minimize any amount of organic residues on the sample. Then, the ZEP 520A electron beam resist was spin-coated and baked for 10 min at 150 °C. The resist was exposed to a 10 kV acceleration voltage with a 30 µm aperture size, resulting in a beam current of ≈200 pA. To remove the resist in the exposed structures on top of the NWs, the sample was developed for 90 s in N-amyl acetate and subsequently immersed for 30 s in isopropanol (IPA) to stop the development. An e-beam evaporator Creavac CREAMET 600 (CREAVAC) was used to evaporate a 5 nm thick titanium adhesion layer initially and then a 90 nm gold deposit with $2\,\text{Å}\,\text{s}^{-1}$ and $5\,\text{Å}\,\text{s}^{-1}$ rates, respectively. The lithography was completed with a lift-off process that included 90 s in N, N-dimethylacetamide (ZDMAC), 30 s in the stopper (Isopropyl Alcohol), and dried by a N_2 flow for 30 s. Due to the arbitrary distribution of NWs on the SiO_x/Si substrate, EBL initally produced arrays of 8 μm markers. The alignment markers' positions determined the placement of the NWs. They were subsequently contacted by two electrodes separated by a wide range of distances depending on the length of the NWs using the same lithography procedure as previously described. Two probes were used for the electrical characterization through two contacts on the NWs using two different experimental arrangements: "ambient setup" and "vacuum setup". The former uses a Semiconductor Characterization System with the Keithley Interactive Test Environment; the system is located in a gray room environment at atmospheric conditions in pressure and temperature. The latter is based on an Agilent 4156C Precision Semiconductor Parameter Analyzer; the samples were placed on a substrate holder in a vacuum chamber at a base pressure of 1×10^{-6} mbar and at room temperature. The I-V measurements for both systems were taken in the -20 mV to 20 mV input voltage range in bright light using the power supplies for each system described above. The resistance of two-probe devices was determined by linear fitting of the I-V curves at room temperature.

4. Conclusions

We have demonstrated that the self-assembly of Au NWs inside quasi-circular molds synthesized using the DNA Origami method leads to the successful formation of continuous metal lines. The percentage of NWs exhibiting very low resistance values was increased substantially compared to earlier studies on metalizing six-helix-bundles [41] or rectangular molds [19]. This result indicates that strain induced in the metallic layers during the growth of the NWs may play an essential role during the formation of continuous metal lines. Some of the wires have large initial resistances that dropped by orders of magnitude after initial measurements; these wires behaved very similarly to wires with low initial resistance values in all further measurements. We explain this significant change in resistance with current-induced annealing of the NWs caused by electromigration of the Au atoms on the substrate surface. These results pave the way for the reliable formation of metal NWs and thus the creation of interconnects in nanoelectronic circuits using self-assembly. Such an approach to nanolithography may offer possibilities for energy-saving fabrication of future nanoelectronic components.

Supplementary Materials: The following supporting information can be downloaded at: https://www.mdpi.com/article/10.3390/ijms241713549/s1.

Author Contributions: Conceptualization, E.C.S. and A.E.; methodology, E.C.S., S.J.G., and D.D.R.A.; validation, E.C.S. and A.E.; formal analysis, S.J.G.; investigation, D.D.R.A. and S.J.G.; writing—original draft preparation, D.D.R.A., S.J.G., A.E., and E.C.S.; writing—review and editing, E.C.S. and A.E.; visualization, D.D.R.A.; supervision, E.C.S. and A.E.; project administration, E.C.S. and A.E.; funding acquisition, E.C.S. and A.E. All authors have read and agreed to the published version of the manuscript.

Funding: This research was funded by DFG grant number GRK 2767 and ER341/19-2.

Data Availability Statement: Data can be made available on request by the authors.

Acknowledgments: We are thankful to T. Schönherr and C. Neisser for their help in the fabrication and measurement processes. The nanofabrication facilities (NanoFaRo) at the Ion Beam Center at the HZDR and TEM Center at North Carolina State University in Raleigh, NC, USA, are also gratefully acknowledged. Enrique C. Samano and David Ruiz Arce are grateful to CONACYT and

UNAM-DGAPA for their financial support by means of the CB-176352 and PAPIIT-IG100417 projects, respectively.

Conflicts of Interest: The authors declare no conflicts of interest.

References

1. *International Roadmap for Devices and Systems: 2020*; Technical Report; IEEE: Manhattan, NY, USA , 2020.
2. Hornyak, G.L.; Tibbals, H.F.; Dutta, J.; Moore, J.J. *Introduction to Nanoscience and Nanotechnology*; CRC Press: Boca Raton, FL, USA, 2008. [CrossRef]
3. Zhang, S. Building from the Bottom Up. *Mater. Today* **2003**, *6*, 20–27. [CrossRef]
4. Xu, A.; Harb, J.N.; Kostiainen, M.A.; Hughes, W.L.; Woolley, A.T.; Liu, H.; Gopinath, A. DNA Origami: The Bridge from Bottom to Top. *MRS Bull.* **2017**, *42*, 943–950. [CrossRef]
5. Xavier, P.L.; Chandrasekaran, A.R. DNA-based Construction at the Nanoscale: Emerging Trends and Applications. *Nanotechnology* **2018**, *29*, 062001. [CrossRef]
6. Madsen, M.; Gothelf, K.V. Chemistries for DNA Nanotechnology. *Chem. Rev.* **2019**, *119*, 6384–6458. [CrossRef] [PubMed]
7. Zhang, L.; Ma, X.; Wang, G.; Liang, X.; Mitomo, H.; Pike, A.; Houlton, A.; Ijiro, K. Non-Origami DNA for Functional Nanostructures: From Structural Control to Advanced Applications. *Nano Today* **2021**, *39*, 101154. [CrossRef]
8. Tapio, K.; Bald, I. The Potential of DNA Origami to Build Multifunctional Materials. *Multifunct. Mater.* **2020**, *3*, 032001. [CrossRef]
9. Bathe, M.; Chrisey, L.A.; Herr, D.J.C.; Lin, Q.; Rasic, D.; Woolley, A.T.; Zadegan, R.M.; Zhirnov, V.V. Roadmap on Biological Pathways for Electronic Nanofabrication and Materials. *Nano Futur.* **2019**, *3*, 012001. [CrossRef]
10. Rothemund, P.W.K. Folding DNA to Create Nanoscale Shapes and Patterns. *Nature* **2006**, *440*, 297–302. [CrossRef]
11. Seeman, N.C. Nanomaterials Based on DNA. *Annu. Rev. Biochem.* **2010**, *79*, 65–87. [CrossRef]
12. Samanta, A.; Medintz, I.L. Nanoparticles and DNA—A Powerful and Growing Functional Combination in Bionanotechnology. *Nanoscale* **2016**, *8*, 9037–9095. [CrossRef]
13. Chen, Z.; Liu, C.; Cao, F.; Ren, J.; Qu, X. DNA Metallization: Principles, Methods, Structures, and Applications. *Chem. Soc. Rev.* **2018**, *47*, 4017–4072. [CrossRef] [PubMed]
14. Bayrak, T.; Martinez-Reyes, A.; Arce, D.D.R.; Kelling, J.; Samano, E.C.; Erbe, A. Fabrication and Temperature-Dependent Electrical Characterization of a C-shape Nanowire Patterned by a DNA Origami. *Sci. Rep.* **2021**, *11*, 1922. [CrossRef] [PubMed]
15. Nguyen, N.V.; Yang, C.H.; Liu, C.J.; Kuo, C.H.; Wu, D.C.; Jen, C.P. An Aptamer-Based Capacitive Sensing Platform for Specific Detection of Lung Carcinoma Cells in the Microfluidic Chip. *Biosensors* **2018**, *8*, 98. [CrossRef] [PubMed]
16. Rojo, M.M.; Calero, O.C.; Lopeandia, A.F.; Rodriguez-Viejo, J.; Martín-Gonzalez, M. Review on Measurement Techniques of Transport Properties of Nanowires. *Nanoscale* **2013**, *5*, 11526–11544. [CrossRef]
17. Watson, S.M.D.; Pike, A.R.; Pate, J.; Houlton, A.; Horrocks, B.R. DNA-templated Nanowires: Morphology and Electrical Conductivity. *Nanoscale* **2014**, *6*, 4027–4037. [CrossRef]
18. Šponer, J.; Šponer, J.E.; Mládek, A.; Jurečka, P.; Banáš, P.; Otyepka, M. Nature and Magnitude of Aromatic Base Stacking in DNA and RNA: Quantum Chemistry, Molecular Mechanics, and Experiment. *Biopolymers* **2013**, *99*, 978–988. [CrossRef]
19. Bayrak, T.; Helmi, S.; Ye, J.; Kauert, D.; Nano, J.K. DNA-mold Templated Assembly of Conductive Gold Nanowires. *ACS Nano Lett.* **2018**, *18*, 2116–2123. [CrossRef]
20. Pilo-Pais, M.; Goldberg, S.; Samano, E.; LaBean, T.H.; Finkelstein, G. Connecting the Nanodots: Programmable Nanofabrication of Fused Metal Shapes on DNA Templates. *Nano Lett.* **2011**, *11*, 3489–3492. [CrossRef]
21. Schreiber, R.; Kempter, S.; Holler, S.; Schüller, V.; Schiffels, D.; Simmel, S.S.; Nickels, P.C.; Liedl, T. DNA Origami-Templated Growth of Arbitrarily Shaped Metal Nanoparticles. *Small* **2011**, *7*, 1795–1799. [CrossRef]
22. Bayrak, T.; Jagtap, N.S.; Erbe, A. Review of the Electrical Characterization of Metallic Nanowires on DNA Templates. *Int. J. Mol. Sci.* **2018**, *19*, 3019. [CrossRef]
23. Pang, C.; Aryal, B.R.; Ranasinghe, D.R.; Westover, T.R.; Ehlert, A.E.F.; Harb, J.N.; Davis, R.C.; Woolley, A.T. Bottom-Up Fabrication of DNA-Templated Electronic Nanomaterials and Their Characterization. *Nanomaterials* **2021**, *11*, 1655. [CrossRef] [PubMed]
24. Aryal, B.R.; Westover, T.R.; Ranasinghe, D.R.; Calvopiña, D.G.; Uprety, B.; Harb, J.N.; Davis, R.C.; Woolley, A.T. Four-Point Probe Electrical Measurements on Templated Gold Nanowires Formed on Single DNA Origami Tiles. *Langmuir* **2018**, *34*, 15069–15077. [CrossRef] [PubMed]
25. Westover, T.R.; Aryal, B.R.; Ranasinghe, D.R.; Uprety, B.; Harb, J.N.; Woolley, A.T.; Davis, R.C. Impact of Polymer-Constrained Annealing on the Properties of DNA Origami-Templated Gold Nanowires. *Langmuir* **2020**, *36*, 6661–6667. [CrossRef]
26. Karim, S.; Toimil-Molares, M.E.; Balogh, A.G.; Ensinger, W.; Cornelius, T.W.; Khan, E.U.; Neumann, R. Morphological Evolution of Au Nanowires Controlled by Rayleigh Instability. *Nanotechnology* **2006**, *17*, 5954. [CrossRef]
27. Helmi, S.; Ziegler, C.; Kauert, D.J.; Seidel, R. Shape-Controlled Synthesis of Gold Nanostructures Using DNA Origami Molds. *Nano Lett.* **2014**, *14*, 6693–6698. [CrossRef]
28. Lienig, J.; Thiele, M. *Fundamentals of Electromigration-Aware Integrated Circuit Design*; Springer International Publishing: Cham, Switzerland, 2018. [CrossRef]
29. Hoffmann-Vogel, R. Electromigration and the Structure of Metallic Nanocontacts. *Appl. Phys. Rev.* **2017**, *4*, 031302. [CrossRef]
30. Kilgore, S. Electromigration in Gold Interconnects. Ph.D. Thesis, Arizona State University, Tempe, AZ, USA, 2013.

31. Cervantes-Salguero, K.; Kawamata, I.; Nomura, S.i.M.; Murata, S. Unzipping and Shearing DNA with Electrophoresed Nanoparticles in Hydrogels. *Phys. Chem. Chem. Phys.* **2017**, *19*, 13414–13418. [CrossRef]
32. Sawtelle, S.D.; Kobos, Z.A.; Reed, M.A. Electromigration in Gold Nanowires under AC Driving. *Appl. Phys. Lett.* **2018**, *113*, 193104. [CrossRef]
33. Heersche, H.B.; Lientschnig, G.; O'Neill, K.; van der Zant, H.S.J.; Zandbergen, H.W. In Situ Imaging of Electromigration-Induced Nanogap Formation by Transmission Electron Microscopy. *Appl. Phys. Lett.* **2007**, *91*, 072107. [CrossRef]
34. Durkan, C.; Schneider, M.A.; Welland, M.E. Analysis of Failure Mechanisms in Electrically Stressed Au Nanowires. *J. Appl. Phys.* **1999**, *86*, 1280–1286. [CrossRef]
35. Sun, W.; Boulais, E.; Hakobyan, Y.; Wang, W.L.; Guan, A.; Bathe, M.; Yin, P. Casting Inorganic Structures with DNA Molds. *Science* **2014**, *346*, 1258361. [CrossRef] [PubMed]
36. Majikes, J.M.; Alexander Liddle, J. Synthesizing the Biochemical and Semiconductor Worlds: The Future of Nucleic Acid Nanotechnology. *Nanoscale* **2022**, *14*, 15586–15595. [CrossRef] [PubMed]
37. Douglas, S.M.; Marblestone, A.H.; Teerapittayanon, S.; Vazquez, A.; Church, G.M.; Shih, W.M. Rapid Prototyping of 3D DNA-origami Shapes with caDNAno. *Nucleic Acids Res.* **2009**, *37*, 5001–5006. [CrossRef] [PubMed]
38. Kim, D.N.; Kilchherr, F.; Dietz, H.; Bathe, M. Quantitative Prediction of 3D Solution Shape and Flexibility of Nucleic Acid Nanostructures. *Nucleic Acids Res.* **2012**, *40*, 2862–2868. [CrossRef]
39. Kuzyk, A.; Schreiber, R.; Fan, Z.; Pardatscher, G.; Roller, E.M.; Högele, A.; Simmel, F.C.; Govorov, A.O.; Liedl, T. DNA-based Self-Assembly of Chiral Plasmonic Nanostructures with Tailored Optical Response. *Nature* **2012**, *483*, 311–314. [CrossRef]
40. Benson, E.; Mohammed, A.; Gardell, J.; Masich, S.; Czeizler, E.; Orponen, P.; Högberg, B. DNA Rendering of Polyhedral Meshes at the Nanoscale. *Nature* **2015**, *523*, 441–444. [CrossRef]
41. Teschome, B.; Facsko, S.; Schoenherr, T.; Kerbusch, J.; Keller, A.; Erbe, A. Temperature-Dependent Charge Transport through Individually Contacted DNA Origami-Based Au Nanowires. *Langmuir* **2016**, *32*, 10159–10165. [CrossRef]

Disclaimer/Publisher's Note: The statements, opinions and data contained in all publications are solely those of the individual author(s) and contributor(s) and not of MDPI and/or the editor(s). MDPI and/or the editor(s) disclaim responsibility for any injury to people or property resulting from any ideas, methods, instructions or products referred to in the content.

Article

In Vitro Study of Composite Cements on Mesenchymal Stem Cells of Palatal Origin

Alina Ioana Ardelean [1], Madalina Florina Dragomir [1], Marioara Moldovan [2,*], Codruta Sarosi [2], Gertrud Alexandra Paltinean [2], Emoke Pall [3], Lucian Barbu Tudoran [4,5], Ioan Petean [6,*] and Liviu Oana [1]

[1] Department of Veterinary Surgery, Faculty of Veterinary Medicine, University of Agricultural Sciences and Veterinary Medicine, 3-5 Manastur Street, 400372 Cluj-Napoca, Romania
[2] Raluca Ripan Institute for Research in Chemistry, Babeș-Bolyai University, 30 Fantanele Street, 400294 Cluj-Napoca, Romania
[3] Department of Veterinary Reproduction, Obstetrics and Gynecology, Faculty of Veterinary Medicine, University of Agricultural Sciences and Veterinary Medicine, 3-5 Manastur Street, 400372 Cluj-Napoca, Romania
[4] Faculty of Biology and Geology, Babes-Bolyai University, 44 Gheorghe Bilaşcu Street, 400015 Cluj-Napoca, Romania
[5] National Institute for Research and Development of Isotopic and Molecular Technologies, 65-103 Donath Street, 400293 Cluj-Napoca, Romania
[6] Faculty of Chemistry and Chemical Engineering, Babes-Bolyai University, 11 Arany Janos Street, 400028 Cluj-Napoca, Romania
* Correspondence: marioara.moldovan@ubbcluj.ro (M.M.); ioan.petean@ubbcluj.ro (I.P.)

Abstract: Uniform filler distribution in composites is an important requirement. Therefore, BaO glass, nano hydroxyapatite and quartz filler distribution was realized through PCL microcapsules which progressively release filler during matrix polymerization. Two composites were realized based on a complex matrix containing BisGMA, UDMA, HEMA and PEG400 mixed with a previously described mineral filler: 33% for C1 and 31% for C2. The spreading efficiency was observed via SEM, revealing a complete disintegration of the microcapsules during C1 polymerization, while C2 preserved some microcapsule parts that were well embedded into the matrix beside BaO filler particles; this was confirmed by means of the EDS spectra. Mesenchymal stem cells of palatal origin were cultured on the composites for 1, 3, 5 and 7 days. The alkaline phosphatase (ALP) level was measured at each time interval and the cytotoxicity was tested after 3, 5 and 7 days of co-culture on the composite samples. The SEM investigation showed that both composites allowed for robust proliferation of the cells. The MSC cell pluripotency stage was observed from 1 to 3 days with an average level of ALP of 209.2 u/L for C1 and 193.0 u/L for C2 as well as a spindle cell morphology. Cell differentiation occurred after 5 and 7 days of culture, implied by morphological changes such as flattened, star and rounded shapes, observed via SEM, which were correlated with an increased ALP level (279.4 u/L for C1 and 284.3 u/L for C2). The EDX spectra after 7 days of co-culture revealed increasing amounts of P and Ca close to the hydroxyapatite stoichiometry, indicating the stimulation of the osteoinductive behavior of MSCs by C1 and C2. The MTT assay test showed a cell viability of 98.08% for C1 and 97.33% for C2 after 3 days, proving the increased biocompatibility of the composite samples. The cell viability slightly decreased at 5 and 7 days but the results were still excellent: 89.5% for C1 and 87.3% for C2. Thus, both C1 and C2 are suitable for further in vivo testing.

Keywords: dental composite cement; mesenchymal stem cells; biocompatibility

Citation: Ardelean, A.I.; Dragomir, M.F.; Moldovan, M.; Sarosi, C.; Paltinean, G.A.; Pall, E.; Tudoran, L.B.; Petean, I.; Oana, L. In Vitro Study of Composite Cements on Mesenchymal Stem Cells of Palatal Origin. *Int. J. Mol. Sci.* **2023**, *24*, 10911. https://doi.org/10.3390/ijms241310911

Academic Editor: Bruce Milthorpe

Received: 18 May 2023
Revised: 21 June 2023
Accepted: 28 June 2023
Published: 30 June 2023

Copyright: © 2023 by the authors. Licensee MDPI, Basel, Switzerland. This article is an open access article distributed under the terms and conditions of the Creative Commons Attribution (CC BY) license (https://creativecommons.org/licenses/by/4.0/).

1. Introduction

The restoration or replacement of damaged and diseased parts of body tissues and organs has led to the continuous development of biomaterials in regenerative medicine. This interdisciplinary and revolutionary field has gained major importance due to the contact of new biomaterials with biological systems in recent years. A biomaterial needs to be biocompatible and nontoxic so as to have a positive interaction with the biological system, to be resistant over time and to have mechanical and physicochemical stability [1,2]. A biomaterial must be designed in order to induce specific biological activity [3,4]. The goal of biomaterials consists in generating efficient interactions to treat, repair or replace damaged tissue. The literature characterizes biomaterials as follows: bioinert or biostable materials are named the first generation, biocompatible and bioactive materials are considered the second generation and biodegradable or bioresorbable materials are the third generation while biomimetic or bioinspired materials are the newest fourth generation [5,6].

The biodegradation of a biomaterial is very important, and it should be designed in a way that assures easy absorption by the body without any harmful effects and substitution of the affected tissues while considering sustainability and ecological responsibility [7–9]. Some of the newest materials are able to influence the bone healing process and osteogenesis and promote cell differentiation and proliferation of osteoblasts according to previous studies [10–12].

The new trend in biomaterial research is their use in dental materials. Their biocompatibility and high chemical stability assure optimal treatment of patients' pathologies. Scientific research in this field facilitates technological development for biomaterials and assures their successful clinical testing in vivo. Such results help dentists to be trained in new procedures and increase their treatment options to fulfill the particular needs of each patient. Thus, a smart combination of dental biomaterials is required to fulfill these purposes.

Bis-GMA (bisphenol A-glycidyl methacrylate) is a resin used as a base matrix in dental sealant composites due to its good mechanical properties, low volatility, diffusivity in tissues and low polymerization shrinkage [13–16]. Its adhesion properties strongly depend on its viscosity which is often adjusted using fluidizer additives such as resins containing phosphoric acid residues with methacrylate groups [17]. It is relatively cytotoxic [18] and thus it requires a smart combination with other polymers to enhance the composite's biocompatibility. Bis-GMA must be substituted with other monomers such as urethane dimethacrylate (UDMA) due to its high viscosity and low degree of conversion. The UDMA substitution increases Bis-GMA's mechanical properties such as flexural strength, flexural modulus of elasticity and polymerization shrinkage in dental restorative composites [19–21].

PEG-400 (polyethylene glycol 400) is a polymer with a low molecular weight that is widely used in a variety of medical formulations such as parenteral, topical, ophthalmic, oral and rectal formulations, mainly due to its low toxicity [22,23]. Glasses such as BaO and BaF_2 have been utilized due to their capacity to form strong chemical bonds with both soft tissues and hard tissues [15,21], assuring good osteogenic and angiogenic characteristics [24,25] and improving the structural stability and bioactivity of the composite materials [26].

Biomaterial functionalization is a very complex subject requiring the stimulation of an enhanced cell response for osteogenic differentiation [27–29]. Size is an important issue for biocompatibility. Since biological structures are nanostructured, the functionalization materials must also be nanostructured for a better physical match. Therefore, particles intended for functionalization must be situated in the size range of 1 to 100 nm [30] and have the ability to increase the material's biocompatibility [31]. A wide variety of nanoparticles have the potential to inhibit some of the cytotoxic effects of materials destined for biological purposes. Consequently, evaluation of the nanoparticles' cytotoxic effects during biomaterial development would increase their efficacy and the final products' safety [32,33].

Dental restorative composites have an important role in dental fissure sealing and enamel surface rehabilitation as well as cavity filling. Their composition might be adapted using colored fillings for mimicking natural teeth color and as a base for glass ionomer cements. Cytotoxic behavior and biocompatibility might be properly adjusted via strict control of their composition, assuring an increase in healing rate [34–36].

The current research is focused on the in vitro investigation of the biocompatibility of two dental composite cements. The organic matrix is based on Bis-GMA(2,2-

bis[p-(2′-hydroxy-3′-metacryloxypropoxy)phenyl]-propane) doped with UDMA (urethane dimethacrylate) and moderated by HEMA (Hydroxyethyl methacrylate) and PEG 400 (polyethylene glycol). The inorganic matrix consists of mineral filler particles such as BaO glass, nanostructured hydroxyapatite nanoparticles (nHA) and silanized quartz nanoparticles. The novelty element within current research is the controlled release of the mineral filler into the polymer matrix during the polymerization process. Filler-targeted delivery is achieved through polycaprolactone microcapsules containing mineral filler particles assisted by buffalo whey as an anti-agglomerate agent and bioactive promoter. Each of the investigated cements has the same polymeric matrix, and the filler amount differs between C1 and C2 samples.

Mesenchymal stem cells (MSCs) of palatal origin are an optimal testing medium for the biocompatibility of dental materials [37–39]. Therefore, both composites were subjected to MSC proliferation aiming at the assessment of biocompatibility and evidence of cytotoxic effects. It is assumed that small differences in the mineral filler ratio might significantly influence the composites' biologic behavior, in good agreement with the literature [40,41]. The null hypothesis states that a small difference in mineral filler has no influence on the composites' bioactivity in vitro.

2. Results

One of the main goals of the present research is to study the targeted delivery of mineral filler within composite cement samples and its influence on the materials' biocompatibility regarding MSCs. Therefore, experimental results were grouped according to their specific aspects.

2.1. Composite Cement Characterization

The microstructural aspect of the composite cement samples is very important and, thus, it was investigated with Scanning Electron Microscopy (SEM). The obtained images and corresponding EDX spectra are presented in Figure 1.

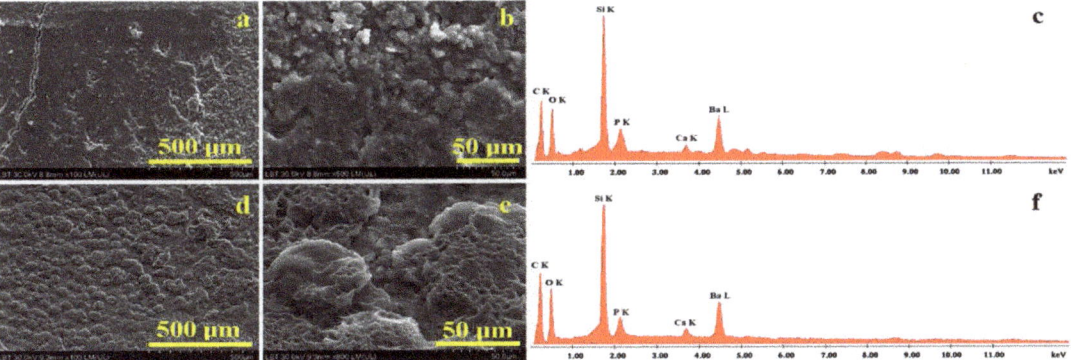

Figure 1. SEM images of composites before cytotoxicity testing: (**a**) C1 overview, (**b**) C1 microstructural details, (**c**) EDX spectrum for details in (**b**), (**d**) C2 overview, (**e**) C2 microstructural details and (**f**) EDX spectrum for details in (**e**).

Figure 1a evidences the overview aspect of the C1 composite. The surface topography is not uniform; two morphologies are observed. On the left side, a smooth area with pores and cracks in the composite is observed, and on the right side, an agglomeration of particles is very well defined. The left side crack occurs due to the sample's superficial excoriation and did not penetrate the composite bulk in depth. High-magnification observation of the C1 composite, Figure 1b, evidences the constituent phases of the sample. In the upper side of the SEM image, the overlapping of particles is evident; barium oxide glass is more evident in this side of the image. They have a boulder shape, and sizes vary from about 1 to 10 µm in diameter. The bottom of the SEM image presents a smoother area where fine particles are more visible and corresponds to nanostructural hydroxyapatite and silanized quartz clusters with rounded shapes and submicron sizes. The elemental composition which resulted for the C1 sample is presented in Table 1. The dominant elements in the surface of the C1 composite are C and O due to the organic matter within the polymer matrix. Si and Ba belong to the nanostructural silica and barium glass filler particles, respectively. Small amounts of P and Ca also occur due to nanostructured hydroxyapatite; their amounts are close to the stoichiometry proportion of nHA.

Table 1. Elemental composition obtained via SEM images for composite samples.

Samples	Identified Elements, Wt.%.					
	C	O	Si	P	Ca	Ba
C1	34.05	16.96	24.13	3.07	2.98	18.81
C2	33.98	21.48	23.91	2.32	2.26	16.05

The microstructural aspects of the C2 composite, Figure 1d, shows the overall morphology of the filler component distribution in the mesoporous structure; it is most likely caused by polycaprolactone microcapsule remains, which are further covered by the polymeric matrix after the composite's polymerization. The mesoporous microcapsules have an average diameter between 40–120 µm (as observed in Figure 1c) and are evenly placed on the surface of the sample. Microstructural details observed in Figure 1e show with accuracy the filler microcapsule's disintegration, releasing filler particles that are embedded into the polymeric matrix. The microcapsules have an average diameter around of 55–60 µm. BaO glass particles have a boulder shape and diameter situated around 2–10 µm, while nHA and quartz nanoparticles clusters are less visible except for the smoother area in the upper central side of the SEM image in Figure 1e. Their aspect is rounded, and sizes are situated in the submicron domain. The elemental composition was investigated via EDX spectrometry, and the obtained values are presented in Table 1. The dominant elements are carbon and oxygen due to the polymer matrix. The filler particles' characteristic elements such as Si, Ba, P and Ca are evidenced, but their amount is less than the one observed in C1, a fact which is in good agreement with the material preparation receipt.

The chemical bonds and molecular interaction within the polymer and those within the mineral filler were investigated via Fourier Transform Infrared Spectroscopy (FTIR), and the obtained results are presented in Figure 2.

Mineral filler particles are embedded well into the polymer matrix, and FTIR spectra evidence their chemical bonds' specific vibrations. The absorption band at 586 cm^{-1} belongs to the Ba–O stretching vibration and the band at 730 cm^{-1} also belongs to the Ba-O bond [42,43]. BaO-related absorption bands are stronger for the C1 sample due to a higher filler amount than in the C2 sample. Hydroxyapatite presence is evidenced by intense absorption bands around 560–600 cm^{-1} and 1000–1100 cm^{-1} belonging to the PO$_4$ chemical bond, and the absorption band at 3300 cm^{-1} corresponds to adsorbed water. Hence, hydroxyapatite is a calcium carbonate mineral. The absorption band related to the CO$_3$ chemical bond appears at 1455 cm^{-1} in good agreement with data in the literature [44].

Silanized quartz particles are evidenced by silicate-specific absorption bands at 1034 cm^{-1}, belonging to in-plane Si-O stretching; 815 cm^{-1} belongs to the Si-O stretching of quartz [45–47]. The polymer matrix is very well evidenced by specific chemical bonds such as 1162 cm^{-1}, related to C-O stretching; 1455 cm^{-1}, related to C-H bending; 1367 cm^{-1}, related to C=C stretching; 1720 cm^{-1}, related to C=O stretching within carboxylic units of Bis-GMA. Broad and less intense bands at 2863 cm^{-1} belong to C-H stretching, and those at and 2935 cm^{-1} belong to O-H stretching within carboxylic groups. The evidenced absorption bands correspond to our previous observation regarding similar compositions [48,49].

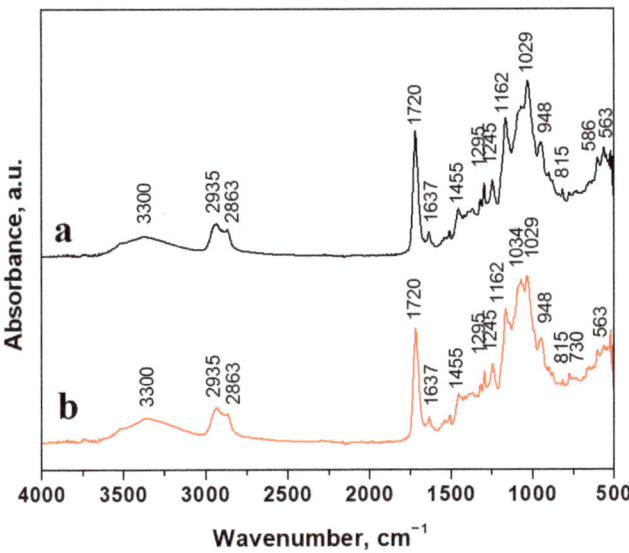

Figure 2. FTIR spectra for composite samples: (**a**) C1 and (**b**) C2.

2.2. MSC Proliferation Morphology

Cell proliferation evolution over time on the composite samples exposed to a culture medium was first investigated via optical microscopy. The obtained images are presented in Figure 3.

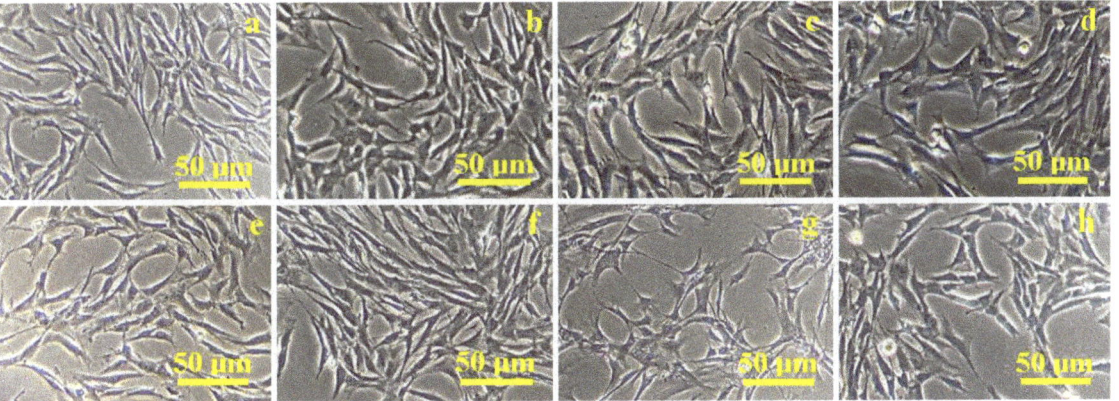

Figure 3. Optical microscopy images of mesenchymal stem cells proliferation on the composite sample C1: (**a**) 1 day, (**b**) 3 days, (**c**) 5 days, (**d**) 7 days; sample C2: (**e**) 1 day, (**f**) 3 days, (**g**) 5 days, (**h**) 7 days.

Cells derived from the palatal tissue have spindle-shaped, fibroblast-like morphology with pronounced bipolarity as observed after the first day of proliferation, shown in Figure 3a,e. Changes in their shape are observed after 3 days from the addition of the composites to the cell co-culture, as shown in Figure 3b,f. Some of the cells in the monolayer are flattened and a small proportion of the cells became round, but a high percentage of the cells in the culture retained their spindle-shaped morphology. This stage indicates cellular differentiation. Thus, the cells showed a marked heterogeneity after 5 days of culture: round cells and a high percentage of starred-shape cells are identified beside spindle-shaped ones (Figure 3c,g). They are predominantly organized in local clusters of 10–15 cells. The number of cell clusters is higher for the C2 composite (Figure 3g) than for C1 composite (Figure 3c). Cell differentiation is more evident; after 7 days of co-culture, they are more flattened, and the starred and round shapes are more evident than spindle shapes, as observed in Figure 3d for the C1 sample and in Figure 3h for C2. Both composites prove their ability to induce cell differentiation starting with 5 days and continuing to 7 days of co-culture. It further requires a more enhanced microscopic technique to investigate cell body morphologic aspects regarding the composite surface.

The aspect of C1 and C2 composites' microstructure visualized via SEM microscopy after MTT assay cytotoxicity testing is given in Figure 4.

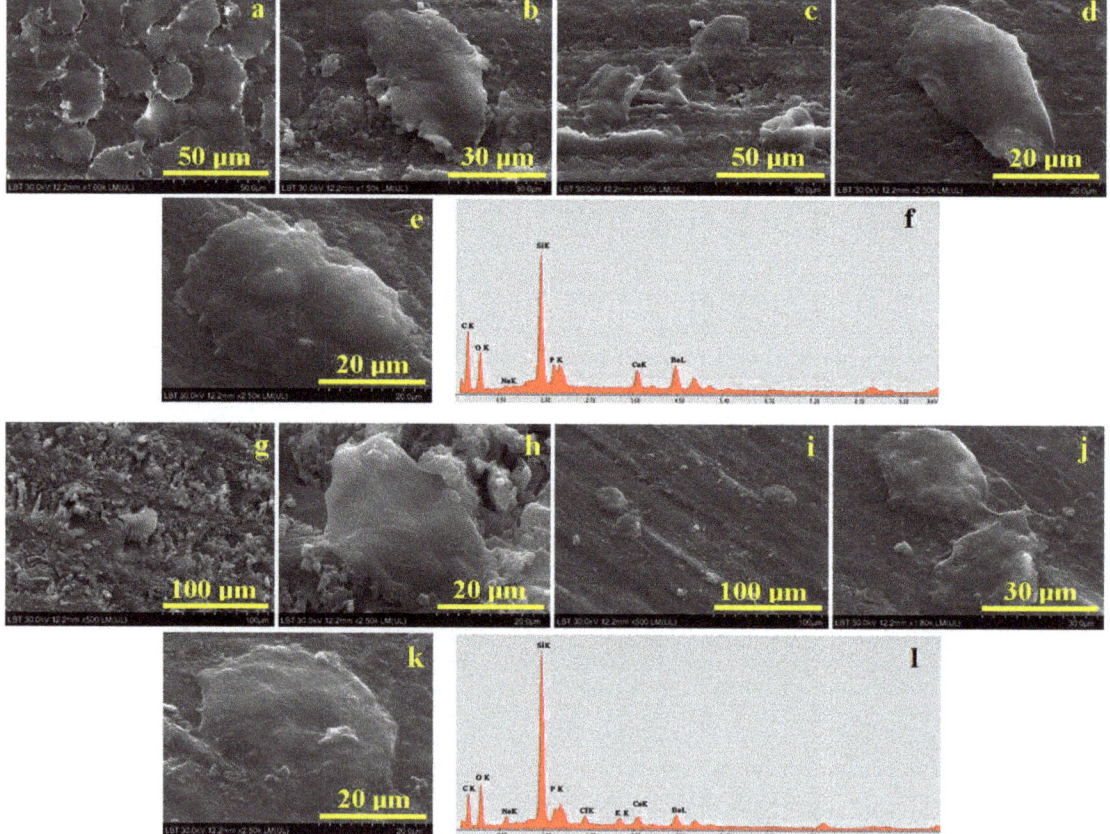

Figure 4. SEM images of composites after cytotoxicity testing for C1: (**a**) 3-day overview, (**b**) 3-day cell details, (**c**) 5-day overview, (**d**) 5-day cell details, (**e**) 7-day cell details and (**f**) EDX spectrum for C1 with cells at 7 days; C2: (**g**) 3-day overview, (**h**) 3-day cell details, (**i**) 5-day overview, (**j**) 5-day cell details, (**k**) 7-day cell details and (**l**) EDX spectrum for C2 with cells at 7 days.

Some representative microphotographs of the obtained results are given after 3, 5 and 7 days of MSC cell proliferation. SEM investigation was effectuated in the shortest possible time from the composite samples' removal from the co-culture medium, and they were not gold sputtered and thus examined in low-vacuum mode. Even with these precautions, only the cell body is predominantly observed, and their terminal pseudopods are almost invisible in Figure 4.

The SEM image from Figure 4a shows the change that occurs in the structure of the C1 composite after 3 days of exposure in co-culture. It can be observed that a group of cells with different sizes is placed randomly and with a flattened structure and rounded or starred shapes. It most likely represents the central body of the MSCs, and the prolonged terminals are not visible due to their thin consistency. The average width of the cells is situated between 15–30 μm and their average length between 20–40 μm. At higher magnification (Figure 3b) a single elongated cell was focalized in the image observation field; its width is about 33 μm and its length is about 52 μm.

After 5 days of exposure, as shown in Figure 4c, only a few cells are observed. However, it was possible to highlight a single cell at higher magnification (Figure 4d) that has an elongated shape, a width of 20 μm and a length of 36 μm, a fact sustaining cell differentiation. We remark an increase of the viable cell body and a significant reduction of the cell number in the image's observation field. Cell morphology is significantly changed after 7 days of co-culture on the C1 surface. Figure 4e shows how its length is about 58 μm and its width is 35 μm, a fact which is in good agreement with the optical microscopy in Figure 3d. It sustains a strong differentiation of the cells on the C1 surface after 7 days of proliferation along with potential osteogenic enhancement. This fact is sustained by a small increase in the Ca and P amounts detected in the EDX spectrum in Figure 4f compared to the initial composition of the C1 sample in Figure 1c. The identified amounts of P and Ca correspond to the hydroxyapatite stoichiometry. Significant amounts of Si and Ba were detected as a consequence of filler presence. The complete elemental composition evidenced by the EDX spectrum in Figure 4f is presented in Table 2.

Table 2. Elemental composition obtained via SEM images after 7 days of co-culture.

Samples	Identified Elements, Wt.%.					
	C	O	Si	P	Ca	Ba
C1—7 days	31.01	12.64	25.46	6.11	5.04	19.74
C2—7 days	27.54	18.32	26.63	5.16	5.22	17.13

A rough structure of the C2 composite surface is evident after 3 days of exposure, as shown in Figure 4g. It deals with a complex morphological convolution between rough aspects of the C2 composite surface, generated by the polycaprolactone fragments that are still visible on the surface, and cell proliferation, as cells prefer to adhere on average rough surfaces. Thus, several cell clusters are visible in Figure 4g, especially the one most evident in the center of the image. This aspect of the sample surface may be due to cell clusters that overlapped with each other. However, at high magnification, as shown in Figure 4h, a single cell was evidenced. The central body of the cell is clearly visible, having a width of 27 μm and a length of 45 μm. The prolonged terminations of the cell are not visible due to their thin consistency and to the particular investigation conditions within the SEM device.

The morphological aspect of the C2 composite after 5 days of exposure in the culture medium (Figure 4i) shows about three cells surrounded by a smooth surface of the sample compared to the one observed after 3 days of exposure. Thus, two well-defined cells attached to the surface of the composite can be observed at high magnification in Figure 4j, proving the cell differentiation is in good agreement with the optical microscopy observation. Their body is well defined and presents a length of about 30 μm and a width of about 25 μm.

Cell differentiation is more evident after 7 days of co-culture, as shown in Figure 4k. It shows a small increase of the cell body, having a slightly elongated shape with dimensions of 48 μm length and 30 μm width. The corresponding EDX composition is presented in Table 1. Identified carbon and oxygen amounts belong to the organic matter within the composite sample and the cell's living body. Filler presence is evidenced by the significant amounts of Si and Ba. There is a significant increase in the P and Ca amounts noticed, respecting the stoichiometry proportion within hydroxyapatite. It indicates a strong cell differentiation after 7 days of co-culture induced by the composite sample that facilitates the osteogenic behavior of the cells.

2.3. Biocompatibility Assessment

Biocompatibility is a relatively abstract term describing the ability of a material to be accepted by living tissue without generating adverse reactions. It must be assessed according to specific indicators that allow a proper quantification of the biological effect.

Therefore, the cytotoxic effect of the investigated composites was assessed via MTT assay using mesenchymal stem cells of palatal origin. This colorimetric assay is based on the reduction of tetrazolium salts to purple formazan crystals by cells that are metabolically active. Viable cells contain NAD(P)H-dependent oxidoreductase enzymes that reduce MTT to formazan. Insoluble formazan crystals were dissolved with dimethylsulfoxide (DMSO), and the absorbance of the chromogenic reaction was evaluated via spectrophotometry at a wavelength of 450 nm. The control sample evidences total cell viability, as shown in Figure 5a. Composite sample C1 evidences an average cell viability of 98.08%, and the C2 sample evidences a mean cell viability of 97.33%. Both values indicate good biocompatibility of the experimental composite cements after 3 days of co-culture, and the statistical analysis shows no significant differences. Cell viability was also tested after 5 and 7 days of co-culture, showing a slow decreased tendency that indicates significant statistical differences. However, the values of 89.5 for C1 and 87.3 for C2 are excellent from the cytotoxic point of view.

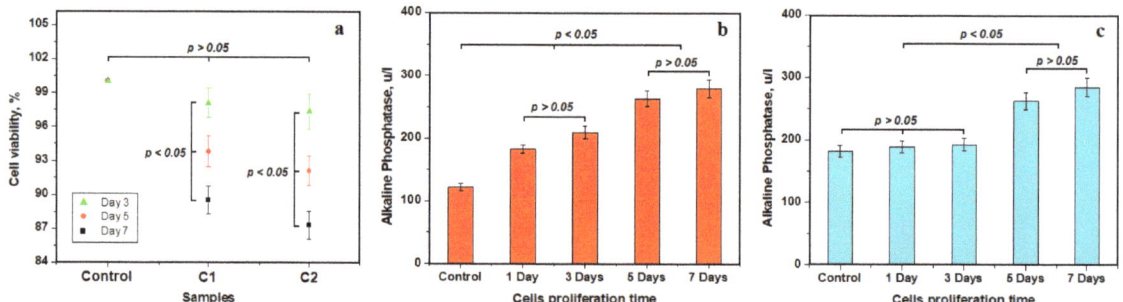

Figure 5. Biocompatibility test results—mean values of three distinct determinations: (**a**) MTT assay cell viability, (**b**) alkaline phosphatase level for C1 sample and (**c**) alkaline phosphatase level for C2 sample.

Alkaline phosphatase (ALP) is another important indicator of mesenchymal stem cells' differentiation potential regarding materials designed to interact in a friendly manner with osseous hard tissues like teeth. It indicates cell bioactivity regarding the composite material surface, which is able to promote bio-integration into the host tissue.

Cell proliferation on the C1 composite surface reveals an average alkaline phosphatase control level of 121.7 u/L, which increases after 1 and 3 days of proliferation to 209.2 u/L. The statistical analysis in Figure 5b reveals an important significance of these values indicating the incipient stage before cell differentiation. The low value of ALP at this early stage is in good agreement with stem cell pluripotency, which agrees with the literature [50,51]. Optical and SEM microstructural investigations reveal physical cell differentiation in the presence of composite samples after 5 and 7 days of proliferation. This fact correlates with the significant increase in ALP level to 279.4 u/L. It indicates an osteoinductive behavior of the differentiated cells that was induced by the C1 sample. Statistical analysis evidence significant difference between the two stages observed.

On the other side, Figure 5c evidences an average alkaline phosphatase control level of 182 u/L that slightly increases after 1 and 3 days of proliferation to a value of about 193 u/L for the C2 sample. Statistical analysis does not reveal any relevant differences in this group indicating an initial stage of MSCs' pluripotency. The ALP level significantly increases after 5 and 7 days of co-culture, respectively, having the same statistical relevance. It indicates that the C2 sample promotes cell differentiation observed via SEM and optical microscopy as well as corresponding osteoinductive behavior. Statistical observation reveals a clear separation between the observed stages of cell evolution.

3. Discussion

SEM microscopy reveals that the filler amount has an important influence on the composite's microstructure (Figure 1a,b). A higher mineral filler amount within the C1 sample ensures better disintegration of the polycaprolactone core shell and a good distribution of the particles within the polymeric matrix. The importance of filler distribution is also reported in the literature [52,53]. Targeted filler delivery assures optimal lamination of the BaO filler particles and proper embedding of the hydroxyapatite and silanized quartz nanoparticles. EDX spectra evidenced filler elements such as Si, Ba, P and Ca on the sample surface, proving that they were properly released onto the polymeric matrix, ensuring the optimal desired distribution. Some superficial cracks are observed for C1 sample, but they do not penetrate the composite bulk in depth.

Less mineral filler within the C2 sample reduces polycaprolactone envelope braking during composite polymerization, assuring a more progressive embedding of the mineral filler particles into a very well-mixed structure without local segregations and superficial cracks. The fact is sustained by the EDX spectrum that evidences the characteristic elements of the filler: mainly Si and Ba due to nano silica and barium glass, but also P and Ca due to the nHA contents. SEM images in Figure 1c,d also reveal the presence of significant fragments belonging to the polycaprolactone envelope which are well embedded into the polymer matrix, assuring the good cohesion of the composite.

FTIR spectra within Figure 2 evidenced effective incorporation of the mineral filler in both C1 and C2 samples and reveals Bis-GMA stabilization due to the presence of UDMA monomers, which stabilize the carboxylic bonds within the polymeric matrix, controlling viscosity during the polymerization process to assure an optimal and uniform distribution of filler particles in the composite bulk. This fact sustained by the homogeneous aspect of microstructural details.

Hydroxyapatite is one of the most important constituents of human hard tissues. Nanostructured crystals are embedded with collagen and other protein materials forming different bone tissues [54,55]. It is also reported as a bone implant surface biofunctionalization agent enhancing the bio-response and implant integration in the living body [56,57]. Thus, the uniform distribution of nanostructured hydroxyapatite clusters within C1 and C2 composites leads to a significant cell proliferation after 1 and 3 days of culture as revealed through SEM and optical microscopy. The general aspect of the cells is spindle-like, and central body size is of moderate width around 15–30 μm and a length of about 20–40 μm. Cell density decreases after 5 days of culture due to their natural growth process. Microscopy evidences significantly increased sizes: width of about 27 μm and length of 45 μm, as well as morphological differentiations such local flattening and shape modifications; e.g., elongated, starred and some rounded shapes are observed. Cell differentiation is induced by the co-culture medium's exposure to the composite samples that stimulated their osteoinductive behavior, a fact sustained by the significantly increased level of P and Ca in the elemental composition measured after 7 days of co-culture. Both composites present good cell proliferation, but C2 proves to be more effective than C1 due to the better distribution of nano-hydroxyapatite clusters within its structure.

The proliferation of mesenchymal stem cells of palatal origin on the composite surface reveals excellent cell viability, proving low cytotoxicity. Cell viability presents a slow decrease along with proliferation time, but the measured values are still excellent, as observed in Figure 5a. Literature data indicate a moderate generation of alkaline phosphatase (ALP) due to the pluripotent state of the MSCs associated with their spindle aspect [50,51]. Statistical analysis of the data in Figure 5b,c proves that the incipient stage lasts up to the third day of co-culture. Cell differentiation is observed from 5 days of co-culture, implying significant generation of ALP. It represents the second stage of ALP variation, and their increase was regarded as a distinct stage via the statistical analysis of the variation in Figure 5b,c. The increase in P and Ca amounts after 7 days of co-culture correlated with an increase in ALP level, indicating that the composite samples stimulate the osteoinductive behavior of MSCs trough the differentiation process. This fact indicates that these experimental materials are compatible with the living body. Data in the literature report an increased level of alkaline phosphatase when the implant bio-integrates; the ALP peak usually occurs at 14 days after in vivo implantation [58,59]. It might facilitate the high-efficacy integration of composite cement into the teeth structure, although this requires further in vivo investigation for our composite samples.

The correlation of experimental results shows that cell proliferation density Is favored by the uniform distribution of the nano-hydroxyapatite clusters within the composite, which induces the differentiation process within MSCs from 5 to 7 days of co-culture and respectively stimulates their osteoinductive behavior. Thus, C2 proves to be more effective than C1, but the relatively high amount of bioactive minerals within C1 samples conducts a significantly greater cell viability and alkaline phosphatase response. Therefore, mineral filler distribution and amount play a key role in the composites' restorative success, and in consequence, the null hypothesis is completely rejected.

4. Materials and Methods

4.1. Composite Cement Sample Preparation

The experimental composite cements were prepared in the laboratories of Raluca Ripan Institute for Research in Chemistry (ICCRR, Cluj-Napoca, Romania).

The polymeric matrix contains Bis-GMA (2,2-bis[p-(2'-hydroxy-3'-metacryloxy propoxy) phenyl]-propane), UDMA (urethane dimethacrylate), HEMA (Hydroxyethyl methacrylate) and PEG 400 (polyethylene glycol); all substances were analytical grade and purchased from Sigma Aldrich, Darmstadt, Germany.

The mineral filler contains BaO glass, silanized quartz nanoparticles and hydroxyapatite nanoparticles purchased from Sigma Aldrich, Darmstadt, Germany. They were encapsulated into polycaprolactone (analytical grade Sigma Aldrich, Darmstadt, Germany) microcapsules produced in our laboratory using buffalo whey (Advanced Materials, Inc., Westerlo Oevel, Belgium) as a nanoparticle dispersion environment.

We prepared two composite cements using the same polymer matrix and filler. The difference consists in the matrix/filler ratio: C1 has a ratio of 67/33% and C2 has a ratio of 69/31%. The composition was mixed using a mechanical homogenizer for 10 min. The photo-polymerization process was initiated using a camphorquinone photoinitiator (CQ) and 2-(Dimethylamino)ethyl methacrylate (DMAEM) (Sigma Aldrich, Darmstadt, Germany). A photo-polymerization lamp was used (Elipar Deep Cure-L 3 M produced by 3M Oral Care Company, Saint Paul, MN, USA). The photo-polymerization lamp was placed as close as possible to the disks' surface. Both composites were molded into disk-shaped samples with a diameter of 5 mm and a thickness of 0.5 mm.

4.2. Palatal Mesenchymal Stem Cell Co-Cultivation and MTT Assay

The palatal mesenchymal stem cells used in the current research were obtained from the Biotechnology Laboratory of the University of Agricultural Sciences and Veterinary Medicine (Cluj-Napoca, Romania) and has the approval of Bioethical Commission No. 352 from 12 December 2022. The stem cell line was stabilized by our team and was previously published [60].

Palatal-tissue-derived mesenchymal cells were cultured in a simple propagation medium of DMEM/F12 (Sigma-Aldrich, Darmstadt, Germany) supplemented with 10% FCS (fetal bovine serum, Sigma-Aldrich), 1% antibiotic-antimycotic (Gibco), 1% glutamax $1\times$ (Sigma-Aldrich, Darmstadt, Germany) and 1% NEA (non-essential amino acids, Gibco). To evaluate the biomaterials, 12-well plates (Thermo Fischer Scientific, Waltham, MA, USA) were used. The concentration of the palatal cells used was 1×10^5. The bioactive potential of bioglass-based composites (C1 and C2) was evaluated through co-culture techniques using inserts in which the composite samples were properly placed, and afterwards, the cells were added on their surface in simple propagation medium. After the desired proliferation time (3, 5 and 7 days of co-culture), the MTT (4,5-dimethylthiazol-2-yl)-2,5-diphenyltetrazolium) assay was used to assess cytotoxicity and degree of proliferation according to the ISO 10993-5:2009 protocols. The MTT (4,5-dimethylthiazol-2-yl)-2,5-diphenyltetrazolium) assay (Sigma-Aldrich, Darmstadt, Germany) measures cellular metabolic activity as an indicator of cell viability, proliferation and cytotoxicity. This colorimetric assay is based on the reduction of tetrazolium salts to purple formazan crystals by cells that are metabolically active. Viable cells contain NAD(P)H-dependent oxidoreductase enzymes that reduce MTT to formazan. Insoluble formazan crystals were dissolved with dimethylsulfoxide (DMSO), and the absorbance of the chromogenic reaction was evaluated using a spectrophotometer at a wavelength of 450 nm.

Alkaline phosphatase (ALP) level was measured on request in the clinical laboratory of the Veterinary Medicine Faculty in Cluj-Napoca, Romania.

Both MTT assay and ALP data were statistically processed using ANOVA one-way followed by Tukey post-hoc tests using Origin 2019b software (Microcal Corporation, Northampton, MA, USA). The significance level was set at 0.05; thus, $p < 0.05$ indicates that significant statistical differences occur and $p > 0.05$ means that there are no significant differences between tested group elements.

4.3. Microscopy Techniques

Optical microscopy investigation was effectuated using a transmitted light biological microscope (ZEISS Axioscope 5, Oberocken, Germany).

Scanning Electron Microscopy was effectuated using a Hitachi SU8230 Scanning Electron Microscope, Tokyo, Japan, at an acceleration voltage of 30 kV in low-vacuum mode without sample metallization. Elemental analysis was effectuated with an Energy-Dispersive Spectroscopy (EDS) detector: X-Max 1160 EDX (Oxford Instruments, Oxford, UK). The culture medium within wells was removed before SEM investigation and afterwards rinsed with ultrapure water and stored under a desiccator for 1 h until their surface was completely dry.

Fourier-transform infrared spectroscopy (FTIR) was performed using an FTIR 610 spectrometer (Jasco Corporation, Tokyo, Japan) in the wavenumber range of 4000–400 cm^{-1}, using the potassium bromide pellet technique. Each spectrum was registered at a resolution of 4 cm^{-1} and represents the average of 100 scans.

5. Conclusions

Controlled filler release through photo-polymerization was properly achieved using polycaprolactone microcapsules. The uniformity of nanostructured filler particles is slightly increased for C2 samples, which is more favorable for palatal mesenchymal stem cell cluster proliferation. Cell viability strongly depends on the amount of bioactive filler. The best results of the MTT assay show that C1 has a greater value of MSC viability. The alkaline phosphatase (ALP) level also was increased for the higher amount of bioactive filler. Two stages occur: the initial stage up to 3 days of co-culture is characterized by an average generation of ALP associated with the pluripotency of the MSCs, and the second stage (from 5 to 7 days of co-culture) is characterized by a significant increase in the ALP level due to MSC differentiation induced by the composite samples that stimulate their osteoinductive behavior. It indicates that both C1 and C2 samples are viable candidates for further in vivo tests.

Author Contributions: Conceptualization, A.I.A. and L.O.; methodology, M.M. and L.O.; validation, M.M. and E.P.; formal analysis, G.A.P. and A.I.A.; investigation, E.P., C.S., L.B.T., M.F.D. and A.I.A.; resources, M.M. and L.O.; data curation, G.A.P. and I.P.; writing—original draft preparation, I.P. and G.A.P.; writing—review and editing, I.P. and M.M.; visualization, L.B.T., C.S. and E.P.; supervision, M.M.; project administration, L.O. All authors have read and agreed to the published version of the manuscript.

Funding: This research received no external funding.

Institutional Review Board Statement: The study was conducted in accordance with the Declaration of Helsinki and approved by the Institutional Bioethics Committee of the University of Agricultural Sciences and Veterinary Medicine (Cluj-Napoca, Romania) No. 352 from 12 December 2022.

Informed Consent Statement: Not applicable.

Data Availability Statement: The research data are available on request from the corresponding author.

Conflicts of Interest: The authors declare no conflict of interest.

References

1. Jurak, M.; Wiącek, A.E.; Ładniak, A.; Przykaza, K.; Szafran, K. What affects the biocompatibility of polymers? *Adv. Colloid Interface Sci.* **2021**, *294*, 102451. [CrossRef] [PubMed]
2. Ratner, B.D.; Schoen, F.J.; Lemons, J.E. (Eds.) *Biomaterials Science: An Introduction to Materials in Medicine*, 1st ed.; Academic Press: San Diego, CA, USA, 1996; p. 484, ISBN 0125824602.
3. Vert, M.; Doi, Y.; Hellwich, K.H.; Hess, M.; Hodge, P.; Kubisa, P.; Rinaudo, M.; Schue, F. Terminology for biorelated polymers and applications (IUPAC Recommendations 2012). *Pure Appl. Chem.* **2012**, *84*, 377–410. [CrossRef]
4. Azevedo, H.S.; Mata, A. Embracing complexity in biomaterials design. *Biomater. Biosyst.* **2022**, *6*, 100039. [CrossRef] [PubMed]
5. Simionescu, B.C.; Ivanov, D. Natural and synthetic polymers for designing composite materials. In *Handbook of Bioceramics and Biocomposites*; Antoniac, I.V., Ed.; Springer: Cham, Switzerland, 2015; pp. 233–286. [CrossRef]
6. Bonferoni, M.C.; Caramella, C.; Catenacci, L.; Conti, B.; Dorati, R.; Ferrari, F.; Genta, I.; Modena, T.; Perteghella, S.; Rossi, S.; et al. Biomaterials for Soft Tissue Repair and Regeneration: A Focus on Italian Research in the Field. *Pharmaceutics* **2021**, *13*, 1341. [CrossRef]
7. Conte, R.; Di Salle, A.; Riccitiello, F.; Petillo, O.; Peluso, G.; Calarco, A. Biodegradable polymers in dental tissue engineering and regeneration. *AIMS Mater. Sci.* **2018**, *5*, 1073–1101. [CrossRef]
8. Calori, I.R.; Braga, G.; de Jesus, P.D.C.C.; Bi, H.; Tedesco, A.C. Polymer scaffolds as drug delivery systems. *Eur. Polym. J.* **2020**, *129*, 109621. [CrossRef]
9. Anju, S.; Prajitha, N.; Sukanya, V.S.; Mohanan, P.V. Complicity of degradable polymers inhealth-care applications. *Mater. Today Chem.* **2020**, *16*, 100236. [CrossRef]
10. Collignon, A.M.; Lesieur, J.; Vacher, C.; Chaussain, C.; Rochefort, G.Y. Strategies Developed to Induce, Direct, and Potentiate Bone Healing. *Front. Physiol.* **2017**, *8*, 927. [CrossRef]
11. Chisnoiu, R.; Moldovan, M.; Pastrav, O.; Delean, A.; Chisnoiu, A.M. The influence of three endodontic sealers on bone healing–An experimental study. *Folia. Morphol.* **2016**, *75*, 14–20. [CrossRef]
12. Lee, S.-Y.; Wu, S.-C.; Chen, H.; Tsai, L.-L.; Tzeng, J.-J.; Lin, C.-H.; Lin, Y.-M. Synthesis and Characterization of Polycaprolactone-Based Polyurethanes for the Fabrication of Elastic Guided Bone Regeneration Membrane. *BioMed Res. Int.* **2018**, *2018*, 3240571. [CrossRef]
13. Luo, S.; Zhu, W.; Liu, F.; He, J. Preparation of a Bis-GMA-Free Dental Resin System with Synthesized Fluorinated Dimethacrylate Monomers. *Int. J. Mol. Sci.* **2016**, *17*, 2014. [CrossRef] [PubMed]
14. Ahovuo-Saloranta, A.; Hiiri, A.; Nordblad, A.; Mäkelä, M.; Worthington, H.V. Pit and fissure sealants for preventing dental decay in the permanent teeth of children and adolescents. *Cochrane Database Syst. Rev.* **2017**, *7*, CD001830. [CrossRef] [PubMed]
15. Haugen, H.J.; Marovic, D.; Par, M.; Le Thieu, M.K.; Reseland, J.E.; Johnsen, G.F. Bulk Fill Composites Have Similar Performance to Conventional Dental Composites. *Int. J. Mol. Sci.* **2020**, *21*, 5136. [CrossRef]
16. Barszczewska-Rybarek, I.M.; Chrószcz, M.W.; Chladek, G. Novel Urethane-Dimethacrylate Monomers and Compositions for Use as Matrices in Dental Restorative Materials. *Int. J. Mol. Sci.* **2020**, *21*, 2644. [CrossRef]
17. Raszewski, Z.; Brząkalski, D.; Derpeński, Ł.; Jałbrzykowski, M.; Przekop, R.E. Aspects and Principles of Material Connections in Restorative Dentistry—A Comprehensive Review. *Materials* **2022**, *15*, 7131. [CrossRef]
18. Limberger, K.M.; Westphalen, G.H.; Menezes, L.M.; Medina-Silva, R. Cytotoxicity of orthodontic materials assessed by survival tests in Saccharomyces cerevisiae. *Dent. Mater.* **2011**, *27*, e81–e86. [CrossRef] [PubMed]
19. Yoshinaga, K.; Yoshihara, K.; Yoshida, Y. Development of new diacrylate monomers as substitutes for Bis-GMA and UDMA. *Dent. Mater.* **2021**, *37*, e391–e398. [CrossRef]
20. Szczesio-Wlodarczyk, A.; Domarecka, M.; Kopacz, K.; Sokolowski, J.; Bociong, K. An Evaluation of the Properties of Urethane Dimethacrylate-Based Dental Resins. *Materials* **2021**, *14*, 2727. [CrossRef]

21. Moszner, N.; Fischer, U.K.; Angermann, J.; Rheinberger, V. A partially aromatic urethane dimethacrylate as a new substitute for Bis-GMA in restorative composites. *Dent. Mater.* **2008**, *24*, 694–699. [CrossRef]
22. Ma, T.Y.; Hollander, D.; Krugliak, P.; Katz, K. PEG 400, a hydrophilic molecular probe for measuring intestinal permeability. *Gastroenterology* **1990**, *98*, 39–46. [CrossRef]
23. Mohl, S.; Winter, G. Continuous release of rh-interferon alpha-2a from triglyceride matrices. *J. Control. Release* **2004**, *97*, 67–78. [CrossRef] [PubMed]
24. Hench, L.L. The future of bioactive ceramics. *J. Mater. Sci. Mater. Med.* **2015**, *26*, 86. [CrossRef] [PubMed]
25. Crush, J.; Hussain, A.; Seah, K.T.M.; Khan, W.S. Bioactive Glass: Methods for Assessing Angiogenesis and Osteogenesis. *Front. Cell Dev. Biol.* **2021**, *9*, 643781. [CrossRef]
26. Gloria, A.; Russo, T.; Rodrigues, D.F.L.; D'Amora, U.; Colella, F.; Improta, G.; Triassi, M.; De Santis, R.; Ambrosio, L. From 3D hierarchical scaffolds for tissue engineering to advanced hydrogel-based and complex devices for in situ cell or drug release. *Procedia CIRP* **2016**, *49*, 72–75. [CrossRef]
27. Rasouli, R.; Barhoum, A.; Uludag, H. A review of nanostructured surfaces and materials for dental implants: Surface coating, patterning and functionalization for improved performance. *Biomater. Sci.* **2018**, *6*, 1312–1338. [CrossRef]
28. Kumar, S.; Nehra, M.; Kedia, D.; Dilbaghi, N.; Tankeshwar, K.; Kim, K.-H. Nanotechnology based biomaterials for orthopaedic applications: Recent advances and future prospects. *Mater. Sci. Eng. C Mater. Biol. Appl.* **2020**, *106*, 110154. [CrossRef] [PubMed]
29. Liu, X.; Chu, P.K.; Ding, C. Surface nano-functionalization of biomaterials. *Mater. Sci. Eng. R* **2010**, *70*, 275–302. [CrossRef]
30. Yang, P.; Ren, J.; Yang, L. Nanoparticles in the New Era of Cardiovascular Therapeutics: Challenges and Opportunities. *Int. J. Mol. Sci.* **2023**, *24*, 5205. [CrossRef]
31. Undin, J.; Finne-Wistrand, A.; Albertsson, A.-C. Adjustable degradation properties and biocompatibility of amorphous and functional poly(ester-acrylate)-based materials. *Biomacromolecules* **2014**, *15*, 2800–2807. [CrossRef]
32. Gu, Q.; Cuevas, E.; Ali, S.F.; Paule, M.G.; Krauthamer, V.; Jones, Y.; Zhang, Y. An Alternative In Vitro Method for Examining Nanoparticle-Induced Cytotoxicity. *Int. J. Toxicol.* **2019**, *38*, 385–394. [CrossRef]
33. Sanità, G.; Carrese, B.; Lamberti, A. Nanoparticle surface functionalization: How to improve biocompatibility and cellular internalization. *Front. Mol. Biosci.* **2020**, *7*, 587012. [CrossRef]
34. Mousavinasab, S.M. Biocompatibility of composite resins. *Dent. Res. J.* **2011**, *8* (Suppl. 1), S21–S29.
35. Pătroi, D.; Gociu, M.; Prejmerean, C.; Colceriu, L.; Dumitrescu, L.S.; Moldovan, M.; Naicu, V. Assessing the biocompatibility of a dental composite product. *Rom. J. Morphol. Embryol.* **2013**, *54*, 321–326.
36. Wataha, J.C. Principles of biocompatibility for dental practitioners. *J. Prosthet. Dent.* **2001**, *86*, 203–209. [CrossRef] [PubMed]
37. Kawai, M.Y.; Ozasa, R.; Ishimoto, T.; Nakano, T.; Yamamoto, H.; Kashiwagi, M.; Yamanaka, S.; Nakao, K.; Maruyama, H.; Bessho, K.; et al. Periodontal Tissue as a Biomaterial for Hard-Tissue Regeneration following *bmp-2* Gene Transfer. *Materials* **2022**, *15*, 993. [CrossRef]
38. Xu, X.Y.; Li, X.; Wang, J.; He, X.T.; Sun, H.H.; Chen, F.M. Concise review: Periodontal tissue regeneration using stem cells: Strategies and translational considerations. *Stem Cells Transl. Med.* **2019**, *8*, 392–403. [CrossRef]
39. Han, J.; Menicanin, D.; Gronthos, S.; Bartold, P.M. Stem cells, tissue engineering and periodontal regeneration. *Aust. Dent. J.* **2014**, *59*, 117–130. [CrossRef] [PubMed]
40. Lawrence, L.M.; Salary, R.; Miller, V.; Valluri, A.; Denning, K.L.; Case-Perry, S.; Abdelgaber, K.; Smith, S.; Claudio, P.P.; Day, J.B. Osteoregenerative Potential of 3D-Printed Poly ε-Caprolactone Tissue Scaffolds In Vitro Using Minimally Manipulative Expansion of Primary Human Bone Marrow Stem Cells. *Int. J. Mol. Sci.* **2023**, *24*, 4940. [CrossRef]
41. Zhao, Y.; Zhang, H.; Hong, L.; Zou, X.; Song, J.; Han, R.; Chen, J.; Yu, Y.; Liu, X.; Zhao, H.; et al. A Multifunctional Dental Resin Composite with Sr-N-Doped TiO_2 and n-HA Fillers for Antibacterial and Mineralization Effects. *Int. J. Mol. Sci.* **2023**, *24*, 1274. [CrossRef] [PubMed]
42. Ansari, M.A.; Jahan, N. Structural and Optical Properties of BaO Nanoparticles Synthesized by Facile Co-precipitation Method. *Mater. Highlights* **2021**, *2*, 23–28. [CrossRef]
43. Osipov, A.A.; Osipova, L.M.; Hruška, B.; Osipov, A.A.; Liška, M. FTIR and Raman spectroscopy studies of ZnO-doped BaO 2B_2O_3 glass matrix. *Vib. Spectrosc.* **2019**, *103*, 102921. [CrossRef]
44. Gheisari, H.; Karamian, E.; Abdellahi, M. A novel hydroxyapatite–Hardystonite nanocomposite ceramic. *Ceram. Int.* **2015**, *41*, 5967–5975. [CrossRef]
45. Koupadi, K.; Boyatzis, S.C.; Roumpou, M.; Kalogeropoulos, N.; Kotzamani, D. Organic remains in early Christian Egyptian metal vessels: Investigation with fourier transform infrared spectroscopy and gas chromatography–mass spectrometry. *Heritage* **2021**, *4*, 3611–3629. [CrossRef]
46. Farcas, I.A.; Dippong, T.; Petean, I.; Moldovan, M.; Filip, M.R.; Ciotlaus, I.; Tudoran, L.B.; Borodi, G.; Paltinean, G.A.; Pripon, E.; et al. Material Evidence of Sediments Recovered from Ancient Amphorae Found at the Potaissa Roman Fortress. *Materials* **2023**, *16*, 2628. [CrossRef] [PubMed]
47. Rusca, M.; Rusu, T.; Avram, S.E.; Prodan, D.; Paltinean, G.A.; Filip, M.R.; Ciotlaus, I.; Pascuta, P.; Rusu, T.A.; Petean, I. Physicochemical Assessment of the Road Vehicle Traffic Pollution Impact on the Urban Environment. *Atmosphere* **2023**, *14*, 862. [CrossRef]
48. Moldovan, M.; Dudea, D.; Cuc, S.; Sarosi, C.; Prodan, D.; Petean, I.; Furtos, G.; Ionescu, A.; Ilie, N. Chemical and Structural Assessment of New Dental Composites with Graphene Exposed to Staining Agents. *J. Funct. Biomater.* **2023**, *14*, 163. [CrossRef]

49. Mazilu, A.; Popescu, V.; Sarosi, C.; Dumitrescu, R.S.; Chisnoiu, A.M.; Moldovan, M.; Dumitrescu, L.S.; Prodan, D.; Carpa, R.; Gheorghe, G.F.; et al. Preparation and In Vitro Characterization of Gels Based on Bromelain, Whey and Quince Extract. *Gels* **2021**, *7*, 191. [CrossRef]
50. Stefkova, K.; Prochazkova, J.; Pachernik, J. Alkaline Phosphatase in Stem Cells. *Stem Cells Int.* **2015**, *2015*, 628368. [CrossRef]
51. Domínguez, L.M.; Bueloni, B.; Cantero, M.J.; Albornoz, M.; Pacienza, N.; Biani, C.; Luzzani, C.; Miriuka, S.; García, M.; Atorrasagasti, C.; et al. Chromatographic Scalable Method to Isolate Engineered Extracellular Vesicles Derived from Mesenchymal Stem Cells for the Treatment of Liver Fibrosis in Mice. *Int. J. Mol. Sci.* **2023**, *24*, 9586. [CrossRef]
52. Tiskaya, M.; Shahid, S.; Gillam, D.; Hill, R.G. The use of bioactive glass (BAG) in dental composites: Critical review. *Dent. Mater.* **2021**, *37*, 296–310. [CrossRef]
53. Bin-Jardan, L.I.; Almadani, D.I.; Almutairi, L.S.; Almoabid, H.A.; Alessa, M.A.; Almulhim, K.S.; AlSheikh, R.N.; Al-Dulaijan, Y.A.; Ibrahim, M.S.; Al-Zain, A.O.; et al. Inorganic Compounds as Remineralizing Fillers in Dental Restorative Materials: Narrative Review. *Int. J. Mol. Sci.* **2023**, *24*, 8295. [CrossRef] [PubMed]
54. Rashid, U.; Becker, S.K.; Sponder, G.; Trappe, S.; Sandhu, M.A.; Aschenbach, J.R. Low Magnesium Concentration Enforces Bone Calcium Deposition Irrespective of 1,25-Dihydroxyvitamin D_3 Concentration. *Int. J. Mol. Sci.* **2023**, *24*, 8679. [CrossRef] [PubMed]
55. Lv, S.; Yuan, X.; Xiao, J.; Jiang, X. Hemostasis-osteogenesis integrated Janus carboxymethyl chitin/hydroxyapatite porous membrane for bone defect repair. *Carbohydr. Polym.* **2023**, *313*, 120888. [CrossRef] [PubMed]
56. Kazimierczak, P.; Wessely-Szponder, J.; Palka, K.; Barylyak, A.; Zinchenko, V.; Przekora, A. Hydroxyapatite or Fluorapatite—Which Bioceramic Is Better as a Base for the Production of Bone Scaffold?—A Comprehensive Comparative Study. *Int. J. Mol. Sci.* **2023**, *24*, 5376. [CrossRef]
57. Makishi, S.; Watanabe, T.; Saito, K.; Ohshima, H. Effect of Hydroxyapatite/β-Tricalcium Phosphate on Osseointegration after Implantation into Mouse Maxilla. *Int. J. Mol. Sci.* **2023**, *24*, 3124. [CrossRef]
58. Chai, Q.; Xu, H.; Xu, X.; Li, Z.; Bao, W.; Man, Z.; Li, W. Mussel-inspired alkaline phosphatase-specific coating on orthopedic implants for spatiotemporal modulating local osteoimmune microenvironment to facilitate osseointegration. *Colloids Surf. B Biointerfaces* **2023**, *225*, 113284. [CrossRef]
59. Zhuravleva, I.Y.; Karpova, E.V.; Dokuchaeva, A.A.; Titov, A.T.; Timchenko, T.P.; Vasilieva, M.B. Calcification of Various Bioprosthetic Materials in Rats: Is It Really Different? *Int. J. Mol. Sci.* **2023**, *24*, 7274. [CrossRef]
60. Páll, E.; Florea, A.; Soriţău, O.; Cenariu, M.; Petruţiu, A.S.; Roman, A. Comparative Assessment of Oral Mesenchymal Stem Cells Isolated from Healthy and Diseased Tissues. *Microsc. Microanal.* **2015**, *21*, 1249–1263. [CrossRef]

Disclaimer/Publisher's Note: The statements, opinions and data contained in all publications are solely those of the individual author(s) and contributor(s) and not of MDPI and/or the editor(s). MDPI and/or the editor(s) disclaim responsibility for any injury to people or property resulting from any ideas, methods, instructions or products referred to in the content.

Communication

Impact of Chronic Oral Administration of Gold Nanoparticles on Cognitive Abilities of Mice

Alexandra L. Ivlieva [1,2], Elena N. Petritskaya [1], Dmitriy A. Rogatkin [1], Inga Zinicovscaia [3,4,5,*], Nikita Yushin [3] and Dmitrii Grozdov [3]

[1] Institute of Higher Nervous Activity and Neurophysiology of RAS, 5A Butlerova St., 117485 Moscow, Russia; ivlieva@medphyslab.com (A.L.I.); medphys@monikiweb.ru (E.N.P.); rogatkin@medphyslab.com (D.A.R.)
[2] Moscow Regional Research and Clinical Institute named after M. F. Vladimirsky, Str. Schepkina 61/2, 129110 Moscow, Russia
[3] Joint Institute for Nuclear Research, Str. Joliot-Curie 6, 141980 Dubna, Russia; ynik_62@mail.ru (N.Y.); dsgrozdov@rambler.ru (D.G.)
[4] Horia Hulubei National Institute for R&D in Physics and Nuclear Engineering, 30 Reactorului Str., MG-6, RO-76900 Bucharest-Magurele, Romania
[5] Institute of Chemistry, Academiei Str. 3, MD-2028 Chisinau, Moldova
* Correspondence: zinikovskaia@mail.ru

Abstract: The influence of gold nanoparticles after their prolonged oral administration to mice (during pregnancy and lactation) on spatial memory and anxiety levels in offspring was investigated. Offspring were tested in the Morris water maze and in the elevated Plus-maze. The average specific mass content of gold which crossed the blood–brain barrier was measured using neutron activation analysis and constituted 3.8 ng/g for females and 1.1 ng/g for offspring. Experimental offspring showed no differences in spatial orientation and memory compared to the control, while their anxiety levels increased. Gold nanoparticles influenced the emotional state of mice exposed to nanoparticles during prenatal and early postnatal development, but not their cognitive abilities.

Keywords: gold nanoparticles; neutron activation analysis; anxiety; cognitive functions; development

1. Introduction

For each new and potentially widely used medical compound, the question about its safety for patients, especially for pregnant women and children arises. Gold nanoparticles (AuNPs) already found wide applications in medicine—in optical visualization, target photosensitization, as sensitizers in radiotherapy of tumors, and as carriers for different drug molecules (antioxidants, immunomodulators, etc.) [1–6]. The ability of AuNPs to cause oxidative stress reported in several studies [7,8] allows their applications as sensitizers [9]. Because of the chemical inertness of gold [10], AuNPs are among the least toxic nanoparticles for mammals [3].

However, in several studies, it was shown that AuNPs are able to penetrate the blood–brain barrier and accumulate in the brain [8,11–13]. They can cause oxidative stress in neurons and glial cells [7,8,14], and increase expression of the markers of inflammation in the brain [11]. The ability of AuNPs to bind to DNA was shown in [15,16]. As it was detected applying the methods for estimation of genotoxicity samples treatment with AuNPs increases the damage of DNA, both, in in vitro and in vivo studies [17]. The ability of small nanoparticles to cross biological barriers is one of their most dangerous properties, despite their usefulness for different medical applications, especially for brain disease treatment [18]. High penetration ability makes not only the brain but also gonads, fetus (through the blood–placenta barrier) [19], and infants (through the milk of mother who had contact with AuNPs) vulnerable to AuNPs' impact.

It should be noted that the behavioral effects caused by contact with AuNPs are described in only several studies, mainly for adult animals after single or several injections [13,20]. Therefore, the present study aimed to examine the influence of AuNPs administrated to mice during pregnancy and lactation on spatial orientation, memory, and anxiety levels in offspring. The accumulation of gold in different tissues of female mice and their offspring was assessed using neutron activation analysis. To our knowledge, it is the first study in which the prolonged effect of the AuNPs on the offspring was studied.

2. Results

2.1. Distribution of Gold in the Organs and Tissues

After oral intake, AuNPs were transported to different organs via the bloodstream. It should be mentioned that the content of gold in the blood of females and offspring was lower than in analyzed organs (Table 1). A slightly higher content of gold was detected in mothers' blood compared to offspring'. In control mice, the content of gold in all analyzed organs was below the detection threshold of the NAA technique.

Table 1. Content of gold in organs and tissues of experimental females and their offspring.

Sample	Organ or Tissue	Mean ± SD (µg/g Dry Weight)	Range (µg/g Dry Weight)
Females (5)	Blood	0.012 ± 0.007	0.006–0.023
	Liver	0.239 ± 0.117	0.131–0.410
	Kidneys	0.323 ± 0.128	0.182–0.516
	Lungs	0.079 ± 0.049	0.038–0.145
Offspring (10)	Blood	0.007 ± 0.005	0.002–0.018
	Liver	0.028 ± 0.022	0.008–0.083
	Kidneys	0.161 ± 0.116	0.053–0.418
	Lungs	0.017 ± 0.010	0.005–0.036

Accumulation of gold in different organs and blood was similar in mothers and offspring: the lowest content was accumulated in the brain (Table 2), then, ten times higher, in blood, followed by lungs, liver, and the highest—in kidneys (Table 1). The content of gold in the liver and kidneys of offspring was ten and two times, respectively lower than in females.

Table 2. Gold accumulated in the brain of female mice and their offspring, excluding its content in blood vessels in the brain.

Sample	Specific Mass in a Sample (ng)		Gold Content (ng/g Dry Weight)
	Mean ± SD	Range	
Females (5)	0.25 ± 0.10	0.14–0.36	3.79
Offspring (10)	0.08 ± 0.03	0.04–0.13	1.10

2.2. Elevated Plus-Maze Results

All mice (SHK line, 20 experimental, 10 control animals) spent most of the trial time in the closed arms, less time—in the center, and visited the open arms rarely or not at all (except one experimental animal which sat in the open arm longer than other animals) (Supplementary Material). In comparison with control mice, experimental animals run more rarely (number of events (running) $p = 0.008$), but longer (mean length (running) $p = 0.043$). These mice entered the closed arm quicker (latency to the first event (closed arms) $p = 0.022$), run into the arm's end, and sat there longer (summary length (closed arms) $p = 0.18$, mean length (closed arms) $p = 0.019$). In addition, experimental animals more rarely run from one arm to another (number of events (closed arms) $p = 0.015$, number of events (center) $p = 0.014$, summary length (center) $p = 0.002$), and less frequently looked into the center while sitting in the arm (summary length (stretch-attend posture) $p = 0.023$, mean length (stretch-attend posture) $p = 0.009$). Experimental mice groomed less often, but longer (mean length (grooming) $p = 0.004$) and reared longer and more frequently (mean length (rearing) $p = 0.0003$). A similar pattern was observed when only males (10) or females (10) of the experimental group were compared with control mice.

2.3. MWM Results

After three days the learning curves typical for MWM were obtained for experimental and control mice and each group. As it can be seen the daily medians and means of all three parameters were declining (Figures 1–3, Tables S1–S6). Values of the parameters from Day 1 and Day 3 were significantly different for all three parameters (all $p < 0.0033$), both in experimental and control mice. Similar significant differences were found in capable and intermediate experimental animals (all $p < 0.039$). However, in control mice differences were significant only in the capable group (all $p < 0.0077$).

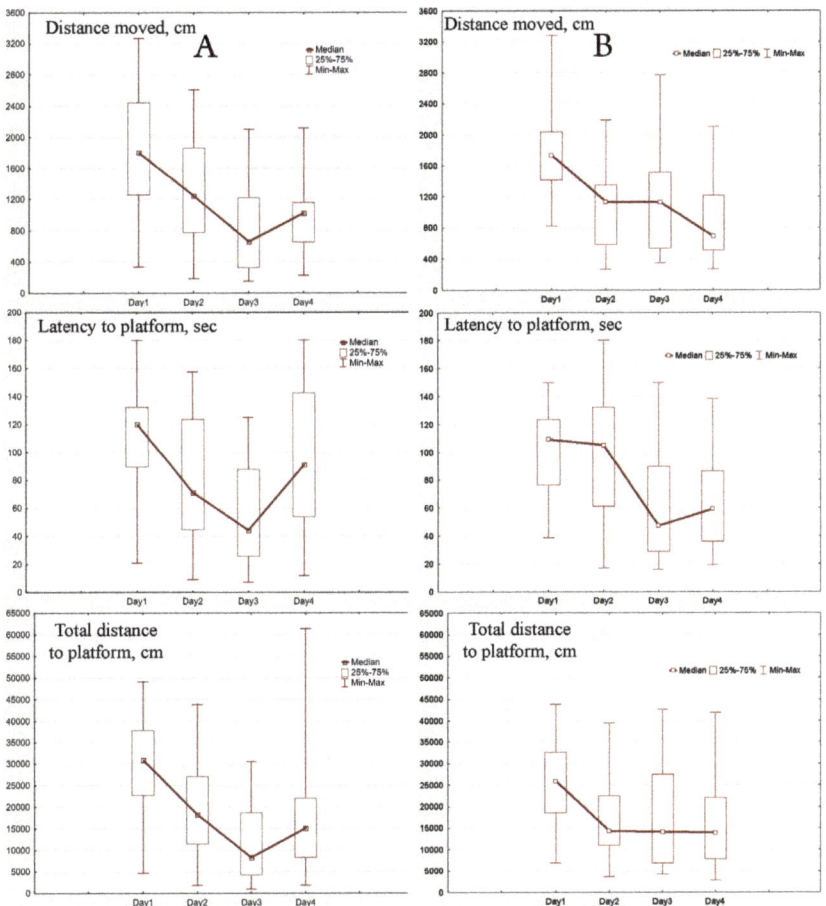

Figure 1. Learning curves obtained for studied groups of animals: (**A**), experimental mice (SHK line, 24 animals), (**B**), control mice (ICR and SHK lines, 18 animals). The central point represents the median; boxes, the lower and upper quartiles; intervals, the ranges.

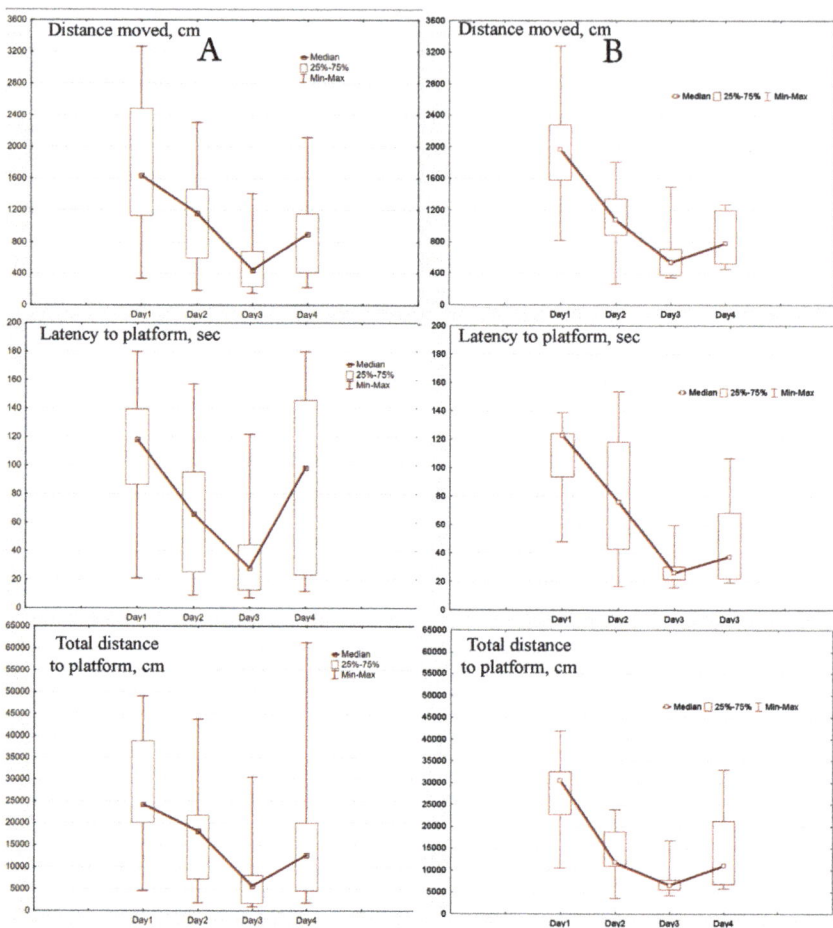

Figure 2. Learning curves obtained in the capable group: (**A**), experimental mice (SHK line, 15 animals), (**B**), control mice (ICR and SHK lines, 9 animals). The central point represents the median; boxes, the lower and upper quartiles; intervals, the ranges.

After a change of platform position in the pool on Day 4 in capable experimental mice parameters' values significantly increased compared to Day 3 (all $p < 0.013$). At the same time, in intermediate experimental mice and in control one they did not change or slightly decrease (Figures 2 and 3). In pairwise comparison between daily values of parameters for experimental and control animals the only differences, the total distance to the platform in the whole sample and the intermediate mice ($p = 0.038$ and $p = 0.034$, respectively), were observed on Day 3.

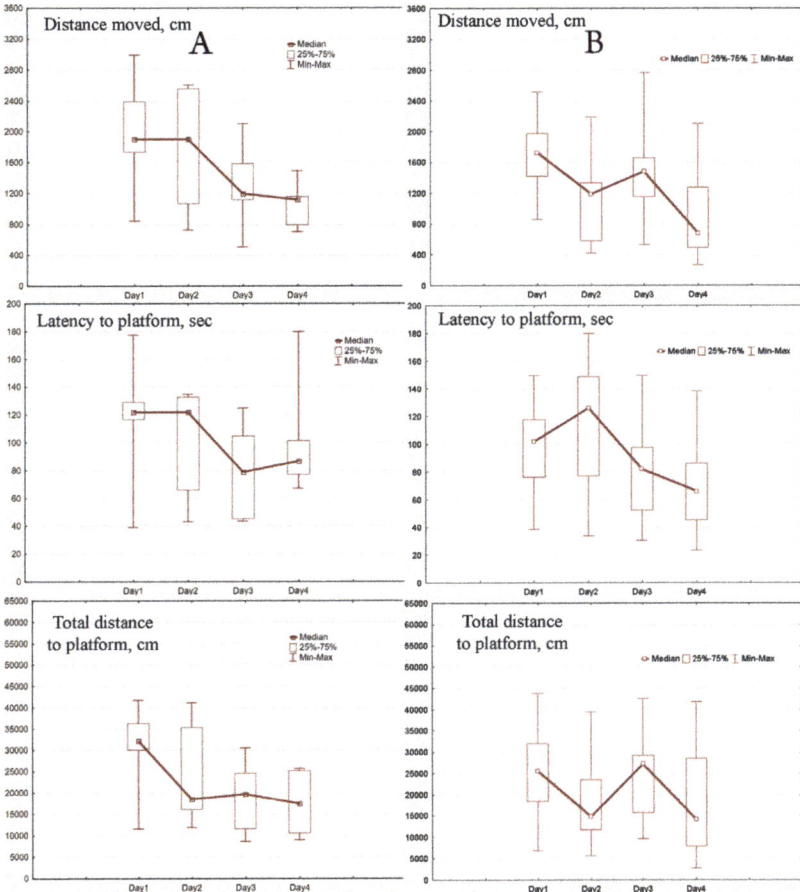

Figure 3. Learning curves obtained in the intermediate group: (**A**), experimental mice (SHK line, 9 animals), (**B**), control mice (ICR and SHK lines, 9 animals). The central point represents the median; boxes, the lower and upper quartiles; intervals, the ranges.

3. Discussion

A significant amount of gold was determined in all tissues of experimental mice and their offspring. The latter contradicts the data that were obtained after acute contact of animals with bare or citrate-coated AuNPs introduced by several intravenous injections to pregnant mice [21]. No traces of gold were found in tissue sections of fetuses (and placenta) by TEM [22], while ICP-MS did not detect gold in fetal organs, and only in significant amount in placenta [21]. It may indicate a slow rate of AuNPs transfers through the blood–placenta barrier. In the abovementioned studies, tissues were taken 24 h after two- or three-day injections. Thus, it can be suggested that AuNPs were not transferred to the fetus 3–4 days after the first dose.

The low content of gold in the blood of experimental animals indicates the rapid transfer of AuNPs from blood to different organs, which confirms the high permeability of tissues for very small (<20 nm) nanoparticles [18]. The gold content in mothers' blood and liver was higher than in offspring, which is consistent with the assumption about the slow rate of AuNPs' transfer through the blood-placenta barrier. This finding contradicts the data obtained in a similar experiment with silver nanoparticles, where the amount of silver in the offspring's blood was four times higher than in females [23].

The highest content of gold was determined in the kidneys of both females and offspring. In [24] it was shown that after 28 days of daily intraperitoneal injections of PEG-coated AuNPs with the size of 5 and 10 nm in adult mice, the highest amounts of gold were accumulated in the liver and spleen. It should be noted that the predominant accumulation of gold in the liver and spleen was detected 24 h after the single intravenous injection of 20 nm sized citrate-coated AuNPs [12], and after 8 days of daily intraperitoneal injections of 12.5 nm sized citrate-coated AuNPs [25]. Although in the latter study, the gold content in the kidneys was higher than in the spleen. The total daily intake of AuNPs per animal in [4] was comparable with the present study, and while the content of gold in kidneys was close to whose determined in the present work, in the liver its content was eight times higher. Although in study [4] the high accumulation of gold in organs can be explained by direct injection of a whole daily dose, the comparatively low amount of gold in the mother's liver in the present study is consistent with considerable amounts of gold found in offspring. It is suggested that nanoparticles were transferred to offspring instead of accumulating in the mother's liver. Thus, it can be assumed that the excretion of AuNPs was mostly done by kidneys, while the liver eliminated AuNPs from the bloodstream and stored them, although it could not prevent AuNPs' transfer to the fetus and milk. The opposite results were obtained for silver nanoparticles in [23], the content of silver in kidneys was the lowest compared to other organs.

The content of gold in the liver was the second largest in the present study and the highest in [4,12]. The liver has a high density of blood vessels, and the phagocytosis of nanoparticles, including citrate- or PEG-coated AuNPs, by Kupffer cells after acute contact through intravenous (quick uptake) or intraperitoneal (slower uptake) injection was described [22,26,27]. It should be noted that macrophages in other tissues (in mesenteric lymph nodes, spleen, and small intestine) also accumulated AuNPs, but in smaller amounts than liver Kupffer cells [22]. The high content of gold was also determined in the spleen [4,12,27]. However, the content of gold in the offspring's liver was lower than in the liver of females, and it can correlate with the lower content of gold in the offspring's blood. In contrast, the silver content in the liver did not differ between mothers and offspring [23]. Lungs have very dense vascularization, and it can be suggested that gold content in the lungs most likely correlates with the content of the metal in blood. However, in [25], the gold content in the lungs of adult mice was comparable with the present study, even though the total daily intake of AuNPs was approximately ten times lower.

After long-term exposure, the accumulation of gold in the brain was the lowest both in females and offspring, which is in line with the data obtained after 24 h [12] and 8 days [25] for adult mice exposure to AuNPs. The known relative inertness of AuNPs, especially in contrast to the high chemical activity of nanosilver [10], could be responsible for the low accumulation of AuNPs in the brain despite the high permeability of the blood–brain barrier for very small nanoparticles.

In an elevated plus-shaped maze all young mice demonstrated a preference for dark and secluded places over open areas, which is typical for rodents. Compared with control animals, experimental mice demonstrated signs of higher levels of anxiety: prolonged stays in closed arms with accompanying increase in long grooming, and decreased running. The exploratory behavior of experimental mice was unevenly changed: stretch-attend posture was shorted, but rearing with support was increased. It could be related to the amount of body exposure to open spaces that were necessary to perform both elements of behavior: for rearing it was enough to expose the nose and part of the head, while the whole half of the body was in the open arm in stretch-attend posture. Thus, the exploratory behavior of experimental animals could not be reduced as a whole but adapted for better body shielding, which is consistent with heightened anxiety. These data contradict the results on the estimation of anxiety levels in an open field test that was done on adult mice after several consecutive injections of bare or PEG-coated AuNPs [13], where no differences were found in comparison to control animals. Such discrepancies could be influenced by the difference in the design of the two tests, as an open field does not have a dark hiding place.

A similar pattern was observed in the experiments in which the influence of silver nanoparticles on anxiety levels in offspring was estimated [28,29]. It can also indicate increased vulnerability of the developing brain to AuNPs' effects. Additionally, the coating could influence the anxiety levels by enabling different mechanisms of AuNPs' accumulation in brain. In both tests, open field and elevated plus-shaped maze, intraperitoneal injection of glycoprotein transferrin coated AuNPs to the adult mice enabled the carrier-mediated transport through the blood–brain barrier and demonstrated increased anxiety (without changes in locomotion). It should be mentioned that after similar contact with bare AuNPs, no changes in anxiety were detected [20].

In the Morris water maze, no signs of impairments in spatial orientation and memory were found; animals successfully learned to find the platform. The differences in parameters' dynamic on Day 4 between experimental capable and intermediate mice correlate with the following types of behavior characteristic for these groups: capable animals with directed search or "scanning" were initially confused, which had raised the parameters' values, but for intermediate mice with random search, the probability of finding the platform did not change or slightly increased. The only difference between experimental and control animals in pairwise comparison was the total distance to the platform on Day 3, which can be attributed to the different types of behavior, which allowed capable animals to find the platform quicker. It should be noted that the small number of the control sample of capable and intermediate mice influenced the statistical significance of comparisons in the control group, but the parameter's dynamic typical for successful learning was clearly shown in the control samples. The lack of differences between experimental and control mice agrees with the data reported in [30] for adult mice after a single injection of citrate-coated AuNPs. The lack of deficiencies in spatial learning in the Barns maze after regular intraperitoneal injections of citrate-coated AuNPs to adult mice was shown in [31].

4. Materials and Methods

4.1. Animals

Outbred white mice (SHK) of the age 1.5–2 months (with an average mass of 20 g), 15 females and 5 males, were purchased from the Stolbovaya Farm (Moscow region, Russia). Mice are heterozygous by an undefined number of genes and are used to assess the safety of medical and cosmetic products, and dietary supplements. The animals and their offspring were maintained in the vivarium of M.F. Vladimirskiy Moscow Regional Research and Clinical Institute. Females were kept in steel cages with sizes of $31.5 \times 23 \times 15.7$ cm, and each cage contained five mice, with a natural cycle of illumination (by daylight through the window in the period from 8 am to 5 pm daily) and the temperature of 22–24 °C; each cage was cleaned once per day. The methodology of the experiments and the maintenance of the animals in the vivarium of the institute were performed according to the principles of Directive 2010/63/EU of the European Parliament and of the Council of 22 September 2010 on the protection of animals used for scientific purposes [32].

4.2. Nanoparticles

Small, 3–5 nm in diameter, AuNPs were selected since the accumulation of smaller particles is usually higher and more widespread than that of larger ones [18]. High-purity polyethylene-glycol-coated gold nanoparticles (PEG-AuNPs) were purchased from the M9 Company (Tolyatti, Russia). According to the manufacturer, the nanoparticle solution is stable for two months. The experimental solution with a concentration of 25 µg/mL was prepared by diluting the concentrated solution with pure water in a ratio of 1:500. The bottles in cages were filled with the solution when its level was low, but at least once a week. The experimental solution was prepared anew every two weeks.

4.3. Experiment

To study the uptake of AuNPs by both mothers and by offspring, experimental females drank the experimental solution with a concentration of 25 µg/mL a week before pregnancy and to the end of lactation (1 month after birth). The total daily intake per animal was 125 μg—the daily liquid consumption per mouse was 5 mL [33]. Thus, the offspring of experimental females received AuNPs from mothers—both through the placental barrier during prenatal development and with milk during lactation. Data for control females and their offspring were taken from the similar experiment performed previously with silver nanoparticles [23]. After the end of the lactation, young experimental animals, 15 females and 15 males, were transferred to their own cages, 3 mice per cage, where they drank tap water and were raised to 2 months of age. Then, animals participated in behavioral tests.

4.4. Measuring of Gold Content

At the end of the lactation, females and some of the offspring were euthanized by intraperitoneal injection of urethane solution. The concentration of the water solution is based on body weight and a dosage of 1.2 g of dry urethane per kg of body weight, proposed by a veterinarian. Brain, liver, lungs, kidneys, and blood (with measured volume) were extracted from each mouse, weighed, and freeze-dried.

The gold content in the isolated tissues was determined using neutron activation analysis at the IBR-2 reactor in Dubna, Russia. The description of irradiation channels and pneumatic transport system of the REGATA installation can be found elsewhere [34].

The analyzed samples were packed in aluminum foil cups, irradiated for 3 days at a neutron flux 1.2×10^{11} cm^{-2} s^{-1}, and measured for 30 min after 4 days of irradiation. The samples were irradiated simultaneously with two reference materials: SRM 2710 (Montana Soil, Highly Elevated Trace Element Concentrations (NIST, USA) and the liquid gold standard (Merck, Darmstadt, Germany). Reference materials and blanks were placed in each container.

Gamma spectra of induced activity were measured using spectrometers based on HPGe detectors with an efficiency of 40–55% and resolution of 1.8–2.0 keV for total-absorption peak 1332 keV of the isotope ^{60}Co and Canberra spectrometric electronics. The spectra analysis was performed using the Genie2000 software (version 3.4.1) (Canberra Industries, Inc.; Toledo, OH, USA) and the software "Concentration" version 6.13.3 (JINR, Dubna, Russia).

In the present study, with lower net inaccuracy the isotope 198Au from the reference material liquid standard was used as a calibrator, and SRM 2710 was used for quality control. The obtained values for concentrations for SRM 2710 in all irradiated containers differed from the certified values in the range (0.2–5.5%).

4.5. Behavioral Tests

Assessment of anxiety levels. Experimental (10 males, 10 females) and control (10 females) mice entered elevated Plus-maze (Open Science, Russia): basic test protocol for estimation of anxiety levels was applied [35], with only the trial duration taken from verification protocol given by the manufacturer of test installation (Verification of elevated Plus-maze by Open Science company's scientific division [36] Test was done in one day for every 10 animals, with one test trial per animal; mice were not previously acquainted with the test. The plus-shaped construction consists of four arms, connected to the common center at 90 degrees angle, and it is elevated at 40 cm height above the table. Two arms are open walkways without walls, and the other two arms have walls from three sides, except the central side. Near the maze, two lamps were positioned opposite to each other from the open arms' sides, so the pronounced contrast lighting of open and closed arms was created: open arms were bright, and closed arms were dark.

Animals were consequently tested; before each trial floor and wall of all arms and centers were cleaned with ethanol to prevent smell marks' influence on mice's behavior. According to the protocol of preparation for this test [35], before the first animal's trial one "zero" trial was done to equalize test conditions for the first mouse with others', one mouse that was not included in this experiment was set in the test for 3 min without collection of data. The animal was placed in the maze center and then had 3 min to roam freely. After that the mouse was gently extracted from the maze and placed in its cage; the maze was cleaned with 95% ethanol before the placement of the next animal.

Video records of elevated plus-maze trials were analyzed in the RealTimer program (OpenScience, Moscow, Russia) (free distribution). Seven aspects of behavior were counted and measured; when a mouse was in closed arms, open arms, or center, running and rearing (with support—when forelegs were placed on the wall), grooming and the stretch-attend posture, when an animal stands with forelegs in one zone and hind legs in other zone and stretches its body to sniff and look. For each aspect four quantitative parameters were measured, including number of events, summary length (s), mean length (s), and latency to the first event (s).

Assessment of spatial learning and memory. One cycle of testing (three days of primary learning) was performed in the Morris water maze (MWM), according to the early developed protocol of the testing [37,38] with additional assessment of the flexibility of developed spatial memory by subsequent re-positioning of the platform into the other part of the swimming pool (fourth day) [39]. The circular white pool (1.5 m in diameter) was filled with water, and various visible clues were positioned around the pool. Water was made opaque with dry milk to hide the underwater platform. Each animal had to find this invisible underwater platform to escape from the water. If the animal did not find the platform during a trial, it was gently placed on the platform manually.

The cycle consisted of four consecutive days with three daily trials: the platform's location was unchanged in the first three days of primary learning, but on the fourth day, the platform was relocated to the other part of the pool for assessment of formed spatial memory. During each trial, a mouse was allowed to swim freely in the pool for a maximum of 180 s. When the animal either found the platform or was placed on it after the time was up, it was left to sit on the platform for 20 s, and then it was picked up, dried with a clean cloth, and returned to its cage until its next trial. Therefore, when 15 mice entered the test, each animal had 20–30 min of rest between the trials.

For each animal was determined the behavioral type was demonstrated during trials in three days of primary learning. Similarly to the previously developed protocol [40], the general sample was divided into three groups, "incapable", "capable", and "intermediate" animals, based on the type of behavior most frequently shown on Day 3. Incapable mice were excluded from the MWM as being non-responsive to potential changes induced by AuNPs that could be detected in this test; they demonstrated types of behavior, which did not include exploration of the pool. Therefore, they did not settle in the paradigm of MWM. All other individuals demonstrated various types of behavior that included an exploration of the pool, so they were assigned either to the capable (directed search, "scanning" [41] or to the intermediate (random search) group.

Thus, 30 young experimental mice, 15 females and 15 males, entered MWM, and data from 24 animals were included in the statistical analysis, with both genders present in each group. After MWM the offspring aged out at 2 months; therefore, to minimize the influence of aging, the data from control female offspring (their mothers drank clear water) from the previous experiment with a similar design (with silver nanoparticles; [33]) were taken into the analysis as the control, and also the control sample included the data from female offspring from the previous experiment with a similar design in which titanium dioxide nanoparticles were too big to pass through the blood–brain barrier so titanium was not detected in the brain at all [42].

Video records of MWM trials were analyzed by the Ethovision program (Noldus Information Technology, Wageningen, The Netherlands). Trajectories (tracks) of all trials were visualized and based on those three parameters were calculated for each trial: the distance moved (track length (cm), the latency to the platform (s), and the total distance to the platform (cm). The last parameter is a sum of distances to the platform from all points of the track; if the value is low, this means that a mouse was swimming near the platform. Means of the parameters per day were calculated for each animal, with subsequent comparison of the daily samples of the means.

4.6. Statistics

The behavioral tests' results were analyzed using Statistica 13.2 software (Dell Inc., Round Rock, TX, USA), at the significance level of $p < 0.05$; all tests were two-tailed. The elevated plus-maze results were compared between experimental and control animals for each parameter of each behavioral aspect by Mann–Whitney U test. Additionally, the results for experimental males or females were compared with the control group.

For MWM results distribution of quantitative variables was checked for normality by Shapiro–Wilk test; this hypothesis was rejected for all three parameters. The daily changes of all three parameters (learning curves) for each group in the experimental and control samples were assessed by comparison between Day 1 and Day 3 in Wilcoxon matched pairs test. Differences between the experimental and control data of each group were assessed by each day comparison with Mann–Whitney U-test.

5. Conclusions

Gold was present in all examined organs of experimental females and their offspring. The highest gold content was found in the kidneys, followed by the liver, which suggests that in case of prolonged exposure excretion of AuNPs was done by the kidneys. In contrast, the liver eliminated AuNPs from the bloodstream and accumulated them. The content of gold in the offspring's liver and blood was lower than in the mother's indicating a relatively low rate of AuNPs' transfer to offspring. Accumulation of gold in the brain was the lowest among analyzed organs both in females and offspring, which could be attributed to the chemical inertness of gold. No significant differences in spatial orientation and memory were found between experimental and control offspring in MWM. Still, the experimental young mice demonstrated increased anxiety levels in an elevated Plus-shaped maze. Thus, the AuNPs influenced the emotional state of mice that were exposed to nanoparticles during prenatal and early postnatal development, but not on their cognitive abilities.

Supplementary Materials: The following supporting information can be downloaded at: https://www.mdpi.com/article/10.3390/ijms24108962/s1.

Author Contributions: Conceptualization, A.L.I., D.A.R. and I.Z.; methodology, E.N.P.; software, A.L.I.; validation, N.Y. and D.G.; formal analysis, A.L.I. and I.Z.; investigation, A.L.I., E.N.P., D.G. and N.Y.; data curation, A.L.I. and I.Z.; writing—original draft preparation, A.L.I.; writing—review and editing, I.Z.; supervision, I.Z. and D.A.R. All authors have read and agreed to the published version of the manuscript.

Funding: This research received no external funding.

Institutional Review Board Statement: The study was conducted in accordance with the Declaration of Helsinki and approved by the Independent Ethics Committee of Moscow Region State Budgetary Healthcare Institution "M.F. Vladimirsky Moscow Regional Clinical Research Institute named" (protocol code 1 January 2019).

Informed Consent Statement: Not applicable.

Data Availability Statement: All data are presented in Supplementary Files.

Conflicts of Interest: The authors declare no conflict of interest.

References

1. Sukhorukov, G.; Fery, A.; Möhwald, H. Intelligent Micro- and Nanocapsules. In *Progress in Polymer Science, Proceedings of the World Polymer Congress, 40th IUPAC International Symposium on Macromolecules, Paris, Francw, 4–9 July 2005*; Elsevier: Amsterdam, The Netherlands, 2005; Volume 30, pp. 885–897.
2. Hou, K.; Zhao, J.; Wang, H.; Li, B.; Li, K.; Shi, X.; Wan, K.; Ai, J.; Lv, J.; Wang, D.; et al. Chiral Gold Nanoparticles Enantioselectively Rescue Memory Deficits in a Mouse Model of Alzheimer's Disease. *Nat. Commun.* **2020**, *11*, 4790. [CrossRef] [PubMed]
3. Kurapov, Y.A.; Litvin, S.; Belyavina, N.N.; Oranskaya, E.I.; Romanenko, S.M.; Stelmakh, Y. Synthesis of Pure (Ligandless) Titanium Nanoparticles by EB-PVD Method. *J. Nanopart. Res.* **2021**, *23*, 20. [CrossRef]
4. Zhang, X.D.; Wu, D.; Shen, X.; Liu, P.X.; Yang, N.; Zhao, B.; Zhang, H.; Sun, Y.M.; Zhang, L.A.; Fan, F.Y. Size-Dependent In Vivo Toxicity of PEG-Coated Gold Nanoparticles. *Int. J. Nanomed.* **2011**, *6*, 2071–2081. [CrossRef] [PubMed]

5. Abu-Dief, A.M.; Alsehli, M.; Awaad, A. A Higher Dose of PEGylated Gold Nanoparticles Reduces the Accelerated Blood Clearance Phenomenon Effect and Induces Spleen B Lymphocytes in Albino Mice. *Histochem. Cell Biol.* **2022**, *157*, 641–656. [CrossRef]
6. Abu-Dief, A.M.; Salaheldeen, M.; El-Dabea, T. Recent Advances in Development of Gold Nanoparticles for Drug Delivery Systems. *J. Mod. Nanotechnol.* **2021**, *1*, 1. [CrossRef]
7. Ferreira, G.K.; Cardoso, E.; Vuolo, F.S.; Galant, L.S.; Michels, M.; Gonçalves, C.L.; Rezin, G.T.; Dal-Pizzol, F.; Benavides, R.; Alonso-Núñez, G.; et al. Effect of Acute and Long-Term Administration of Gold Nanoparticles on Biochemical Parameters in Rat Brain. *Mater. Sci. Eng. C* **2017**, *79*, 748–755. [CrossRef] [PubMed]
8. Lee, U.; Yoo, C.J.; Kim, Y.J.; Yoo, Y.M. Cytotoxicity of Gold Nanoparticles in Human Neural Precursor Cells and Rat Cerebral Cortex. *J. Biosci. Bioeng.* **2016**, *121*, 341–344. [CrossRef]
9. Schuemann, J.; Bagley, A.F.; Berbeco, R.; Bromma, K.; Butterworth, K.T.; Byrne, H.L.; Chithrani, B.D.; Cho, S.H.; Cook, J.R.; Favaudon, V.; et al. Roadmap for Metal Nanoparticles in Radiation Therapy: Current Status, Translational Challenges, and Future Directions. *Phys. Med. Biol.* **2020**, *65*, 21RM02. [CrossRef]
10. Behra, R.; Sigg, L.; Clift, M.J.D.; Herzog, F.; Minghetti, M.; Johnston, B.; Petri-Fink, A.; Rothen-Rutishauser, B. Bioavailability of Silver Nanoparticles and Ions: From a Chemical and Biochemical Perspective. *J. R. Soc. Interface* **2013**, *10*, 20130396. [CrossRef]
11. Khan, H.A.; Alamery, S.; Ibrahim, K.E.; El-Nagar, D.M.; Al-Harbi, N.; Rusop, M.; Alrokayan, S.H. Size and Time-Dependent Induction of Proinflammatory Cytokines Expression in Brains of Mice Treated with Gold Nanoparticles. *Saudi J. Biol. Sci.* **2019**, *26*, 625–631. [CrossRef]
12. Takeuchi, I.; Nobata, S.; Oiri, N.; Tomoda, K.; Makino, K. Biodistribution and Excretion of Colloidal Gold Nanoparticles after Intravenous Injection: Effects of Particle Size. *Biomed. Mater. Eng.* **2017**, *28*, 315–323. [CrossRef]
13. Tuna, B.G.; Yesilay, G.; Yavuz, Y.; Yilmaz, B.; Culha, M.; Maharramov, A.; Dogan, S. Electrophysiological Effects of Polyethylene Glycol Modified Gold Nanoparticles on Mouse Hippocampal Neurons. *Heliyon* **2020**, *6*, e05824. [CrossRef] [PubMed]
14. Zhang, X.; Guo, X.; Kang, X.; Yang, H.; Guo, W.; Guan, L.; Wu, H.; Du, L. Surface Functionalization of Pegylated Gold Nanoparticles with Antioxidants Suppresses Nanoparticle-Induced Oxidative Stress and Neurotoxicity. *Chem. Res. Toxicol.* **2020**, *33*, 1195–1205. [CrossRef] [PubMed]
15. Liu, Y.; Meyer-Zaika, W.; Franzka, S.; Schmid, G.; Tsoli, M.; Kuhn, H. Gold-Cluster Degradation by the Transition of B-DNA into A-DNA and the Formation of Nanowires. *Angew. Chemie Int. Ed.* **2003**, *42*, 2853–2857. [CrossRef] [PubMed]
16. Tsoli, M.; Kuhn, H.; Brandau, W.; Esche, H.; Schmid, G. Cellular Uptake and Toxicity of Au55 Clusters. *Small* **2005**, *1*, 841–844. [CrossRef] [PubMed]
17. Wang, Y.; Zhang, H.; Shi, L.; Xu, J.; Duan, G.; Yang, H. A Focus on the Genotoxicity of Gold Nanoparticles. *Nanomedicine* **2020**, *15*, 319–323. [CrossRef]
18. Lewinski, N.; Colvin, V.; Drezek, R. Cytotoxicity of Nanoparticles. *Small* **2008**, *4*, 26–49. [CrossRef] [PubMed]
19. Takeda, K.; Suzuki, K.I.; Ishihara, A.; Kubo-Irie, M.; Fujimoto, R.; Tabata, M.; Oshio, S.; Nihei, Y.; Ihara, T.; Sugamata, M. Nanoparticles Transferred from Pregnant Mice to Their Offspring Can Damage the Genital and Cranial Nerve Systems. *J. Health Sci.* **2009**, *55*, 95–102. [CrossRef]
20. Yavuz, Y.; Yesilay, G.; Guvenc Tuna, B.; Maharramov, A.; Culha, M.; Erdogan, C.S.; Garip, G.A.; Yilmaz, B. Investigation of Effects of Transferrin-Conjugated Gold Nanoparticles on Hippocampal Neuronal Activity and Anxiety Behavior in Mice. *Mol. Cell. Biochem.* **2022**. [CrossRef]
21. Rattanapinyopituk, K.; Shimada, A.; Morita, T.; Sakurai, M.; Asano, A.; Hasegawa, T.; Inoue, K.; Takano, H. Demonstration of the Clathrin- and Caveolin-Mediated Endocytosis at the Maternal-Fetal Barrier in Mouse Placenta after Intravenous Administration of Gold Nanoparticles. *J. Vet. Med. Sci.* **2014**, *76*, 377–387. [CrossRef]
22. Sadauskas, E.; Wallin, H.; Stoltenberg, M.; Vogel, U.; Doering, P.; Larsen, A.; Danscher, G. Kupffer Cells Are Central in the Removal of Nanoparticles from the Organism. *Part. Fibre Toxicol.* **2007**, *4*, 10. [CrossRef] [PubMed]
23. Zinicovscaia, I.; Grozdov, D.; Yushin, N.; Ivlieva, A.; Petritskaya, E.; Rogatkin, D. Neutron Activation Analysis as a Tool for Tracing the Accumulation of Silver Nanoparticles in Tissues of Female Mice and Their Offspring. *J. Radioanal. Nucl. Chem.* **2019**, *322*, 1079–1083. [CrossRef]
24. Zhang, X.D.; Wu, D.; Shen, X.; Chen, J.; Sun, Y.M.; Liu, P.X.; Liang, X.J. Size-Dependent Radiosensitization of PEG-Coated Gold Nanoparticles for Cancer Radiation Therapy. *Biomaterials* **2012**, *33*, 6408–6419. [CrossRef] [PubMed]
25. Lasagna-Reeves, C.; Gonzalez-Romero, D.; Barria, M.A.; Olmedo, I.; Clos, A.; Sadagopa Ramanujam, V.M.; Urayama, A.; Vergara, L.; Kogan, M.J.; Soto, C. Bioaccumulation and Toxicity of Gold Nanoparticles after Repeated Administration in Mice. *Biochem. Biophys. Res. Commun.* **2010**, *393*, 649–655. [CrossRef] [PubMed]
26. Nghiem, T.H.L.; Nguyen, T.T.; Fort, E.; Nguyen, T.P.; Hoang, T.M.N.; Nguyen, T.Q.; Tran, H.N. Capping and In Vivo Toxicity Studies of Gold Nanoparticles. *Adv. Nat. Sci. Nanosci. Nanotechnol.* **2012**, *3*, 015002. [CrossRef]
27. Valentini, X.; Rugira, P.; Frau, A.; Tagliatti, V.; Conotte, R.; Laurent, S.; Colet, J.M.; Nonclercq, D. Hepatic and Renal Toxicity Induced by TiO2 Nanoparticles in Rats: A Morphological and Metabonomic Study. *J. Toxicol.* **2019**, *2019*, 5767012. [CrossRef]
28. Ghaderi, S.; Tabatabaei, S.R.F.; Varzi, H.N.; Rashno, M. Induced Adverse Effects of Prenatal Exposure to Silver Nanoparticles on Neurobehavioral Development of Offspring of Mice. *J. Toxicol. Sci.* **2015**, *40*, 263–275. [CrossRef]

29. Dǎnilǎ, O.O.; Berghian, A.S.; Dionisie, V.; Gheban, D.; Olteanu, D.; Tabaran, F.; Baldea, I.; Katona, G.; Moldovan, B.; Clichici, S.; et al. The Effects of Silver Nanoparticles on Behavior, Apoptosis and Nitro-Oxidative Stress in Offspring Wistar Rats. *Nanomedicine* **2017**, *12*, 1455–1473. [CrossRef]
30. Sanati, M.; Khodagholi, F.; Aminyavari, S.; Ghasemi, F.; Gholami, M.; Kebriaeezadeh, A.; Sabzevari, O.; Hajipour, M.J.; Imani, M.; Mahmoudi, M.; et al. Impact of Gold Nanoparticles on Amyloid β-Induced Alzheimer's Disease in a Rat Animal Model: Involvement of STIM Proteins. *ACS Chem. Neurosci.* **2019**, *10*, 2299–2309. [CrossRef]
31. Muller, A.P.; Ferreira, G.K.; Pires, A.J.; de Bem Silveira, G.; de Souza, D.L.; Brandolfi, J.d.A.; de Souza, C.T.; Paula, M.M.S.; Silveira, P.C.L. Gold Nanoparticles Prevent Cognitive Deficits, Oxidative Stress and Inflammation in a Rat Model of Sporadic Dementia of Alzheimer's Type. *Mater. Sci. Eng. C* **2017**, *77*, 476–483. [CrossRef]
32. The European Parliament; The Council of The European Union. Directive 2010/63/EU of the European Parliament and of the Council of 22 September 2010 on the Protection of Animals Used for Scientific Purposes Text with EEA Relevance. *Off. J. Eur. Union* **2010**, *53*, 33–79.
33. Lopatina, M.; Petritskaya, E.; Ivlieva, A. Dynamics of Body Weight of Laboratory Mice Depending on The Type of Feed and Feeding Regime. *Lab. Zhivotnye Dlya Nauchnych Issled. Lab. Anim. Sci.* **2019**, *2*. [CrossRef]
34. Frontasyeva, M.V. Neutron Activation Analysis in the Life Sciences. *Phys. Part. Nucl.* **2011**, *42*, 332–378. [CrossRef]
35. Walf, A.A.; Frye, C.A. The Use of the Elevated plus Maze as an Assay of Anxiety-Related Behavior in Rodents. *Nat. Protoc.* **2007**, *2*, 322–328. [CrossRef]
36. ООО "НПК Открытая Наука". Available online: https://www.openscience.ru/ (accessed on 13 May 2023).
37. Ivlieva, A.L.; Petritskaya, E.N.; Lopatina, M.V.; Rogatkin, D.A.; Zinkovskaya, I. Evaluation of the Cognitive Abilities of Mice Exposed to Silver Nanoparticles during Prenatal Development and Lactation. In *Cognitive Modeling, Proceedings of the Seventh International Forum on Cognitive Modeling, Retimno, Greece, 5–15 September 2019*; Science and Studies Foundation: Rostov-on-Don, Russia, 2019; pp. 276–282.
38. Ivlieva, A.L.; Petritskaya, E.N.; Rogatkin, D.A.; Demin, V.A. Methodological Characteristics of the Use of the Morris Water Maze for Assessment of Cognitive Functions in Animals. *Neurosci. Behav. Physiol.* **2017**, *47*, 484–493. [CrossRef]
39. Nieto-Escámez, F.A.; Sánchez-Santed, F.; De Bruin, J.P.C. Pretraining or Previous Non-Spatial Experience Improves Spatial Learning in the Morris Water Maze of Nucleus Basalis Lesioned Rats. *Behav. Brain Res.* **2004**, *148*, 55–71. [CrossRef]
40. Ivlieva, A.L.; Petritskaya, E.N.; Rogatkin, D.A.; Demin, V.A.; Glazkov, A.A.; Zinicovscaia, I.; Pavlov, S.S.; Frontasyeva, M.V. Impact of Chronic Oral Administration of Silver Nanoparticles on Cognitive Abilities of Mice. *Phys. Part. Nucl. Lett.* **2021**, *18*, 250–265. [CrossRef]
41. Garthe, A.; Behr, J.; Kempermann, G. Adult-Generated Hippocampal Neurons Allow the Flexible Use of Spatially Precise Learning Strategies. *PLoS ONE* **2009**, *4*, e5464. [CrossRef]
42. Zinicovscaia, I.; Ivlieva, A.L.; Petritskaya, E.N.; Rogatkin, D.A.; Yushin, N.; Grozdov, D.; Vergel, K.; Kutláková, K.M. Assessment of TiO_2 Nanoparticles Accumulation in Organs and Their Effect on Cognitive Abilities of Mice. *Phys. Part. Nucl. Lett.* **2021**, *18*, 378–384. [CrossRef]

Disclaimer/Publisher's Note: The statements, opinions and data contained in all publications are solely those of the individual author(s) and contributor(s) and not of MDPI and/or the editor(s). MDPI and/or the editor(s) disclaim responsibility for any injury to people or property resulting from any ideas, methods, instructions or products referred to in the content.

International Journal of *Molecular Sciences*

Article

Injectable Hydrogel Guides Neurons Growth with Specific Directionality

Yun-Hsiu Tseng [1], Tien-Li Ma [1], Dun-Heng Tan [1], An-Jey A. Su [2,*], Kia M. Washington [2], Chun-Chieh Wang [3], Yu-Ching Huang [4], Ming-Chung Wu [5,6,7,*] and Wei-Fang Su [1,4]

[1] Department of Materials Science and Engineering, National Taiwan University, Taipei 10617, Taiwan
[2] Department of Surgery, University of Colorado Anschutz Medical Campus, Aurora, CO 80045, USA
[3] National Synchrotron Radiation Research Center, Hsinchu 30076, Taiwan
[4] Department of Materials Engineering, Ming Chi University of Technology, New Taipei 24301, Taiwan
[5] Department of Chemical and Materials Engineering, Chang Gung University, Taoyuan 33302, Taiwan
[6] Center for Green Technology, Chang Gung University, Taoyuan 33302, Taiwan
[7] Division of Neonatology, Department of Pediatrics, Chang Gung Memorial Hospital at Linkou, Taoyuan 33305, Taiwan
* Correspondence: an-jey.su@cuanschutz.edu (A.-J.A.S.); mingchungwu@cgu.edu.tw (M.-C.W.)

Citation: Tseng, Y.-H.; Ma, T.-L.; Tan, D.-H.; Su, A.-J.A.; Washington, K.M.; Wang, C.-C.; Huang, Y.-C.; Wu, M.-C.; Su, W.-F. Injectable Hydrogel Guides Neurons Growth with Specific Directionality. *Int. J. Mol. Sci.* **2023**, *24*, 7952. https://doi.org/10.3390/ijms24097952

Academic Editors: Marcel Popa and Thomas Dippong

Received: 23 February 2023
Revised: 19 April 2023
Accepted: 25 April 2023
Published: 27 April 2023

Copyright: © 2023 by the authors. Licensee MDPI, Basel, Switzerland. This article is an open access article distributed under the terms and conditions of the Creative Commons Attribution (CC BY) license (https://creativecommons.org/licenses/by/4.0/).

Abstract: Visual disabilities affect more than 250 million people, with 43 million suffering from irreversible blindness. The eyes are an extension of the central nervous system which cannot regenerate. Neural tissue engineering is a potential method to cure the disease. Injectability is a desirable property for tissue engineering scaffolds which can eliminate some surgical procedures and reduce possible complications and health risks. We report the development of the anisotropic structured hydrogel scaffold created by a co-injection of cellulose nanofiber (CNF) solution and co-polypeptide solution. The positively charged poly (L-lysine)-r-poly(L-glutamic acid) with 20 mol% of glutamic acid (PLLGA) is crosslinked with negatively charged CNF while promoting cellular activity from the acid nerve stimulate. We found that CNF easily aligns under shear forces from injection and is able to form hydrogel with an ordered structure. Hydrogel is mechanically strong and able to support, guide, and stimulate neurite growth. The anisotropy of our hydrogel was quantitatively determined in situ by 2D optical microscopy and 3D X-ray tomography. The effects of PLLGA:CNF blend ratios on cell viability, neurite growth, and neuronal signaling are systematically investigated in this study. We determined the optimal blend composition for stimulating directional neurite growth yielded a 16% increase in length compared with control, reaching anisotropy of 30.30% at 10°/57.58% at 30°. Using measurements of calcium signaling in vitro, we found a 2.45-fold increase vs. control. Based on our results, we conclude this novel material and unique injection method has a high potential for application in neural tissue engineering.

Keywords: hydrogel; polypeptide; cellulose nanofiber; injectable; aligned structure; neuron; tissue engineering; three-dimensional tomography; calcium imaging

1. Introduction

The digital revolution has changed eye-use habits, causing negative consequences for ocular health. Long-term use of electronic products has been reported to increase intraocular pressure (IOP) [1,2], forcing the elongation of the optic axis, which then compresses the cornea and optic nerve. An increase in IOP is often accompanied by a variety of orbital disorders such as keratoconus or glaucoma [3]. Studies have also pointed out that the annual myopia rate is increasing, with a particular effect upon the world's younger generations [4,5]. Clinical studies confirm that high myopia can increase the incidence of glaucoma [6]; the morbidity rate of glaucoma has indeed increased annually in recent surveys [7]. Currently, glaucoma is the most common cause of blindness, second only to cataracts [8]. In primary closed angle glaucoma, the patient is born with an anatomically narrow drainage angle. This can restrict the outflow of the aqueous fluid and elevate IOP. Additionally, the displacement of the lens caused by phubbing and mydriasis resulting from a lack of light or tight ciliary muscles associated with eye tension can cause aberrant aqueous humor drainage. Ischemia, secondary to elevated IOP, causes retinal ganglion cell (RGC) death and degeneration of the optic nerve, which is comprised of RGC axons [9]. Therefore, blindness caused by optic nerve damage is a permanent and irreversible condition with no current treatments available to patients [10,11].

Neural tissue engineering (NTE) was first proposed in 1987 as a promising technology to regenerate damaged CNS neurons and their axons [12]. The concept of tissue engineering combines three main elements: cells, engineering tissue architecture, and engineering materials [13,14]. A potential benefit of using tissue engineering in medical applications is relief from organ or tissue recovery and transplantation. Instead, damaged or missing organs and tissues can be cultivated and used to repair, replace, or reconstruct defects. Engineered 3D tissue architecture that mimics the extracellular matrix can be designed with different functions to match the intrinsic characteristics of the target tissue; for example, the high-frequency contraction/relaxation of cardiomyocytes [15,16] and the specific directional communication of nerves [17,18]. To reestablish proper nervous system signaling, engineered neural must incorporate specific directionality to guide the orderly growth of nerves [19]. Thus, it is the objective of this research to develop injectable hydrogel under shear force that can promote the neurite growth with directionality and signal transport functionality in a biomimetic extracellular.

Previously electrospinning scaffolds, hydrogels, and hybrid scaffolds are commonly used in engineered tissue architecture. Although electrospinning scaffolds can effectively guide the growth of neurites, studies have found that cells grown on stiff surfaces may cause unusual adhesion [20,21]. Thus, we chose hydrogels as our tissue engineering architecture for their mechanical properties, which are similar to those of living organisms. Additionally, hydrogels have a bio-friendly high water-content structure, which can be made for injectable delivery, reducing the risks associated with surgical transplantation [22]. One potential issue is that the high water content of hydrogels causes difficulty in obtaining highly ordered and stable structures. Anisotropic hydrogels can be arranged through self-assembled peptides (SAPs), electromagnetic field induction, or stress–strain induction. The advantage of SAPs hydrogels arises from the utilization of natural amino acids or peptide sequences with beneficial properties such as the IKVAV sequence taken from Laminin, which can improve cell attachment when applied. Unfortunately, its low synthesis yield causes high material costs and difficulties in obtaining long-range arrangements make the SAPs difficult to apply clinically [23,24]. Electromagnetic field-induced hydrogels are limited by high costs despite the technique's efficiency for obtaining high-ordered structures. Additionally, high electromagnetic fields are required to align the molecules that present hidden dangers to cells in 3D culture [25–29].

Instead, we elected to orient our hydrogels using simple and low-cost shear force via syringe injection to align the molecules [30–36]. The repeated monomeric units of glucose can self-arrange cellulose chains into the large and ordered structure of cellulose microfibril [37], which can be easily aligned under shear force. Thus, we selected the easy-to-align nanocellulose fiber (CNF) as one of the components in our hydrogels. We confined the molecules of hydrogels in a long (5.7 cm) and small diameter needle (22 Gauge: 0.168 mm I.D.) and capillary tube (9 cm long and 0.2 mm O.D.) to minimize the stress differences between the wall and center of the syringe device.

CNF contains negative charges which quickly crosslink with cation poly-L-lysine-based crosslinkers through static charge attraction and hydrogen bonding. The poly-L-lysine-based crosslinkers were selected for their anti-inflammatory properties and cell/protein adhesion properties (which create a neuroprotective and neurotrophic environment), and are amenable to neural tissues and nerve-like cells used in research such as the transformed rat cell line, PC12 and human ARPE-19 cells [38–41]. Poly-L-lysine was further modified to copolymerize with the excitatory neurotransmitter, L-glutamic acid, to improve the growth and differentiation of the neurites [17,18,42–44]. Figure 1 shows the schematic diagram of the preparation of anisotropic structured hydrogel.

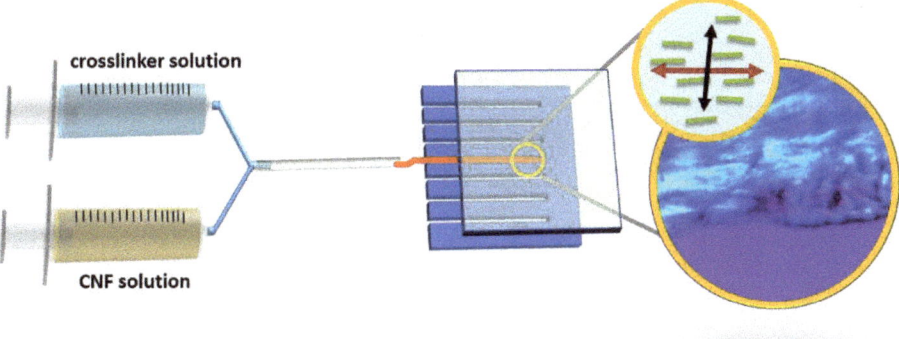

Figure 1. Schematic diagram of the preparation of the anisotropic-structured hydrogel via shear force using co-injection of CNF and crosslinker solutions.

The parallel needle walls of a co-injection syringe were used to align the CNF and crosslinker solutions, respectively. Then, the two-shear aligned individual solutions were combined through a capillary tube to form a hydrogel. The diffusion gradient causes a second stage arrangement from within the hydrogels. The shear stress applied to the hydrogels by the wall of the capillary tube during injection provided a third stage alignment coinciding with the second stage arrangement. A polarized optical microscope (POM) and transmission X-ray microscope (TXM) were used to analyze the alignment structures of our hydrogels in the micro- and nano-scales, respectively. They are the techniques to examine the morphologies of hydrogels in situ without using high vacuum techniques such as SEM or TEM which would distort the wet gel structures [45]. PC12 cells are regularly used as a model of sympathetic neurons and neuronal differentiation [46]. We performed in vitro experiments to demonstrate that PC12 neurites were guided by the ordered hydrogel structures. Ca^{2+} is utilized in this study and others as an intracellular messenger that controls many cellular responses in various excitable and non-excitable cell types [47]. We used calcium imaging in ARPE-19 cells to demonstrate an enhanced responsiveness of ocular tissue derived cells cultivated on our scaffolds [48]. The combination of natural and synthetic materials holds tremendous potential for developing injectable hydrogels with enhanced biocompatibility, reduced cytotoxicity, and improved mechanical properties [49–52]. This study demonstrates the successful crosslinking of negatively charged CNFs with positively charged copolypeptides into hydrogels. The copolypeptides contain a moiety of lysine and glutamic acid that promote nerve axon growth and a water solubility that represents a breakthrough in hydrogel material for neural tissue engineering.

2. Results and Discussion

Hydrogels were prepared by blending aqueous solutions of CNF and polylysine-based crosslinkers PLL and PLLGA through injection methods as depicted in Figure 2 and described above in Materials and Methods. The chemical structures of CNF, poly-L-lysine-based, and the crosslinkers (PLL and PLLGA) are shown in Figure 3. The crosslinkers were synthesized in our laboratory accordingly to established methods [45].

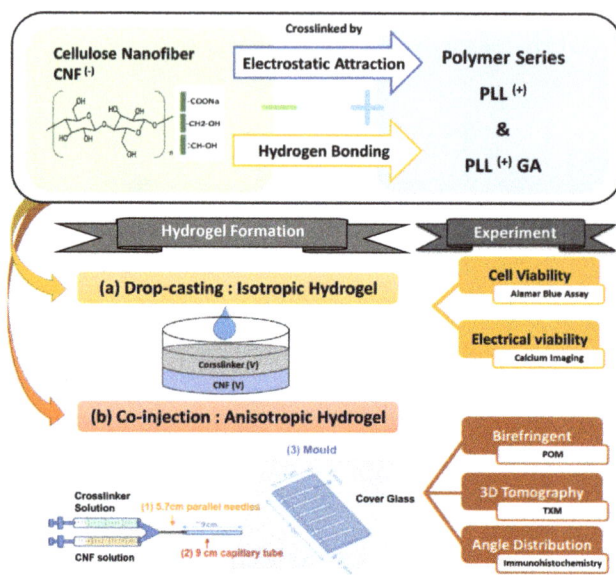

Figure 2. Schematic diagrams of hydrogel preparation (**a**) isotropic gel and (**b**) anisotropic gel.

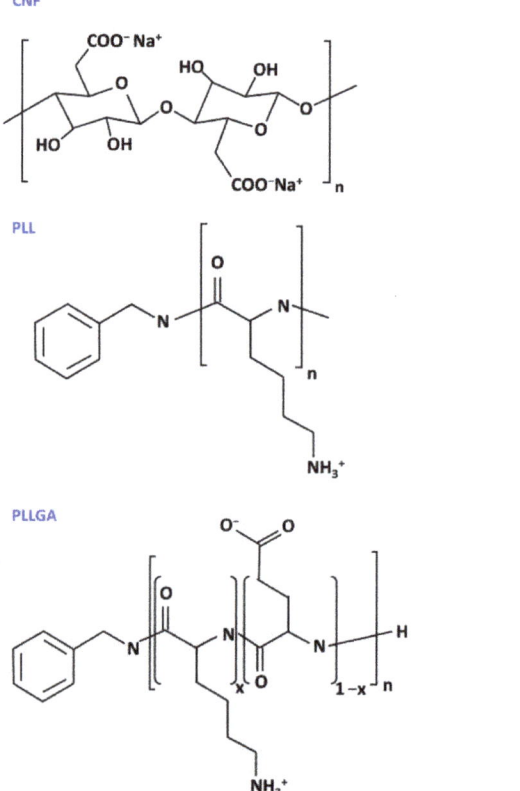

Figure 3. Chemical structures of CNF, PLL, and PLLGA in the hydrogel.

To determine an appropriate hydrogel composition for neurite growth study, we first considered the biotoxicity of the scaffolds by quantifying cell viability using AlamarBlue assay, a redox indicator for cell proliferation. PC12 cells were seeded onto the isotropic hydrogels using two kinds of poly-L-lysine-based, crosslinkers (PLL and PLLGA). For PLL crosslinked CNF hydrogel, on Day 6, the relative cell viability plummeted to less than 20% at 0.5C and 1C samples for all PLL concentrations tested (6.25 mM to 12.5 mM), as shown in Figure 4a. This result demonstrates that PLL exhibits biotoxicity at high dose.

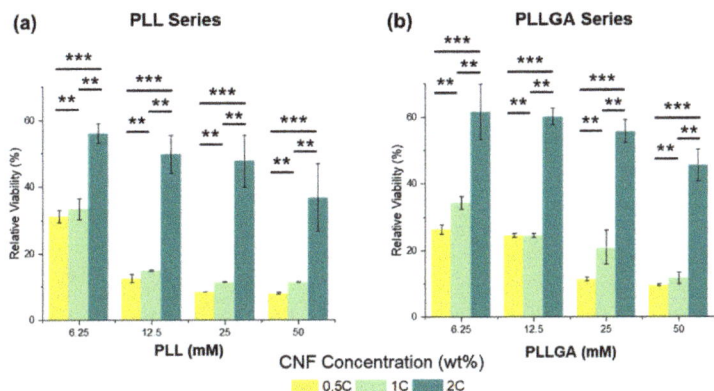

Figure 4. The CNF concentration effect in (a) PLL and (b) PLLGA crosslinker series on the relative viability from AlamarBlue assay on Day 6 PC12 cells. $n = 7$. The Kruskal–Wallis H-test was used to determine significant differences between groups. (** = $p < 0.01$, *** = $p < 0.001$).

However, the negative impact of PLL crosslinker can be offset and cell viability improves when CNF solution concentration is increased to 2C. Previous studies report that glutamate, an endogenous peptide that acts as an excitatory neurotransmitter, can promote not only cell viability, but also neural function [17,53]. To address potential cytotoxicity, PLL was modified by copolymerizing 20 mol% L-glutamic acid to L-lysine to obtain PLLGA (Figure 4b). Of note, this addition of L-glutamic acid reduces the positive charge of L-lysine accordingly. Thus, the actual crosslinker molar concentration of PLLGA was higher than the equivalent molar concentration of the crosslink sites. Despite the higher actual crosslinker molar concentration of PLLGA, we observed increased cell viability at equivalent crosslink sites (Figure 5). Significant differences in viability reduction were seen between PLL and PLLGA under the same effective concentration ($p < 0.001$). Therefore, the modification of PLL by glutamic acid (PLLGA) was an effective method to increase cell viability when using poly-L-lysine-based crosslinkers. Next, in vitro experiments were performed using the optimized 2C-6.25PLL and 2C-6.25PLLGA to represent the PLL series and PLLGA series, respectively.

Neuronal communication, like other secretory cells communication, depends on ligand engagement of receptors on the cell membrane, often coupled to ionic channels whose activation and ability to change voltage potential is conditionally dependent on the linkage between site and gating by ATP occupancy [54,55]. ARPE-19 cells have been used extensively to characterize Ca^{2+} signaling and transport mechanisms [56–58]. Phototransduction by the neural retina rod and cone photoreceptors utilizes Ca^{2+} dynamics by modulating the cGMP-gated channels and cGMP turnover [59]. Moreover, the retinal pigmented epithelium (RPE), which is vital for the nourishment of the retina, depends on Ca^{2+} as a second messenger to maintain homeostasis [60]. To investigate the effects of our hydrogels on neural signaling in the setting of the ocular system, we imaged calcium signaling in adult retinal pigment epithelial cell line-19 (ARPE-19), which have apically oriented cilia that are widely used in researching retinal pathology, particularly macular degeneration [61–64].

RPE cells are known to release ATP, and stimulating adenosine receptors regulate RPE cells interaction with photoreceptors [64]. We found that the intracellular Ca^{2+} concentration increased in ARPE19 cells after the extracellular dosing of ATP. This could result from the direct influx of Ca^{2+} via ATP-dependent calcium channels or a release from intracellular stores, possibly mediated by $P2Y_1$ receptor activation [65]. The fluorescence intensity peak of the calcium indicator can correlate with electroactive neuronal behavior. As shown in Figure 6, hydrogels with polypeptide-based crosslinkers (PLL and PLLGA) had a significant improvement compared with the control cells cultured on the coverslip. We attribute this to the well-established neurotrophic effects of lysine [38–40]. PLLGA with 20 mol% glutamic acid had the larger effect on Ca^{2+} response compared with the PLL series. Glutamic acid is an excitatory neurotransmitter that can additionally activate ionotropic glutamate NMDA receptors [66]. This could account for the drastic enhancement of activation signals associated with PLLGA over PLL. Furthermore, these results suggest ARPE-19 cells are highly electroactive on the glutamate containing hydrogel PLLGA.

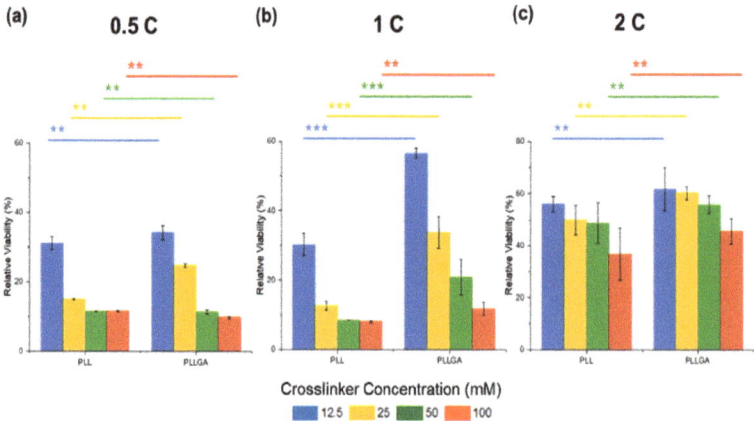

Figure 5. The enhancement of the cell viability on Day 6 through the modification from PLL to PLLGA by varying the CNF concentration (**a**) 0.5 wt.%, (**b**) 1.0 wt.%, and (**c**) 2.0 wt.%. $n = 7$. The Kruskal–Wallis H-test was used to determine significant differences between groups. (** = $p < 0.01$, *** = $p < 0.001$).

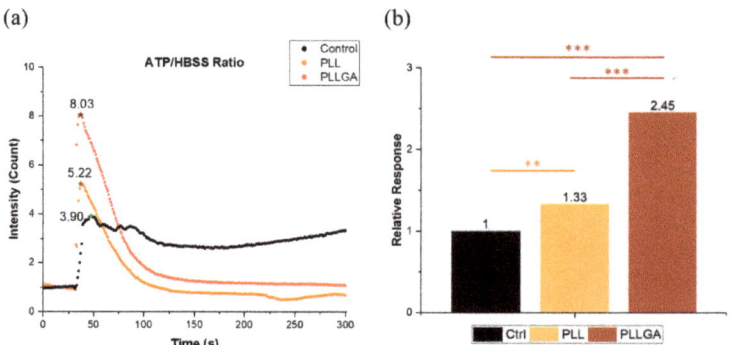

Figure 6. (**a**) The fluorescent intensity profile of the calcium indicator to ATP stimulation and (**b**) the electroactivity of ARPE cultured on 2C-6.25PLL and 2C-PLLGA on Day 1 was represented as the response peak heights which were normalized with the control group. $n = 7$. The Kruskal–Wallis H-test was used to determine significant differences between groups. (** = $p < 0.01$, *** = $p < 0.001$). The star sign marks the highest point of each curve.

After using Ca^{2+} imaging and AlamarBlue viability assays to determine the optimal scaffolds for cell culture on isotropic hydrogels, we examined whether aligned cellulose nanofibers could guide the orderly growth of NGF stimulated PC12 neurites. A scanning electron microscope (SEM) is typically used to confirm engineered nano-scale structures; however, in this case, the high vacuum environment of SEM complicates the analysis of aqueous samples. Hydrogels with over 90% water content need to be dried, which might cause an unexpected deformation or collapse due to the loss of swelling by water. Instead of SEM, POM and TXM were used to characterize the aqueous samples under 2D micro-scale and 3D nanoscale, respectively. Due to the high aspect ratio of CNF, the optical properties of the long axis and the short axis of CNF are significantly different, which makes it an optically anisotropic material. POM, which can analyze the birefringence, is appropriate for characterizing the orientation of fiber in hydrogels. We found the resolution to be insufficient for characterizing the hydrogel at the nanoscale of neurite structure. Freeze-dried samples can be analyzed by SEM, but it will lose their original nanofibrous hydrogel structure. Wet hydrogel samples can be analyzed by TEM, but its procedure is tedious and high cost. Therefore, TXM which can resolve down to 60 nm using high coherence synchrotron radiation was employed to confirm hydrogel directed neurite anisotropy. Samples with very small contrast can be resolved by leveraging the phase shift associated with RuO$_4$ staining. Even so, it is still difficult to distinguish between water and hydrogels composed of light elements of carbon, hydrogen, oxygen, and nitrogen in the hydrogels under TXM. To address these challenges, it was necessary to increase the concentration of crosslinkers to enhance signal contrast. So, for the structure characterizations by POM and TXM, the high concentration crosslinker samples, 2C-50PLL and 2C-50PLLGA were used instead of those determined to be optimal for the in vitro work (2C-6.25PLL and 2C-6.25PLLGA).

A typical CNF hydrogel photo image was shown in Figure 7a. A uniform magenta POM image of the well-dispersed 2.0 wt.% CNF solution was captured under cross-polar mode with the compensator filter inserted as depicted in Figure 7b. Figure 7c,d POM shows that crosslinking the peptide-based crosslinkers (PLL or PLLGA) produces some random anisotropic regions, likely due to shrinkage during gelation. The random anisotropic areas were unable to guide neurite extension with consistent orientation (alignment). Comparatively, the anisotropic hydrogels with persistent directionality successfully guided neurites along their axes over increased distances. The birefringence of ani-2C-50PLL and ani-2C-50PLLGA with the shear stress was applied to align the fibers emerged as a uniform blue along with the 1st and 3rd quadrant. The blue region represents the constructive retardation effect due to two coincident ordinary axes of compensator and specimen. Thus, we verified the creation, functionality, and uniform structural orientation using the co-injection method for CNF hydrogels at a submicron scale, as shown in Figure 7e,f. Despite these data, additional research should be carried out in the future to demonstrate that the hydrogel fibers are orientated along the long axis and capable of supporting regenerative neurite guidance over long distances, such as for peripheral nerves and for optic nerve from the retina, across the optic chiasm, and then to properly target the visual centers within the brain's cortex.

Since CNF has a large amount of hydrogen bonds, it easily aggregates to form a micro- or even larger-scale cellulose bundle. The CNF has to be well dispersed without any aggregation in the solution using an ultrasonic oscillator before crosslinking with polypeptide, as shown in Figure 7b. The ordinary axis of the CNF can be derived from POM images of the hydrogels formed by aggregation CNF solution, as shown in Figure 7g,h. The compensator is a positive crystal. When inserted into the optical path along a northwest-southeast direction, the ordinary axis was along a northwest–southeast direction. CNF was derived to be a positive crystal as the aggregate bundles along with northwest–southeast present blue. The aggregation images also demonstrated that the arrangement applied by the shear stress became less effective when the CNF was poorly dispersed.

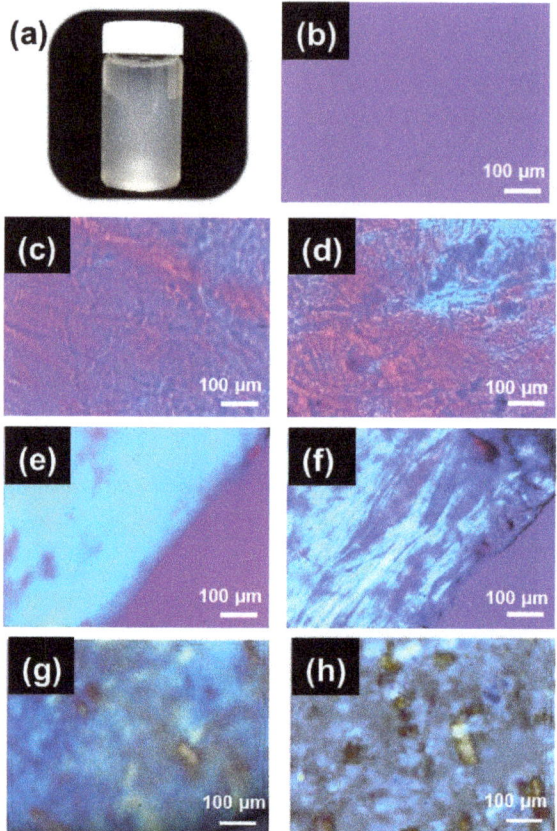

Figure 7. (**a**) Typical photo image of CNF hydrogel. POM images of the well-dispersed (**b**) 2.0 wt.% CNF solution, (**c**) iso-2C-50PLL, (**d**) iso-2C-50 PLLGA, (**e**) ani-2C-50PLL, (**f**) ani-2C-50 PLLGA, and anisotropic hydrogels, (**g**) iso-2C-50PLL, and (**h**) iso-2C-50 PLLGA, which were fabricated with not well-dispersed CNF solution showed the micro-scale aggregation bundle.

We also considered that fibers in the hydrogel were arranged in a 3D space, while the 2D POM images only show the azimuth angle distribution projected on the focus plane. In this case, the fiber distribution at different angles of elevation was ignored. As a result, the reported orientation could be overestimated. That is why using TXM with the capacity for submicron resolution was essential for characterizing the 3D orientation. This in situ analysis for the hydrogels was completed under an ambient environment that prevented any deformation due to dehydration. The structure of the isotropic hydrogels presented sheet-like porous crosslinked features, with multiple groups of small anisotropic regions appearing randomly in the 3D models obtained by TXM images acquired at different angles, as seen in the left column of Figure 8a,b. This result is consistent with the observation in the POM study. The anisotropic hydrogels prepared by the co-injection method presented here were arranged in the long axis direction, as shown in Figure 8c,d. Additionally, the middle of the hydrogels presented larger-scale fiber bundles, which can reach tens to hundreds of nanometers, and lengthen out to 15 μm across the entire observation area (15 μm × 15 μm). The mechanism behind the formation of this structure is speculated to be the driving force of diffusion at the solution interface that occurs when the two solutions in the capillary tube encounter one another. We hypothesize the diffusion gradient forces molecular chains to present an arrangement of parallel interfaces.

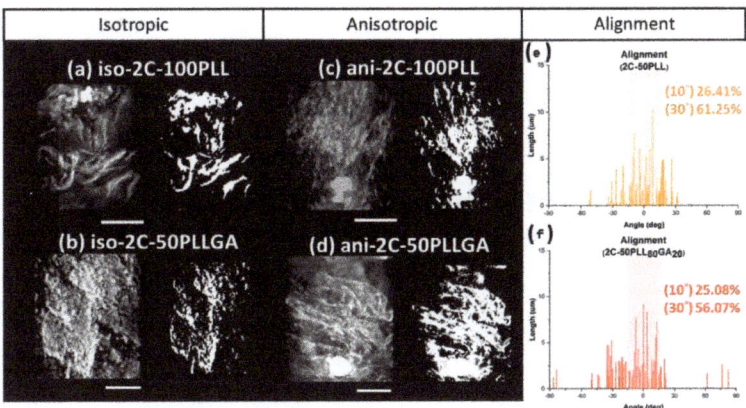

Figure 8. The TXM image of (**a**) iso-2C-50PLL, (**b**) iso-2C-50PLLGA, (**c**) ani-2C-50PLL, and (**d**) ani-2C-50PLLGA, and the alignment of hydrogel molecules was performed by angle distribution analysis from (**e**) ani-2C-50PLL and (**f**) ani-2C-50PLLGA. (n = 50), scale bar 5 micron.

The extent of fiber alignment in the hydrogel was analyzed using TXM images acquired under different rotation angles. The left column of Figure 8a–d shows the 3D images of different hydrogel samples. The alignment structure of the hydrogel gel under shear force can be seen in the left column of Figure 8c,d. The 3D images were transformed into 2D black and white 2D images through Fourier transformation, as shown in the right column of Figure 8a–d. Short segments (down to tens of nanometers) and long segments across the entire observation area coexisted in the TXM images; thus, the angle distribution of fibers was calculated by weighting the length to the angle. The weighted average of the angles was redefined as the axis of orientation. The statistically significant results are shown in Figure 8e,f. The alignment of Ani-2C-50PLL and Ani-2C-50PLLGA was 26.41% at 10°/61.25% at 30° and 25.08% at 10°/56.07% at 30°, respectively, as the degree of alignment was defined as the proportion of angles between −5°~5° and −15°~15°. The average of the alignment is calculated by weighted average due to nanofiber bundle. The single nanofiber under the TXM image may be formed by several CNF nanofibers, so weighted average is more accurate than simple number average.

The extent of alignment was calculated by statistical analysis. We found that our co-injection method aligned fibers on the surface, making contact with the syringe and capillary tube walls. Additionally, we found consistent long-range arrangement in a 3D space. The 3D tomography model exhibits a complex network of pores which will be useful for future 3D cell culture work. Such pores can allow oxygen, nutrients, and waste material diffusion and exchange in biological settings. Importantly, the fiber structural organization is not disjointed, but rather, parallel. It is possible that neuronal dendritic arborization might be enhanced by the complex interlaced fiber features. The highly plexiform high-order branching of neural dendrites arises from complex, nonlinear interactions between neural signals [67]. As the size and complexity of dendrites decrease, the activity in neurons is attenuated [68]. The loss of nerve complexity may be a precursor to neurodegeneration [69]. Thus, an engineered increase in the dendritic complexity might play a crucial role in neural protection.

On Day 5 after seeding, the cytoskeleton of PC12 cells stimulated with 2.5s-NGF was stained to investigate whether anisotropic hydrogels could successfully guide neurites growth with specific directionality as shown in Figure 9. Nerve growth and differentiation were determined by measuring neurite length distribution, as shown in Figure 10a. Neurite length distribution presented a positive Skew distribution, which was consistent with the distribution of normal nerve growth [38]. The average neurites length on the coverslip was 59.62 μm, while the ani-2C-6.25PLL and ani-2C-6.25PLLGA can reach 59.38 μm and 69.37 μm, respectively, as shown in Figure 10b. A 16% increase in neurite length is achieved using ani-2C-6.25PLLGA hydrogel scaffold. Typically, cells have difficulty adhering and growing on soft material substrates [70]. Here, neurites differentiation upon ani-2C-6.25PLL is comparable with growth on the coverslip control group. One explanation for this is that the high mechanical properties of the peptide-crosslinker-based hydrogels enhance cellular adhesion [45]. Another rationale is that the lysine in PLL mimics the endogenous neuroprotective role in the CNS [71–73]. Regardless, the addition of the 20 mol% glutamic acid significantly increased neurite length, presumably due to stimulation by glutamate, a neurotransmitter used by various neurobiological systems including the neural retina [74]. The results of our neuronal cell differentiation study corroborate the results of this study's calcium signal imaging (Figure 6). The low doses of polypeptide crosslinkers effectively promoted the electroactivity and differentiation performance in cultured neuronal cell lines. Among them, the 2C-6.25 PLLGA with 20 mol% glutamic acid copolymerized with lysine exhibits significant and outstanding performance.

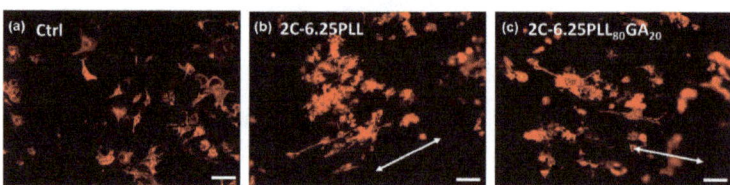

Figure 9. Representative images of immunohistochemical staining results of PC12 cells cultured on (**a**) untreated glass coverslip and (**b**) 2C-6.25PLL, (**c**) 2C-6.25PLL80GA20 anisotropic hydrogel for five days. The arrow bar indicates the directionality of the neurites.

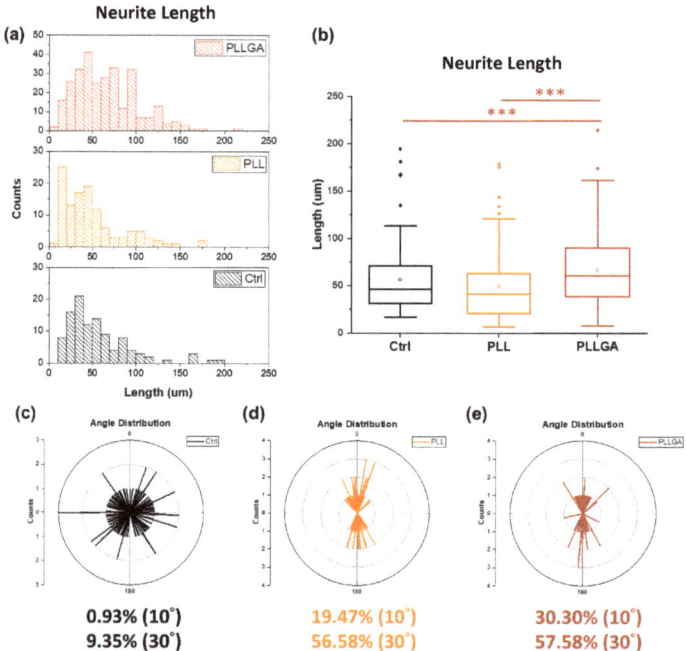

Figure 10. (**a**) The neurite length distribution and (**b**) the box chart of the neurite length. The % alignment of (**c**) control group, (**d**) ani-2C-6.25PLL, (**e**) ani 2C-6.25PLLGA. The counts (x-axis) represent the "number of the neurites". Additionally, 10 degrees represent neurite growth between +10 degree and −10 degree. The Kruskal–Wallis H-test was used to determine significant differences between groups. (*** = $p < 0.001$).

The extent of alignment of neurites guided by the anisotropic hydrogels was determined using polar plot (detailed in Section 3.5.3). Cultured cells on the untreated glass coverslips were typically radial in morphology (Figure 9a). The alignment was only 0.93% in the range of ±5°, and 9.35% in the range of ±15° (Figure 10c). In contrast, neurites grown on an ani-2C-6.25 PLL and an ani-2C-6.25 PLLGA were 19.74% at 10°/56.58% at 30° and 30.30% at 10°/57.58% at 30°, respectively, as shown in Figure 10d,e. Thus, the anisotropic hydrogels fabricated by the co-injection method show great potential for use in effectively guiding nerve growth in a bipolar direction in both crosslinker systems.

3. Materials and Methods

3.1. Materials

The polylysine-based crosslinkers, PLL and PLLGA, were synthesized by ring opening reaction of their corresponding N-carboxyanhydrides [45]. CNF was synthesized as previously described [75]. Chemicals for the characterization of the hydrogel structure are ruthenium tetroxide, 0.5% stabilized aqueous solution RuO_4(aq), (20427-56-9, Polysciences Inc., Warrington, DC, USA), and Au nanoparticles (AC11-500-CIT-DIH-100-1-EP, Nanopartz, Loveland, CO, USA). Reagents for in vitro experiments are as follows: AlamarBlue™ cell viability reagent (DAL1025, Thermofisher, Waltham, MA, USA), nerve growth factor-2.5S from the murine submaxillary gland (NGF, N6009-4X25UG, Sigma, St. Louis, MI, USA), rhodamine phalloidin reagent (AB125138, Abcam, Cambridge, MA, USA), 4′,6-diamidino-2-phenylindole (DAPI, D1847, Sigma-Aldrich, St. Louis, MI, USA), Hank's balanced salt solution, (1×), with calcium, magnesium, without phenol red (HBSS, SH30588, HyClone, Logan, UT, USA), Fluo-2 AM, green fluorescent Ca^{2+} binding dye (ab142775, Abcam, Cambridge, MA, USA), adenosine triphosphate (ATP, A2383, Sigma-Aldrich, St. Louis, MI, USA), Dulbecco's modified eagle medium (DMEM/F12 1:1, SH30023.02, HyClone, Logan, UT, USA), and RPMI 1640 with L-glutamine (SH30011.02, HyClone, Logan, UT, USA).

3.2. Nomenclature

Cellulose nanofiber is abbreviated as (CNF). The CNF solution is abbreviated as C; the prefix before "C" represents the weight concentration of the CNF solution. Abbreviations for the two kinds of polypeptide crosslinker, poly-L-lysine, poly-(L-lysine)-γ-poly-(L-glutamic acid) are PLL and PLLGA, respectively. The prefix before each crosslinker indicates the effective molar concentration of the crosslinker. The nomenclature of the hydrogels is denoted by the composition of CNF solution and the crosslinker. The prefixes iso- and ani- represent the isotropic and anisotropic hydrogels, respectively. For instance, ani-1C-25PLL is an anisotropic hydrogel fabricated with 1.0 wt% CNF and 25 mM PLL.

3.3. Hydrogel Preparation

The CNF has to be well dispersed without any aggregation in solution using ultrasonic oscillator before crosslinking with polypeptide. Isotropic hydrogels were fabricated by drop-casting a mixture of equal volume of crosslinker solution (concentration ranged from 6.25–50 mM PLL, PLLGA or Ca^{2+}) and CNF solution (~1.5 wt%), as shown in Figure 2a. After 24 h of aging, isotropic hydrogels were then prepared. Similar solution concentrations were used for anisotropic hydrogel preparation. The anisotropic hydrogel was placed in the capillary tube for 30 min in order to prevent chain relaxation before crosslinked. While anisotropic hydrogels were prepared by co-injecting the crosslinker solution and CNF solution through the applicator tip, dual cannula (SA-3605, FibriJet, Micromedics, Ventura, CA, USA), and gelled in the capillary tube (34507, Kimble, Vineland, NJ, USA), which was connected at the front end of the co-injector, as shown in Figure 2b. Aligned hydrogels were placed in a mold, then soaked in DI water to avoid water evaporation, preventing changes in the swelling ratio of hydrogels.

3.4. Characterization of Hydrogel Morphology

3.4.1. Polarized Optical Microscopy

A polarized optical microscope (POM, DM 2500 M, Leica, Wetzlar, Germany) was used to study the micro-scale birefringence of all hydrogels. Hydrogels were placed on a glass slide (22-310397, Fisher Scientific, Waltham, MA, USA). The samples were observed under cross polar mode at 100× with the compensator filter inserted into the optical path. Hydrogels were kept moist by applying 1–2 drops of DI water during the measurement to prevent any structural changes due to water evaporation.

3.4.2. Transmission X-ray Microscope

Hydrogels were stained with RuO_4 (aq.) for 24 h, then any residual dye was removed with 1X DI water wash, and then Kapton tape (P-222 AMB, Nitto, Osaka, Japan) that was cut into triangles was used to affix samples onto the sample holder. Long thin samples around 100 μm were preferentially selected using the needle to avoid signals that were perpendicular to the rotation axis. Such signals could block the target imaging area. Next, a 5 μL Au nanoparticle suspension with a particle size around 400–500 nm was dropped on the sample as a calibration object. Samples were then sent into the transmission X-ray microscope (TXM, NSRRC 01B1, Taiwan) for imaging. After rotating samples from $-75°$ to $75°$, a 3D model of the hydrogel nanostructure was created by Fourier transformation of the 151 TXM images.

3.5. In Vitro Studies Using Hydrogel

3.5.1. Cell Viability

AlamarBlue Assay was employed to study PC12 cell viability at 2, 4, and 6 days in vitro (Day 2, 4, 6) when 2D cocultured on top of the different compositions of hydrogels. The hydrogel gel sample was prepared by drop-casting a mixture of 180 μL crosslinker solution and 180 μL CNF solution in 48-well TC cell culture plate. The PC12 cells (Passage 9) were a gift from Prof. Jia-Shing Yu, Department of Chemical Engineering, National Taiwan University. The coverslip was not coated by any ECM molecules, nor was the hydrogel. PC12 cells were cultured in RPMI-1640 with L-glutamine with 10 v/v% HS (16050122, Gibco, Rockville, MD, USA), 5 v/v% FBS (SH30071, HyClone, Logan, UT, USA), 1 v/v% PSA (A5955, Sigma-Aldrich), and 1.0 g sodium bicarbonate (144-55-8, Sigma-Aldrich). Cells were incubated under 5% CO_2 at 37 °C and sub-cultured with trypsin EDTA (0.5%, 03-051-5C, Biological Industries, Haemek, Israel). PC12 cells were seeded into sterile polystyrene (PS) 48-well TC cell culture plate (353078, Corning Falcon, New York, USA) which had been UV sterilized overnight. Cells grown on 12 mm coverslip (0111520, Marienfeld, Germany) were used as the control group, while coverslips sterilized with UV (24 h) without cells served as the blank group. After staining for 4h with 10 v/v% AlamarBlue kit diluted with 37 °C DMEM (21063-029, Gibco, Rockville, MD, USA), stained cell suspensions were transferred to a sterile PS 96-well treated cell culture microplate (351172, Corning Falcon, New York, NY, USA) for detection by a microplate detector (800TS, BioTEK, Winooski, VT, USA) to analyze the absorbance at 570 nm and 600 nm wavelength, respectively. The absorbance was converted into a reduced percent by Equation (1) to represent the cell viability [76].

$$\text{Percent reduced (\%)} = \frac{\varepsilon_{ox}\lambda_{595} A_{\lambda 570} - \varepsilon_{ox}\lambda_{570} A_{\lambda 595}}{\varepsilon_{red}\lambda_{570} A'_{\lambda 595} - \varepsilon_{red}\lambda_{596} A'_{\lambda 570}} \quad (1)$$

$\lambda_1 = 570$ nm; $(\varepsilon_{ox})_{\lambda 1} = 80{,}586$; $(\varepsilon_{RED})_{\lambda 1} = 155{,}677$

$\lambda_2 = 600$ nm; $(\varepsilon_{ox})_{\lambda 2} = 117{,}216$; $(\varepsilon_{RED})_{\lambda 2} = 14{,}652$

A: Absorption of tested sample
A': Absorption of blank

3.5.2. Calcium Imaging

ATP stimulated calcium signaling was imaged 24 h after seeding cells (Day 1) on the 2D hydrogels. ARPE-19 cells (passage 37) were supplied by Dr. Ta-Ching Chen, National Taiwan University Hospital, Taipei, Taiwan. The cells were grown in DMEM/F12 1:1 medium with 10 v/v% FBS and 1 v/v% PSA (SV30010, HyClone, Logan, UT, USA). The cells were incubated under 5% CO_2 at 37 °C and subcultured using trypsin EDTA for seeding at a density of 5×10^5 cells/well into sterile PS 6-Well TC cell culture plate (353046, Corning Falcon, New York, NY, USA) which had been UV sterilized overnight. The hydrogel gel sample was prepared by drop-casting a mixture of 500 µL crosslinker solution and 500 µL CNF solution in a 6-well TC cell culture plate. Cells grown on the 22 mm coverslip (0111520, Marienfeld, Lauda-Konigshofen, Germany) were used as the control group. Calcium imaging was performed as previously described [77,78]. Fluo-2 AM was diluted with 37 °C HBSS to 5 µM and used to stain Day1 ARPE-19 for 30 min. We diluted 20 mM ATP with 37 °C HBSS to a final concentration of 1 mM to stimulate the cellular calcium signal. After removing the dye residues by 1.0 mL HBSS washing per well, an inverted fluorescent microscope (Eclipse Ti, Nikon, Tokyo, Japan) recorded the region of interest selected by Nikon NIS elements per second for 5 min. Next, 100 µL HBSS and 20 mM ATP solution were, respectively, added to the culture plates 30 s after staining to reveal changed induced by adding 1 mM ATP.

3.5.3. Measurement of Length, Distribution, and the Extent of the Alignment of Neurites

NGF (100 ng/mL) was used to induce PC12 neurite growth. PC12 cells were seeded at a density of 1.6×10^4 cells/well on the hydrogels in a 48-well cell culture plate. On Day 5, culture media was removed and then cells washed with 37 °C phosphate buffered saline (PBS, 1 tablet/160 mL H_2O, SLCC8330, Sigma-Aldrich, St.Louis, USA) solution. Cells were fixed with 400 µL/well of 3.7 v/v% formaldehyde (37 wt% in water, 50-00-0, Acros Organics, Branchburg, NJ, USA) diluted with PBS. After 15 min of incubating at 37 °C under 5% CO_2, fixative was removed and cells washed twice with PBS. A total of 400 µL per well of 1% Triton X-100 (X198-07, J.T. Baker, Phillipsburg, NJ, USA) diluted with PBS was used to permeabilize cells at room temperature for 10 min, then removed and washed using PBS. A total of 400 µL per well of 1 w/v% bovine serum albumin (BSA, 9048468, Sigma, St. Louis, MI, USA) was then added to PBS to pre-incubate for 60 min at room temperature to enhance staining. After removing the BSA/PBS pre-incubated solution, 300 units of rhodamine phalloidin dissolved in 1.5 mL methanol (67-56-1, Macron fine chemicals, Center Valley, PA, USA) was diluted 60 times in BSA/PBS. A total of 400 µL rhodamine phalloidin working solution was used to visualize the filamentous actin after 60 min of staining. The dye was removed and washed twice by BSA/PBS. Then, 400 µL of 300 nM DAPI solution diluted with BSA/PBS was used to label PC12 nuclei. After 15 min of staining, the dye was removed and cells washed twice by BSA/PBS. The samples were soaked in BSA/PBS to be imaged and measured. Neurite length and orientation were measured from the cell body to the end of the neurites by ImageJ. The average direction of the neurites was redefined as 0° for analysis of the neurite growth alignment on substrates. Three repeat measurements were performed for each experimental group, and over 50 neurites were measured in each sample.

3.6. Statistical Analysis

The Kruskal–Wallis H-test was used to determine significant differences between groups. The p-values < 0.05, < 0.01, and < 0.001 are represented as significant, highly significant, and extremely significant, respectively, and are marked as *, **, and ***, accordingly.

4. Conclusions

We have successfully developed anisotropic-structured hydrogels by crosslinking negatively charged CNF with positively charged copolypeptides under shear force using a co-injection method. The arrangement of anisotropic hydrogels was characterized in situ by POM and TXM under micro- and nano-scale, respectively. We show the co-injected

CNF hydrogels can enhance the extent of the alignment of NGF-stimulated neurite growth from 9.35% to 57.58% in the range of $\pm 15°$. We show a 2.45-fold increase over control in electroactivity of functional ATP-dependent Ca^{2+} signaling for ARPE-19 cells grown on our hydrogel. The cell viability of hydrogel is dependent on the amount of CNF and polypeptides, as well as the type of polypeptide. The cytotoxicity of hydrogel can not only be reduced, but its neurite growth performance is also increased by the incorporation of 20 molar % of glutamic acid moiety in polylysine as a co-polypeptide crosslinker: PLLGA. The cytotoxicity decreases with increasing amount of CNF and reducing amount of copolypeptide. The best biocompatibility of more than 60% cell viability is observed for the hydrogel sample of 2C-6.25PLLGA (2 wt% CNF and 6.25 mM PLLGA). We demonstrate a 16% increase in neurite length in stimulating directional neurite growth compared with the control using the optimal composition of hydrogel: 2C-6.25PLLGA. With this hydrogel, under shear force via co-injection method, an anisotropy of 30.30% at $10°$ /57.58% at $30°$ is reached. The injectable anisotropic-structured hydrogel reported here is a promising neural tissue engineering strategy for treating various neuropathies including those specifically related to the visual system and vision loss. The clinical use of injectable hydrogel has the potential advantage of avoiding open surgical procedure complications.

Author Contributions: Y.-H.T. and T.-L.M.: Investigation, Writing—Original draft preparation. D.-H.T. and C.-C.W.: Investigation. A.-J.A.S., K.M.W., M.-C.W., and Y.-C.H.: Methodology, Investigation, Writing—Review and Editing. W.-F.S.: Conceptualization, Methodology, Writing—Review and Editing, Funding acquisition, Supervision. All authors have read and agreed to the published version of the manuscript.

Funding: This research is granted by the National Science and Technology Council, Taiwan (NSTC 108-2221-E-002-027-MY3, 110-2628-E-182-001, 111-2221-E-182-040-MY3, 111-2628-E-182-001-MY2 and 111-2221-E-002-029), Department of Defense of USA (W81XWH-16-1-0775) and University of Colorado Anschutz Medical Campus, USA (Su-AEF Seed Grant: 63503960). The authors also thank the Chang Gung University (QZRPD181) and Chang Gung Memorial Hospital at Linkou (CMRPD2L0082 and BMRPC74) for financial support.

Institutional Review Board Statement: Not applicable.

Informed Consent Statement: Not applicable.

Data Availability Statement: The data that support the findings of this study are available on request from the corresponding author.

Acknowledgments: We appreciate the National Synchrotron Radiation Research Center (NSRRC) for supporting Transmission X-ray Microscope (TXM) facility.

Conflicts of Interest: The authors declare no conflict of interest.

References

1. Ha, A.; Kim, Y.K.; Park, Y.J.; Jeoung, J.W.; Park, K.H. Intraocular pressure change during reading or writing on smartphone. *PLoS ONE* **2018**, *13*, e0206061. [CrossRef] [PubMed]
2. Qudsiya, S.M.; Khatoon, F.; Khader, A.A.; Ali, M.A.; Hazari, M.A.H.; Sultana, F.; Farheen, A. Study of intraocular pressure among individuals working on computer screens for long hours: Effect of exposure to computer screens on IOP. *Ann. Med. Physiol.* **2017**, *1*, 22–25. [CrossRef]
3. Nassr, M.A.; Morris, C.L.; Netland, P.A.; Karcioglu, Z.A. Intraocular Pressure Change in Orbital Disease. *Surv. Ophthalmol.* **2009**, *54*, 519–544. [CrossRef] [PubMed]
4. Chen, M.; Wu, A.; Zhang, L.; Wang, W.; Chen, X.; Yu, X.; Wang, K. The increasing prevalence of myopia and high myopia among high school students in Fenghua city, eastern China: A 15-year population-based survey. *BMC Ophthalmol.* **2018**, *18*, 159. [CrossRef]
5. Matsumura, H.; Hirai, H. Prevalence of Myopia and Refractive Changes in Students from 3 to 17 Years of Age. *Surv. Ophthalmol.* **1999**, *44*, S109–S115. [CrossRef]
6. Saw, S.-M.; Gazzard, G.; Shih-Yen, E.C.; Chua, W.-H. Myopia and associated pathological complications. *Ophthalmic Physiol. Opt.* **2005**, *25*, 381–391. [CrossRef]
7. Allison, K.; Patel, D.; Alabi, O. Epidemiology of Glaucoma: The Past, Present, and Predictions for the Future. *Cureus* **2020**, *12*, e11686. [CrossRef]

8. Kingman, S. Glaucoma is second leading cause of blindness globally. *Bull. World Health Organ.* **2004**, *82*, 887–888.
9. Goel, M.; Pacciani, R.G.; Lee, R.K.; Battacharya, S.K. Aqueous Humor Dynamics: A Review. *Open Ophthalmol. J.* **2010**, *4*, 52–59. [CrossRef]
10. Almasieh, M.; Wilson, A.M.; Morquette, B.; Vargas, J.L.C.; Di Polo, A. The molecular basis of retinal ganglion cell death in glaucoma. *Prog. Retin. Eye Res.* **2012**, *31*, 152–181. [CrossRef]
11. Yang, P.; Wei, L.; Tian, H.; Yu, F.; Shi, Y.; Gao, L. Gadolinium chloride protects neurons by regulating the activation of microglia in the model of optic nerve crush. *Biochem. Biophys. Res. Commun.* **2022**, *618*, 119–126. [CrossRef] [PubMed]
12. Wu, J.; Hu, C.; Tang, Z.; Yu, Q.; Liu, X.; Chen, H. Tissue-engineered Vascular Grafts: Balance of the Four Major Requirements. *Colloid Interface Sci. Commun.* **2018**, *23*, 34–44. [CrossRef]
13. Wei, Y.; Baskaran, N.; Wang, H.-Y.; Su, Y.-C.; Nabilla, S.C.; Chung, R.-J. Study of polymethylmethacrylate/tricalcium silicate composite cement for orthopedic application. *Biomed. J.* **2022**. [CrossRef] [PubMed]
14. Deegan, A.J.; Hendrikson, W.J.; El Haj, A.J.; Rouwkema, J.; Yang, Y. Regulation of endothelial cell arrangements within hMSC–HUVEC co-cultured aggregates. *Biomed. J.* **2019**, *42*, 166–177. [CrossRef] [PubMed]
15. Carrier, R.L.; Papadaki, M.; Rupnick, M.; Schoen, F.J.; Bursac, N.; Langer, R.; Freed, L.E.; Vunjak-Novakovic, G. Cardiac tissue engineering: Cell seeding, cultivation parameters, and tissue construct characterization. *Biotechnol. Bioeng.* **1999**, *64*, 580–589. [CrossRef]
16. Chen, Q.-Z.; Harding, S.E.; Ali, N.N.; Lyon, A.R.; Boccaccini, A.R. Biomaterials in cardiac tissue engineering: Ten years of research survey. *Mater. Sci. Eng. R Rep.* **2008**, *59*, 1–37. [CrossRef]
17. Lin, C.-Y.; Luo, S.-C.; Yu, J.-S.; Chen, T.-C.; Su, W.-F. Peptide-Based Polyelectrolyte Promotes Directional and Long Neurite Outgrowth. *ACS Appl. Bio Mater.* **2018**, *2*, 518–526. [CrossRef]
18. Wang, Z.-H.; Chang, Y.-Y.; Wu, J.-G.; Lin, C.-Y.; An, H.-L.; Luo, S.-C.; Tang, T.K.; Su, W.-F. Novel 3D Neuron Regeneration Scaffolds Based on Synthetic Polypeptide Containing Neuron Cue. *Macromol. Biosci.* **2017**, *18*, 1700251. [CrossRef]
19. Steed, M.B.; Mukhatyar, V.; Valmikinathan, C.; Bellamkonda, R.V. Advances in Bioengineered Conduits for Peripheral Nerve Regeneration. *Atlas Oral Maxillofac. Surg. Clin.* **2011**, *19*, 119–130. [CrossRef]
20. Petersen, O.W.; Rønnov-Jessen, L.; Howlett, A.R.; Bissell, M.J. Interaction with basement membrane serves to rapidly distinguish growth and differentiation pattern of normal and malignant human breast epithelial cells. *Proc. Natl. Acad. Sci. USA* **1992**, *89*, 9064. [CrossRef]
21. Caliari, S.R.; Burdick, J.A. A practical guide to hydrogels for cell culture. *Nat. Methods* **2016**, *13*, 405–414. [CrossRef] [PubMed]
22. Niemczyk, B.; Sajkiewicz, P.; Kolbuk, D. Injectable hydrogels as novel materials for central nervous system regeneration. *J. Neural Eng.* **2018**, *15*, 051002. [CrossRef]
23. He, L.; Xiao, Q.; Zhao, Y.; Li, J.; Reddy, S.; Shi, X.; Su, X.; Chiu, K.; Ramakrishna, S. Engineering an Injectable Electroactive Nanohybrid Hydrogel for Boosting Peripheral Nerve Growth and Myelination in Combination with Electrical Stimulation. *ACS Appl. Mater. Interfaces* **2020**, *12*, 53150–53163. [CrossRef] [PubMed]
24. Chetia, M.; Debnath, S.; Chowdhury, S.; Chatterjee, S. Self-assembly and multifunctionality of peptide organogels: Oil spill recovery, dye absorption and synthesis of conducting biomaterials. *RSC Adv.* **2020**, *10*, 5220–5233. [CrossRef] [PubMed]
25. Lu, Q.; Bai, S.; Ding, Z.; Guo, H.; Shao, Z.; Zhu, H.; Kaplan, D.L. Hydrogel Assembly with Hierarchical Alignment by Balancing Electrostatic Forces. *Adv. Mater. Interfaces* **2016**, *3*, 1500687. [CrossRef]
26. Omidinia-Anarkoli, A.; Boesveld, S.; Tuvshindorj, U.; Rose, J.C.; Haraszti, T.; De Laporte, L. An injectable hybrid hydrogel with oriented short fibers induces unidirectional growth of functional nerve cells. *Small* **2017**, *13*, 1702207. [CrossRef]
27. Wang, L.; Song, D.; Zhang, X.; Ding, Z.; Kong, X.; Lu, Q.; Kaplan, D.L. Silk–Graphene Hybrid Hydrogels with Multiple Cues to Induce Nerve Cell Behavior. *ACS Biomater. Sci. Eng.* **2018**, *5*, 613–622. [CrossRef]
28. De France, K.J.; Yager, K.G.; Chan, K.J.W.; Corbett, B.; Cranston, E.D.; Hoare, T. Injectable Anisotropic Nanocomposite Hydrogels Direct in Situ Growth and Alignment of Myotubes. *Nano Lett.* **2017**, *17*, 6487–6495. [CrossRef]
29. Ghaderinejad, P.; Najmoddin, N.; Bagher, Z.; Saeed, M.; Karimi, S.; Simorgh, S.; Pezeshki-Modaress, M. An injectable anisotropic alginate hydrogel containing oriented fibers for nerve tissue engineering. *Chem. Eng. J.* **2021**, *420*, 130465. [CrossRef]
30. Vader, D.; Kabla, A.; Weitz, D.; Mahadevan, L. Strain-Induced Alignment in Collagen Gels. *PLoS ONE* **2009**, *4*, e5902. [CrossRef]
31. Antman-Passig, M.; Levy, S.; Gartenberg, C.; Schori, H.; Shefi, O. Mechanically Oriented 3D Collagen Hydrogel for Directing Neurite Growth. *Tissue Eng. Part A* **2017**, *23*, 403–414. [CrossRef] [PubMed]
32. Kim, S.H.; Im, S.K.; Oh, S.J.; Jeong, S.; Yoon, E.S.; Lee, C.J.; Choi, N.; Hur, E.M. Anisotropically organized three-dimensional culture platform for reconstruction of a hippocampal neural network. *Nat. Commun.* **2017**, *8*, 14346. [CrossRef] [PubMed]
33. Seo, Y.; Jeong, S.; Chung, J.J.; Kim, S.H.; Choi, N.; Jung, Y. Development of an Anisotropically Organized Brain dECM Hydrogel-Based 3D Neuronal Culture Platform for Recapitulating the Brain Microenvironment in Vivo. *ACS Biomater. Sci. Eng.* **2019**, *6*, 610–620. [CrossRef] [PubMed]
34. Du, J.; Liu, J.; Yao, S.; Mao, H.; Peng, J.; Sun, X.; Cao, Z.; Yang, Y.; Xiao, B.; Wang, Y.; et al. Prompt peripheral nerve regeneration induced by a hierarchically aligned fibrin nanofiber hydrogel. *Acta Biomater.* **2017**, *55*, 296–309. [CrossRef] [PubMed]
35. Yao, S.; Yu, S.; Cao, Z.; Yang, Y.; Yu, X.; Mao, H.-Q.; Wang, L.-N.; Sun, X.; Zhao, L.; Wang, X. Hierarchically aligned fibrin nanofiber hydrogel accelerated axonal regrowth and locomotor function recovery in rat spinal cord injury. *Int. J. Nanomed.* **2018**, *13*, 2883–2895. [CrossRef]

36. Perez, R.A.; Kim, M.; Kim, T.-H.; Kim, J.-H.; Lee, J.H.; Park, J.-H.; Knowles, J.C.; Kim, H.-W. Utilizing Core–Shell Fibrous Collagen-Alginate Hydrogel Cell Delivery System for Bone Tissue Engineering. *Tissue Eng. Part A* **2014**, *20*, 103–114. [CrossRef]
37. Martínez-Sanz, M.; Mikkelsen, D.; Flanagan, B.M.; Rehm, C.; de Campo, L.; Gidley, M.J.; Gilbert, E.P. Investigation of the micro- and nano-scale architecture of cellulose hydrogels with plant cell wall polysaccharides: A combined USANS/SANS study. *Polymer* **2016**, *105*, 449–460. [CrossRef]
38. Hong-Ping, G.; Bao-Shan, K. Neuroprotective effect of L-lysine monohydrochloride on acute iterative anoxia in rats with quantitative analysis of electrocorticogram. *Life Sci.* **1999**, *65*, PL19–PL25. [CrossRef]
39. Cheng, J.; Tang, J.C.; Pan, M.X.; Chen, S.F.; Zhao, D.; Zhang, Y.; Liao, H.B.; Zhuang, Y.; Lei, R.X.; Wang, S.; et al. l-lysine confers neuroprotection by suppressing inflammatory response via microRNA-575/PTEN signaling after mouse intracerebral hemorrhage injury. *Exp. Neurol.* **2020**, *327*, 113214. [CrossRef]
40. Kondoh, T.; Kameishi, M.; Mallick, H.N.; Ono, T.; Torii, K. Lysine and arginine reduce the effects of cerebral ischemic insults and inhibit glutamate-induced neuronal activity in rats. *Front. Integr. Neurosci.* **2010**, *4*, 18. [CrossRef]
41. Yavin, E.; Yavin, Z. Attachment and culture of dissociated cells from rat embryo cerebral hemispheres on polylysine-coated surface. *J. Cell Biol.* **1974**, *62*, 540–546. [CrossRef] [PubMed]
42. Low, C.M.; Wee, K.S. New insights into the not-so-new NR3 subunits of N-methyl-D-aspartate receptor: Localization, structure, and function. *Mol. Pharm.* **2010**, *78*, 1–11. [CrossRef] [PubMed]
43. Pin, J.-P.; Acher, F. The metabotropic glutamate receptors: Structure, activation mechanism and pharmacology. *Curr. Drug Target CNS Neurol. Disord.* **2002**, *1*, 297–317. [CrossRef] [PubMed]
44. Ma, T.; Yang, S.; Luo, S.; Chen, W.; Liao, S.; Su, W. Dual-Function Fibrous Co-Polypeptide Scaffolds for Neural Tissue Engineering. *Macromol. Biosci.* **2022**, *23*, 2200286. [CrossRef]
45. Yu, T.; Tseng, Y.; Wang, C.; Lin, T.; Wu, M.; Tsao, C.; Su, W. Three-Level Hierarchical 3D Network Formation and Structure Elucidation of Wet Hydrogel of Tunable-High-Strength Nanocomposites. *Macromol. Mater. Eng.* **2022**, *307*, 2100871. [CrossRef]
46. Shui, G. *5.44-Development of in vitro Neural Models for Drug Discovery and Toxicity Screening, in Comprehensive Biotechnology*, 2nd ed.; Moo-Young, M., Ed.; Academic Press: Burlington, VT, USA, 2011; pp. 565–572.
47. Koizumi, S.; Bootman, M.D.; Bobanović, L.K.; Schell, M.J.; Berridge, M.J.; Lipp, P. Characterization of Elementary Ca^{2+} Release Signals in NGF-Differentiated PC12 Cells and Hippocampal Neurons. *Neuron* **1999**, *22*, 125–137. [CrossRef]
48. Wittig, D.; Wang, X.; Walter, C.; Gerdes, H.-H.; Funk, R.H.W.; Roehlecke, C. Multi-Level Communication of Human Retinal Pigment Epithelial Cells via Tunneling Nanotubes. *PLoS ONE* **2012**, *7*, e33195. [CrossRef]
49. Hasanzadeh, E.; Seifalian, A.; Mellati, A.; Saremi, J.; Asadpour, S.; Enderami, S.E.; Nekounam, H.; Mahmoodi, N. Injectable hydrogels in central nervous system: Unique and novel platforms for promoting extracellular matrix remodeling and tissue engineering. *Mater. Today Bio* **2023**, *20*, 100614. [CrossRef]
50. Khan, J.; Rudrapal, M.; Bhat, E.A.; Ali, A.; Alaidarous, M.; Alshehri, B.; Banwas, S.; Ismail, R.; Egbuna, C. Perspective Insights to Bio-Nanomaterials for the Treatment of Neurological Disorders. *Front. Bioeng. Biotechnol.* **2021**, *9*, 724158. [CrossRef]
51. Grimaudo, M.; Krishnakumar, G.; Giusto, E.; Furlani, F.; Bassi, G.; Rossi, A.; Molinari, F.; Lista, F.; Montesi, M.; Panseri, S. Bioactive injectable hydrogels for on demand molecule/cell delivery and for tissue regeneration in the central nervous system. *Acta Biomater.* **2021**, *140*, 88–101. [CrossRef]
52. Li, S.; Ke, Z.; Peng, X.; Fan, P.; Chao, J.; Wu, P.; Xiao, P.; Zhou, Y. Injectable and fast gelling hyaluronate hydrogels with rapid self-healing ability for spinal cord injury repair. *Carbohydr. Polym.* **2022**, *298*, 120081. [CrossRef] [PubMed]
53. Ma, T.-L.; Yang, S.-C.; Cheng, T.; Chen, M.-Y.; Wu, J.-H.; Liao, S.-L.; Chen, W.-L.; Su, W.-F. Exploration of biomimetic poly(γ-benzyl-l-glutamate) fibrous scaffolds for corneal nerve regeneration. *J. Mater. Chem. B* **2022**, *10*, 6372–6379. [CrossRef] [PubMed]
54. Catterall, W.A. Voltage-gated calcium channels. *Cold Spring Harb. Perspect. Biol.* **2011**, *3*, a003947. [CrossRef] [PubMed]
55. Li, L.; Geng, X.; Yonkunas, M.; Su, A.; Densmore, E.; Tang, P.; Drain, P. Ligand-dependent Linkage of the ATP Site to Inhibition Gate Closure in the KATP Channel. *J. Gen. Physiol.* **2005**, *126*, 285–299. [CrossRef]
56. Duman, J.G.; Chen, L.; Hille, B. Calcium Transport Mechanisms of PC12 Cells. *J. Gen. Physiol.* **2008**, *131*, 307–323. [CrossRef]
57. Li, G.-Y.; Fan, B.; Zheng, Y.-C. Calcium Overload Is A Critical Step in Programmed Necrosis of ARPE-19 Cells Induced by High-Concentration H_2O_2. *Biomed. Environ. Sci.* **2010**, *23*, 371–377. [CrossRef]
58. Strauss, O.; Abdusalamova, K.; Huber, C.; Rohrer, B.; Busch, C. Interactive Ca^{2+} signaling in the RPE by anaphylatoxins. *Investig. Ophthalmol. Vis. Sci.* **2017**, *58*, 2277.
59. Nakatani, K.; Chen, C.; Yau, K.-W.; Koutalos, Y. Calcium and Phototransduction. *Adv. Exp. Med. Biol.* **2002**, *514*, 1–20. [CrossRef]
60. Zhang, L.; Hui, Y.-N.; Wang, Y.-S.; Ma, J.-X.; Wang, J.-B.; Ma, L.-N. Calcium overload is associated with lipofuscin formation in human retinal pigment epithelial cells fed with photoreceptor outer segments. *Eye* **2011**, *25*, 519–527. [CrossRef]
61. Dunn, K.C.; Aotaki-Keen, A.E.; Putkey, F.R.; Hjelmeland, L.M. ARPE-19, A Human Retinal Pigment Epithelial Cell Line with Differentiated Properties. *Exp. Eye Res.* **1996**, *62*, 155–170. [CrossRef]
62. Hazim, R.A.; Volland, S.; Yen, A.; Burgess, B.L.; Williams, D.S. Rapid differentiation of the human RPE cell line, ARPE-19, induced by nicotinamide. *Exp. Eye Res.* **2019**, *179*, 18–24. [CrossRef] [PubMed]
63. Mannino, G.; Cristaldi, M.; Giurdanella, G.; Perrotta, R.E.; Furno, D.L.; Giuffrida, R.; Rusciano, D. ARPE-19 conditioned medium promotes neural differentiation of adipose-derived mesenchymal stem cells. *World J. Stem Cells* **2021**, *13*, 1783–1796. [CrossRef]

64. Mitter, S.K.; Song, C.; Qi, X.; Mao, H.; Rao, H.; Akin, D.; Lewin, A.; Grant, M.; Dunn, W.; Ding, J.; et al. Dysregulated autophagy in the RPE is associated with increased susceptibility to oxidative stress and AMD. *Autophagy* **2014**, *10*, 1989–2005. [CrossRef] [PubMed]
65. Reigada, D.; Lu, W.; Zhang, X.; Friedman, C.; Pendrak, K.; McGlinn, A.; Stone, R.A.; Laties, A.M.; Mitchell, C.H. Degradation of extracellular ATP by the retinal pigment epithelium. *Am. J. Physiol. Physiol.* **2005**, *289*, C617–C624. [CrossRef] [PubMed]
66. Guo, H.; Camargo, L.M.; Yeboah, F.; Digan, M.E.; Niu, H.; Pan, Y.; Reiling, S.; Soler-Llavina, G.; Weihofen, W.A.; Wang, H.-R.; et al. A NMDA-receptor calcium influx assay sensitive to stimulation by glutamate and glycine/D-serine. *Sci. Rep.* **2017**, *7*, 11608. [CrossRef]
67. Jan, Y.-N.; Jan, L. Branching out: Mechanisms of dendritic arborization. *Nat. Rev. Neurosci.* **2010**, *11*, 316–328. [CrossRef]
68. Kirch, C.; Gollo, L.L. Single-neuron dynamical effects of dendritic pruning implicated in aging and neurodegeneration: Towards a measure of neuronal reserve. *Sci. Rep.* **2021**, *11*, 1309. [CrossRef]
69. López-Doménech, G.; Higgs, N.F.; Vaccaro, V.; Roš, H.; Arancibia-Cárcamo, I.L.; MacAskill, A.F.; Kittler, J.T. Loss of Dendritic Complexity Precedes Neurodegeneration in a Mouse Model with Disrupted Mitochondrial Distribution in Mature Dendrites. *Cell Rep.* **2016**, *17*, 317–327. [CrossRef]
70. Guo, W.-H.; Frey, M.T.; Burnham, N.A.; Wang, Y.-L. Substrate Rigidity Regulates the Formation and Maintenance of Tissues. *Biophys. J.* **2006**, *90*, 2213–2220. [CrossRef]
71. Sharma, A.; Goring, A.; Staines, K.A.; Emery, R.J.; Pitsillides, A.A.; Oreffo, R.O.; Mahajan, S.; Clarkin, C.E. Raman spectroscopy links differentiating osteoblast matrix signatures to pro-angiogenic potential. *Matrix Biol. Plus* **2020**, *5*, 100018. [CrossRef]
72. Orlowska, A.; Perera, P.T.; Al Kobaisi, M.; Dias, A.; Nguyen, H.K.D.; Ghanaati, S.; Baulin, V.; Crawford, R.J.; Ivanova, E.P. The Effect of Coatings and Nerve Growth Factor on Attachment and Differentiation of Pheochromocytoma Cells. *Materials* **2017**, *11*, 60. [CrossRef] [PubMed]
73. Yu, P.; Santiago, L.Y.; Katagiri, Y.; Geller, H.M. Myosin II activity regulates neurite outgrowth and guidance in response to chondroitin sulfate proteoglycans. *J. Neurochem.* **2011**, *120*, 1117–1128. [CrossRef] [PubMed]
74. Reiner, A.; Levitz, J. Glutamatergic Signaling in the Central Nervous System: Ionotropic and Metabotropic Receptors in Concert. *Neuron* **2018**, *98*, 1080–1098. [CrossRef]
75. Takaichi, S.; Saito, T.; Tanaka, R.; Isogai, A. Improvement of nanodispersibility of oven-dried TEMPO-oxidized celluloses in water. *Cellulose* **2014**, *21*, 4093–4103. [CrossRef]
76. Jagan, S.; Paganessi, L.A.; Frank, R.R.; Venugopal, P.; Larson, M.; Christopherson, K.W. Bone Marrow and Peripheral Blood AML Cells Are Highly Sensitive to CNDAC, the Active Form of Sapacitabine. *Adv. Hematol.* **2012**, *2012*, 727683. [CrossRef] [PubMed]
77. Hung, H.H.; Kao, L.S.; Liu, P.S.; Huang, C.C.; Yang, D.M.; Pan, C.Y. Dopamine elevates intracellular zinc concentration in cultured rat embryonic cortical neurons through the cAMP-nitric oxide signaling cascade. *Mol. Cell Neurosci.* **2017**, *82*, 35–45. [CrossRef]
78. Hung, H.-H.; Huang, W.-P.; Pan, C.-Y. Dopamine- and zinc-induced autophagosome formation facilitates PC12 cell survival. *Cell Biol. Toxicol.* **2013**, *29*, 415–429. [CrossRef]

Disclaimer/Publisher's Note: The statements, opinions and data contained in all publications are solely those of the individual author(s) and contributor(s) and not of MDPI and/or the editor(s). MDPI and/or the editor(s) disclaim responsibility for any injury to people or property resulting from any ideas, methods, instructions or products referred to in the content.

MDPI
St. Alban-Anlage 66
4052 Basel
Switzerland
www.mdpi.com

International Journal of Molecular Sciences Editorial Office
E-mail: ijms@mdpi.com
www.mdpi.com/journal/ijms

Disclaimer/Publisher's Note: The statements, opinions and data contained in all publications are solely those of the individual author(s) and contributor(s) and not of MDPI and/or the editor(s). MDPI and/or the editor(s) disclaim responsibility for any injury to people or property resulting from any ideas, methods, instructions or products referred to in the content.

www.ingramcontent.com/pod-product-compliance
Lightning Source LLC
LaVergne TN
LVHW070410100526
838202LV00014B/1428